Computer Algebra and Symbolic Computation

Computer Algebra and Symbolic Computation
Mathematical Methods

Joel S. Cohen

Department of Computer Science
University of Denver

CRC Press
Taylor & Francis Group
Boca Raton London New York

CRC Press is an imprint of the
Taylor & Francis Group, an **informa** business

AN A K PETERS BOOK

CRC Press
Taylor & Francis Group
6000 Broken Sound Parkway NW, Suite 300
Boca Raton, FL 33487-2742

First issued in paperback 2020

© 2003 by Taylor & Francis Group, LLC
CRC Press is an imprint of Taylor & Francis Group, an Informa business

No claim to original U.S. Government works

ISBN-13: 978-1-56881-159-8 (hbk)
ISBN-13: 978-0-367-65947-9 (pbk)

Visit the Taylor & Francis Web site at
http://www.taylorandfrancis.com

and the CRC Press Web site at
http://www.crcpress.com

Library of Congress Cataloging-in-Publication Data
Cohen, Joel S.
   Computer algebra and symbolic computation : mathematical methods
/ Joel S. Cohen
      p. cm.
   Includes bibliographical references and index.
   ISBN 1-56881-159-4
   1. Algebra–Data processing. I. Title.

QA155.7.E4 .C6352 2002
512–dc21   2002024315

For my wife Kathryn

Contents

Preface

Computer algebra is the field of mathematics and computer science that is concerned with the development, implementation, and application of algorithms that manipulate and analyze mathematical expressions. This book and the companion text, *Computer Algebra and Symbolic Computation: Elementary Algorithms*, are an introduction to the subject that addresses both its practical and theoretical aspects. *Elementary Algorithms* addresses the practical side; it is concerned with the formulation of algorithms that solve symbolic mathematical problems and with the implementation of these algorithms in terms of the operations and control structures available in computer algebra programming languages. This book, which addresses more theoretical issues, is concerned with the basic mathematical and algorithmic concepts that are the foundation of the subject. Both books serve as a bridge between texts and manuals that show how to use computer algebra software and graduate level texts that describe algorithms at the forefront of the field.

These books have been in various stages of development for over 15 years. They are based on the class notes for a two-quarter course sequence in computer algebra that has been offered at the University of Denver every other year for the past 16 years. The first course, which is the basis for *Elementary Algorithms*, attracts primarily undergraduate students and a few graduate students from mathematics, computer science, and engineering. The second course, which is the basis for *Mathematical Methods*, attracts primarily graduate students in both mathematics and computer science. The course is cross-listed under both mathematics and computer science.

Prerequisites

The target audience for these books includes students and professionals from mathematics, computer science, and other technical fields who would like to know about computer algebra and its applications.

In the spirit of an introductory text, we have tried to minimize the prerequisites. The mathematical prerequisites include the usual two year freshman–sophomore sequence of courses (calculus through multivariable calculus, elementary linear algebra, and applied ordinary differential equations). In addition, an introductory course in discrete mathematics is recommended because mathematical induction is used as a proof technique throughout. Topics from elementary number theory and abstract algebra are introduced as needed.

On the computer science side, we assume that the reader has had some experience with a computer programming language such as Fortran, Pascal, C, C++, or Java. Although these languages are not used in these books, the skills in problem solving and algorithm development obtained in a beginning programming course are essential. One programming technique that is especially important in computer algebra is recursion. Although many students will have seen recursion in a conventional programming course, the topic is described in Chapter 5 of *Elementary Algorithms* from a computer algebra perspective.

Realistically speaking, while these prerequisites suffice in a formal sense for both books, in a practical sense there are some sections as the texts progress where greater mathematical and computational sophistication is required. Although the mathematical development in these sections can be challenging for students with the minimum prerequisites, the algorithms are accessible, and these sections provide a transition to more advanced treatments of the subject.

Organization and Content

Broadly speaking, these books are intended to serve two (complementary) purposes:

- *To provide a systematic approach to the algorithmic formulation and implementation of mathematical operations in a computer algebra programming language.*

Algorithmic methods in traditional mathematics are usually not presented with the precision found in numerical mathematics or conventional computer programming. For example, the algorithm for the expansion of products and powers of polynomials is usually given informally instead of with (recursive) procedures that can be expressed as a computer program.

The material in *Elementary Algorithms* is concerned with the algorithmic formulation of solutions to elementary symbolic mathematical problems. The viewpoint is that mathematical expressions, represented as expression trees, are the data objects of computer algebra programs, and, using a few primitive operations that analyze and construct expressions, we can implement many elementary operations from algebra, trigonometry, calculus, and differential equations. For example, algorithms are given for the analysis and manipulation of polynomials and rational expressions, the manipulation of exponential and trigonometric functions, differentiation, elementary integration, and the solution of first order differential equations. Most of the material in this book is not found in either mathematics textbooks or in other, more advanced computer algebra textbooks.

- *To describe some of the mathematical concepts and algorithmic techniques utilized by modern computer algebra software.*

For the past 35 years, the research in computer algebra has been concerned with the development of effective and efficient algorithms for many mathematical operations including polynomial greatest common divisor (gcd) computation, polynomial factorization, polynomial decomposition, the solution of systems of linear equations and multivariate polynomial equations, indefinite integration, and the solution of differential equations. Although algorithms for some of these problems have been known since the nineteenth century, for efficiency reasons they are not suitable as general purpose algorithms for computer algebra software. The classical algorithms are important, however, because they are much simpler and provide a context to motivate the basic algebraic ideas and the need for more efficient approaches.

The material in *Mathematical Methods* is an introduction to the mathematical techniques and algorithmic methods of computer algebra. Although the material in this book is more difficult and requires greater mathematical sophistication, the approach and selection of topics is designed so that it is accessible and interesting to the intended audience. Algorithms are given for basic integer and rational number operations, automatic (or default) simplification of algebraic expressions, greatest common divisor calculation for single and multivariate polynomials, resultant computation, polynomial decomposition, polynomial simplification with Gröbner bases, and polynomial factorization.

Topic Selection

The author of an introductory text about a rapidly changing field is faced with a difficult decision about which topics and algorithms to include in a text. This decision is constrained by the background of the audience, the mathematical difficulty of the material and, of course, by space limitations. In addition, we believe that an introductory text should really be an introduction to the subject that describes some of the important issues in the field but should not try to be comprehensive or include all refinements of a particular topic or algorithm. This viewpoint has guided the selection of topics, choice of algorithms, and level of mathematical rigor.

For example, polynomial gcd computation is an important topic in *Mathematical Methods* that plays an essential role in modern computer algebra software. We describe classical Euclidean algorithms for both single and multivariate polynomials with rational number coefficients and a Euclidean algorithm for single variable polynomials with simple algebraic number coefficients. It is well known, however, that, for efficiency reasons, these algorithms are not suitable as general purpose algorithms in a computer algebra system. For this reason, we describe the more advanced subresultant gcd algorithm for multivariate polynomials but omit the mathematical justification, which is quite involved and far outside the scope and spirit of these books.

One topic that is not discussed is the asymptotic complexity of the time and space requirements of algorithms. Complexity analysis for computer algebra, which is often quite involved, uses techniques from algorithm analysis, probability theory, discrete mathematics, the theory of computation, and other areas that are well beyond the background of the intended audience. Of course, it is impossible to ignore efficiency considerations entirely and, when appropriate, we indicate (usually by example) some of the issues that arise. A course based on *Mathematical Methods* is an ideal prerequisite for a graduate level course that includes the complexity analysis of algorithms along with recent developments in the field.[1]

Chapter Summaries

A more detailed description of the material covered in these books is given in the following chapter summaries.

Elementary Algorithms

Chapter 1: Introduction to Computer Algebra. This chapter is an introduction to the field of computer algebra. It illustrates both the possibilities and limitations for computer symbolic computation through dialogues with a number of commercial computer algebra systems.

---

[1]A graduate level course could be based on one of the books: Akritas [2], Geddes, Czapor, and Labahn [39], Mignotte [66], Mignotte and Ştefănescu [67], Mishra [68], von zur Gathen and Gerhard [96], Winkler [101], Yap [105], or Zippel [108].

Chapter 2: Elementary Concepts of Computer Algebra. This chapter introduces an algorithmic language called *mathematical pseudo-language* (or simply MPL) that is used throughout the books to describe the concepts, examples, and algorithms of computer algebra. MPL is a simple language that can be easily translated into the structures and operations available in modern computer algebra languages. This chapter also includes a general description of the evaluation process in computer algebra software (including automatic simplification) and a case study which includes an MPL program that obtains the change of form of quadratic expressions under rotation of coordinates.

Chapter 3: Recursive Structure of Mathematical Expressions. This chapter is concerned with the internal tree structure of mathematical expressions. Both the conventional structure (before evaluation) and the simplified structure (after evaluation and automatic simplification) are described. The structure of automatically simplified expressions is important because all algorithms assume that the input data is in this form. Four primitive MPL operators (*Kind*, *Operand*, *Number_of_operands*, and *Construct*) that analyze and construct mathematical expressions are introduced. The chapter also includes a description of four MPL operators (*Free_of*, *Substitute*, *Sequential_substitute*, and *Concurrent_substitute*) which depend only on the tree structure of an expression.

Chapter 4: Elementary Mathematical Algorithms. In this chapter we describe the basic programming structures in MPL and use these structures to describe a number of elementary algorithms. The chapter includes a case study which describes an algorithm that solves a class of first order ordinary differential equations using the separation of variables technique and the method of exact equations with integrating factors.

Chapter 5: Recursive Algorithms. This chapter describes recursion as a programming technique in computer algebra and gives a number of examples that illustrate its advantages and limitations. It includes a case study that describes an elementary integration algorithm which finds the antiderivatives for a limited class of functions using the linear properties of the integral and the substitution method. Extensions of the algorithm to include the elementary rational function integration, some trigonometric integrals, elementary integration by parts, and one algebraic function form are described in the exercises.

Chapter 6: Structure of Polynomials and Rational Expressions. This chapter is concerned with the algorithms that analyze and manipulate polynomials and rational expressions. It includes computational definitions for various classes of polynomials and rational expressions that are based on the internal tree structure of expressions. Algorithms based on the primitive operations introduced in Chapter 3 are given for degree and coefficient computation, coefficient collection, expansion, and rationalization of algebraic expressions.

Chapter 7: Exponential and Trigonometric Transformations. This chapter is concerned with algorithms that manipulate exponential and trigonometric functions. It includes algorithms for exponential expansion and reduction, trigonometric expansion and reduction, and a simplification algorithm that can verify a large class of trigonometric identities.

Mathematical Methods

Chapter 1: Background Concepts. This chapter is a summary of the background material from *Elementary Algorithms* that provides a framework for the mathematical and computational discussions in the book. It includes a description of the mathematical psuedo-language (MPL), a brief discussion of the tree structure and polynomial structure of algebraic expressions, and a summary of the basic mathematical operators that appear in our algorithms.

Chapter 2: Integers, Rational Numbers, and Fields. This chapter is concerned with the numerical objects that arise in computer algebra, including integers, rational numbers, and algebraic numbers. It includes Euclid's algorithm for the greatest common divisor of two integers, the extended Euclidean algorithm, the Chinese remainder algorithm, and a simplification algorithm that transforms an involved arithmetic expression with integers and fractions to a rational number in standard form. In addition, it introduces the concept of a field which describes in a general way the properties of number systems that arise in computer algebra.

Chapter 3: Automatic Simplification. Automatic simplification is defined as the collection of algebraic and trigonometric simplification transformations that are applied to an expression as part of the evaluation process. In this chapter we take an in-depth look at the algebraic component of this process, give a precise definition of an automatically simplified expression, and describe an (involved) algorithm that transforms mathematical expressions to automatically simplified form. Although automatic simplification is essential for the operation of computer algebra software, this is the only detailed treatment of the topic in the textbook literature.

Chapter 4: Single Variable Polynomials. This chapter is concerned with algorithms for single variable polynomials with coefficients in a field. All algorithms in this chapter are ultimately based on polynomial division. It includes algorithms for polynomial division and expansion, Euclid's algorithm for greatest common divisor computation, the extended Euclidean algorithm, and a polynomial version of the Chinese remainder algorithm. In addition, the basic polynomial division and gcd algorithms are used to give algorithms for numerical computations in elementary algebraic number fields. These algorithms are then used to develop division and gcd algorithms for polynomials with algebraic number coefficients. The chapter concludes with an algorithm for partial fraction expansion that is based on the extended Euclidean algorithm.

Chapter 5: Polynomial Decomposition. Polynomial decomposition is a process that determines if a polynomial can be represented as a composition of lower degree polynomials. In this chapter we discuss some theoretical aspects of the decomposition problem and give an algorithm based on polynomial factorization that either finds a decomposition or determines that no decomposition exists.

Chapter 6: Multivariate Polynomials. This chapter generalizes the division and gcd algorithms to multivariate polynomials with coefficients in an integral domain. It includes algorithms for three polynomial division operations (recursive division, monomial-based division, and pseudo-division), polynomial expansion (including an application to the algebraic substitution problem), and the primitive and subresultant algorithms for gcd computation.

Chapter 7: The Resultant. This chapter introduces the resultant of two polynomials, which is defined as the determinant of a matrix whose entries depend on the coefficients of the polynomials. We describe a Euclidean algorithm and a subresultant algorithm for resultant computation and use the resultant to find polynomial relations for explicit algebraic numbers.

Chapter 8: Polynomial Simplification with Side Relations. This chapter includes an introduction to Gröbner basis computation with an application to the polynomial simplification problem. To simplify the presentation, we assume that polynomials have rational number coefficients and use the lexicographical ordering scheme for monomials.

Chapter 9: Polynomial Factorization. The goal of this chapter is the description of a basic version of a modern factorization algorithm for single variable polynomials in $Q[x]$. It includes square-free factorization algorithms (in $Q[x]$ and $Z_p[x]$), Kronecker's classical factorization algorithm for $Z[x]$, Berlekamp's algorithm for factorization in $Z_p[x]$, and a basic version of the Hensel lifting algorithm.

Computer Algebra Software and Programs

We use a procedure style of programming that corresponds most closely to the programming structures and style of the Maple, Mathematica, and MuPAD systems and, to a lesser degree, to the Macsyma and Reduce systems. In addition, some algorithms are described by transformation rules that translate to the pattern matching languages in the Mathematica and Maple systems. Unfortunately, the programming style used here does not translate easily to the structures in the Axiom system.

The dialogues and algorithms in these books have been implemented in the Maple 7.0, Mathematica 4.1, and MuPAD Pro (Version 2.0) systems. The dialogues and programs are found on a CD included with the books. In each book, available dialogues and programs are indicated by the word "Implementation" followed by a system name Maple, Mathematica, or MuPAD. System dialogues are in a notebook format (mws in Maple, nb in Mathematica, and mnb in MuPAD), and procedures are in text (ASCII) format. (For example, see the dialogue in Figure 3.4 on page 72 and the procedure in Figure 2.2 on page 23.) In some examples, the dialogue display of a computer algebra system given in the text has been modified so that it fits on the printed page.

Electronic Version of the Book

These books have been processed in the LATEX 2_ε system with the *hyperref* package, which allows hypertext links to chapter numbers, section numbers, displayed (and numbered) formulas, theorems, examples, figures, footnotes, exercises, the table of contents, the index, the bibliography, and web sites. An electronic version of the book (as well as additional reference files) in the portable document format (PDF), which is displayed with the Adobe Acrobat software, is included on the CD.

Acknowledgements

I am grateful to the many students and colleagues who read and helped debug preliminary versions of this book. Their advice, encouragement, suggestions, criticisms, and corrections have greatly improved the style and structure of the book. Thanks to Norman Bleistein, Andrew Burt, Alex Champion, the late Jack Cohen, Robert Coombe, George Donovan, Bill Dorn, Richard Fateman, Clayton Ferner, Carl Gibbons, Herb Greenberg, Jillane Hutchings, Lan Lin, Zhiming Li, Gouping Liu, John Magruder, Jocelyn Marbeau, Stanly Steinberg, Joyce Stivers, Sandhya Vinjamuri, and Diane Wagner.

I am grateful to Gwen Diaz and Alex Champion for their help with the LATEX document preparation; Britta Wienand, who read most of the text and translated many of the programs to the MuPAD language; Aditya Nagrath, who created some of the figures; and Michael Wester who translated many of the programs to the Mathematica, MuPAD, and Macsyma languages. Thanks to Camie Bates, who read the entire manuscript and made numerous suggestions that improved the exposition, notation, and helped clarify confusing sections of the book. Her careful reading discovered numerous typographical, grammatical, and mathematical errors.

I also acknowledge the long-term support and encouragement of my home institution, the University of Denver. During the writing of the book, I was awarded two sabbatical leaves to develop this material.

Special thanks to my family for encouragement and support: my late parents Elbert and Judith Cohen, Daniel Cohen, Fannye Cohen, and Louis and Elizabeth Oberdorfer.

Finally, I would like to thank my wife, Kathryn, who as long as she can remember, has lived through draft after draft of this book, and who with patience, love, and support has helped make this book possible.

Joel S. Cohen
Denver, Colorado
November 19, 2002

1

Background Concepts

In this chapter we summarize the background material that provides a framework for the mathematical and computational discussions in the book. A more detailed discussion of this material can be found on the CD that accompanies this book and in our companion book, *Computer Algebra and Symbolic Computation, Elementary Algorithms*, (Cohen [24]). Readers who are familiar with this material may wish to skim this chapter and refer to it as needed.

1.1 Computer Algebra Systems

A *computer algebra system* (CAS) or *symbol manipulation system* is a computer program that performs symbolic mathematical operations. In this book we refer to the computer algebra capabilities of the following three systems which are readily available and support a programming style that is most similar to the one used here:

- **Maple** – a very large CAS originally developed by the Symbolic Computation Group at the University of Waterloo (Canada) and now distributed by Waterloo Maple Inc. Information about Maple is found in Heck [45] or at the web site `http://www.maplesoft.com`.

- **Mathematica** – a very large CAS developed by Wolfram Research Inc. Information about Mathematica can be found in Wolfram [102] or at the web site `http://www.wolfram.com`.

- **MuPAD** – a large CAS developed by the University of Paderborn (Germany) and SciFace Software GmbH & Co. KG. Information about MuPAD can be found in Gerhard et al. [40] or at the web site `http://www.mupad.com`.

1.2 Mathematical Pseudo-Language (MPL)

Mathematical pseudo-language (MPL) is an algorithmic language that is used throughout this book to describe the concepts, examples, and algorithms of computer algebra. MPL algorithms are readily expressed in the programming languages of Maple, Mathematica, and MuPAD, and implementations of the dialogues and algorithms in these systems are included on the CD that accompanies this book.

Mathematical Expressions

MPL *mathematical expressions* are constructed with the following symbols and operators:

- *Integers* and *fractions* that utilize infinite precision rational number arithmetic.

- *Identifiers* that are used both as *programming variables* that represent the result of a computation and as *mathematical symbols* that represent indeterminates (or variables) in a mathematical expression.

- The *algebraic operators* $+$, $-$, $*$, $/$, $\wedge$ (power), and ! (factorial). (As with ordinary mathematical notation, we usually omit the $*$ operator and use raised exponents for powers.)

- *Function forms* that are used for mathematical functions ($\sin(x)$, $\exp(x)$, $\arctan(x)$, etc.), mathematical operators (*Expand(u)*, *Factor(u)*, *Integral(u,x)*, etc.), and undefined functions ($f(x)$, $g(x,y)$, etc.).

- The *relational operators* $=$, $\neq$, $<$, $\leq$, $>$, and $\geq$, the *logical constants* **true** and **false**, and the *logical operators*, **and**, **or**, and **not**.

- *Finite sets* of expressions that utilize the set operations $\cup$, $\cap$, $\sim$ (set difference), and $\in$ (set membership). Following mathematical convention, sets do not contain duplicate elements and the contents of a set does not depend on the order of the elements (e.g., $\{a, b\} = \{b, a\}$).

- *Finite lists* of expressions. A list is represented using the brackets [and] (e.g., $[1, x, x^2]$). The empty list, which contains no expressions, is represented by [ ]. Lists may contain duplicate elements, and the order of elements is significant (e.g., $[a, b] \neq [b, a]$).

The MPL set and list operators and the corresponding operators in computer algebra systems are given in Figure 1.1.

MPL mathematical expressions have two (somewhat overlapping) roles as either *program statements* that represent a computational step in a program or as *data objects* that are processed by program statements.

Assignments, Functions, and Procedures

The MPL assignment operator is a colon followed by an equal sign ($:=$) and an assignment statement has the form $f := u$ where u is a mathematical expression.

An MPL *function definition* has the form $f(x_1, \ldots, x_l) \overset{\text{function}}{:=} u$, where $x_1, \ldots, x_l$ is a sequence of symbols called the *formal parameters*, and u is a mathematical expression. MPL *procedures* extend the function concept to *mathematical operators* that are defined by a sequence of statements. The general form of an MPL procedure is given in Figure 1.2. Functions and procedures are invoked with an expression of the form $f(a_1, \ldots, a_l)$, where $a_1, \ldots, a_l$ is a sequence of mathematical expressions called the *actual parameters*.

In order to promote a programming style that works for all languages, we adopt the following conventions for the use of local variables and formal parameters in a procedure:

- An unassigned local variable cannot appear as a symbol in a mathematical expression. In situations where a procedure requires a local (unassigned) mathematical symbol, we either pass the symbol through the parameter list or use a global symbol.

- Formal parameters are used only to transmit data into a procedure and not as local variables or to return data from a procedure. When we need to return more than one expression from a procedure, we return a set or list of expressions.

Decision and Iteration Structures

MPL provides three decision structures: the **if** structure, the **if-else** structure which allows for two alternatives, and the *multi-branch* decision structure which allows for a sequence of alternatives.

MPL contains two iteration structures that allow for repeated evaluation of a sequence of statements, the **while** structure and the **for** structure. Some of our procedures contain **for** loops that include a *Return* statement.

MPL	Maple	Mathematica	MuPAD
set notation $\{a, b, c\}$	$\{a,b,c\}$	$\{a,b,c\}$	$\{a,b,c\}$
$\emptyset$	$\{\}$	$\{\}$	$\{\}$
$A \cup B$ (set union)	A union B	Union[A,B]	A union B
$A \cap B$ (set intersection)	A intersect B	Intersection[A,B]	A intersect B
$A \sim B$ (set difference)	A minus B	Complement[A,B]	A minus B
$x \in A$ (set membership)	member(x, A)	MemberQ[x,A]	contains(A,x)

(a) Sets. (Implementation: Maple (mws), Mathematica (nb), MuPAD (mnb).)

MPL	Maple	Mathematica	MuPAD
list notation $[a, b, c]$	[a,b,c]	$\{a,b,c\}$	[a,b,c]
empty list []	[]	$\{\}$	[]
$First(L)$ (first member of L)	op(1,L)	First[L]	op(L,1)
$Rest(L)$ (a new list with first member of L removed)	[op(2..nops(L),L)]	Rest[L]	[op(L,2..nops(L))]
$Adjoin(x, L)$ (a new list with x adjoined to the beginning of L)	[x,op(L)]	Prepend[L,x]	[x, op(L)]
$Join(L, M)$ (a new list with members of L followed by members of M)	[op(L),op(M)]	Join[L,M]	_concat(L,M)
$x \in L$ (list membership)	member(x,L)	MemberQ[x,L]	contains(L,x)

(b) Lists. (Implementation: Maple (mws), Mathematica (nb), MuPAD (mnb).)

Figure 1.1. MPL set and list operations in CAS languages.

Procedure $f(x_1, \ldots, x_l)$;
Input
    x_1 : description of input to x_1;

     $\vdots$

    x_l : description of input to x_l;
Output
    description of output;
Local Variables
    $v_1, \ldots, v_m$;
Begin
    S_1;

     $\vdots$

    S_n
End

Figure 1.2. The general form of an MPL procedure. (Implementation: Maple (txt), Mathematica (txt), MuPAD (txt).)

In this case, we intend that both the loop and the procedure terminate when the *Return* is encountered.[1]

All computer algebra languages provide decision and iteration structures (Figure 1.3).

1.3 Automatic Simplification and Expression Structure

As part of the evaluation process, computer algebra systems apply some "obvious" simplification rules from algebra and trigonometry that remove extraneous symbols from an expression and transform it to a standard form. This process is called *automatic simplification*. For example,

$$x + 2x + y\,y^2 + z^0 + \sin(\pi/4) \to 3\,x + y^3 + 1 + \sqrt{2}/2$$

where the expression to the right of the arrow gives the automatically simplified form after evaluation.

In MPL (as in a CAS), all expressions in dialogues and computer programs operate in the *context of automatic simplification*. This means: (1)

---

[1] The **for** statements in both Maple and MuPAD work in this way. However, in Mathematica, a **Return** in a **For** statement will only work in this way if the upper limit contains a relational operator (e.g., **i<=N**). (Implementation: Mathematica (nb).)

MPL	Maple	Mathematica	MuPAD
if	if	If	if
if-else	if-else	If	if-else
if-elseif-else (multi-branch)	if-elseif-else	Which	if-elseif-else

(a) Decision Structures. (Implementation: Maple (mws), Mathematica (nb), MuPAD (mnb).)

MPL	Maple	Mathematica	MuPAD
for	for	For	for
while	while	While	while

(b) Iteration Structures. (Implementation: Maple (mws), Mathematica (nb), MuPAD (mnb).)

Figure 1.3. MPL decision structures and iteration structures and the corresponding structures in Maple, Mathematica, and MuPAD.

all input operands to mathematical operators are automatically simplified before the operators are applied; (2) the result obtained by evaluating an expression is in automatically simplified form.

Expression Structure

The *structure* of an expression involves the relationships between the operators and operands that make up the expression. Since mathematical expressions are the data objects in computer algebra programming, an understanding of this structure is essential.

An *expression tree* is a diagram that displays this structure. For example, the expression $c + d * x \wedge 2$ is represented by the expression tree in Figure 1.4. The operator at the root of the tree is called the *main operator* of the expression, a designation that emphasizes that $c + d * x \wedge 2$ is viewed as a sum with two operands c and $d * x \wedge 2$.

Algebraic Expressions

The *algebraic expressions* are constructed using integers, symbols, function forms, and the algebraic operators $(+, -, *, /, \wedge,$ and $!)$. For example, the expression $x^2 + \cos(x) + f(x, y, z)$ is an algebraic expression, while $[a, b, c]$ and $x + y = 2$ are not. The evaluation process modifies the structure of these expressions to a form where the algebraic operators satisfy the following properties:

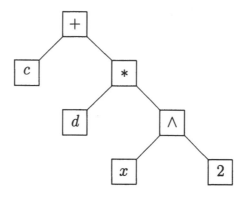

Figure 1.4. The expression tree for $c + d * x \wedge 2$.

- The operator $+$ is an n-ary infix operator with two or more operands and none of its operands is a sum. In addition, at most one operand of $+$ is an integer or fraction.

- The operator $*$ is an n-ary infix operator with two or more operands and none of its operands is a product. In addition, at most one operand of $*$ is an integer or fraction, and when an integer or fraction is an operand of a product, it is the first operand.[2]

- The unary operator $-$ and the binary operator $-$ do not appear in simplified expressions. Unary differences are represented as products (e.g. $-x \to (-1)\,x$) and binary differences as sums (e.g. $(a - b \to a + (-1)\,b)$.

- The binary operator $/$ does not appear in simplified expressions. Quotients are represented as either products (e.g., $a/b \to a\,b^{-1}$), powers (e.g., $1/a^2 \to a^{-2}$), or numerical fractions (described below).

- A quotient that represents a fraction c/d, where c and $d \neq 0$ are integers is represented by an expression tree with root the symbol **fraction**, first operand c, and second operand d.

---

[2]In both Maple and Mathematica, an integer or fraction operand in a product is the first operand. In MuPAD, however, an integer or fraction operand in a product is represented internally as the last operand even though the displayed form indicates it is the first operand. Since some algorithms in later chapters assume that an integer or fraction in a product is the first operand, the MuPAD implementations are modified to account for this difference.

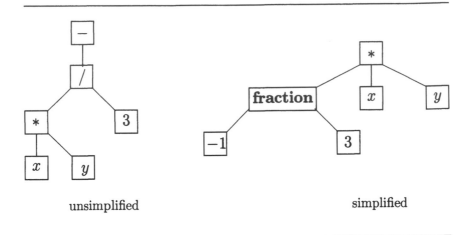

unsimplified   simplified

Figure 1.5. Expression trees for $-x*y/3$ and its simplified form $(-1/3)*x*y$.

- For $u = v \wedge n$, where n is an integer, the expression v is not an integer, fraction, product, or power (e.g., $(x^2)^3 \to x^6$).

- The operand of ! is not a non-negative integer (e.g., $3! \to 6$).

Figure 1.5 shows the expression trees for the expression $-x*y/3$ and its simplified form $((-1)/3)*x*y$. Observe that in simplified form, the operator $*$ is an n-ary operator with three operands, the $-$ is part of the integer -1, and the fraction $(-1)/3$ has main operator **fraction**.

The structure of algebraic expressions is described in detail in Cohen [24], Chapter 3. *Non-algebraic expressions* include relational and logical expressions, lists, and sets. The structure of these expressions in a particular CAS can be determined using the primitive operators in Figure 1.6.

Primitive Operators for Simplified Mathematical Expressions

MPL uses four primitive operators to access the structure of expressions and to construct expressions.

- *Kind(u)*. This operator returns the type of expression (e.g., **symbol, integer, fraction, +, *, $\wedge$, !, =, <, $\leq$, >, $\geq$, $\neq$, and, or, not, set, list**, and function names). For example, $Kind(m*x+b) \to +$.

- *Number_of_operands(u)*. This operator returns the number of operands of the main operator of u. For example,

$$Number\_of\_operands(a*x+b*x+c) \to 3.$$

- *Operand*(u, i). This operator returns the ith operand of the main operator of u. For example, *Operand*$(m * x + b, 2) \rightarrow b$.

- *Construct*(f, L). Let f be an operator ($+$, $*$, $=$, etc.) or a symbol, and let $L = [a, b, \ldots, c]$ be a list of expressions. This operator returns an expression with main operator f and operands $a, b, \ldots, c$. For example, *Construct*$(" + ", [a, b, c]) \rightarrow a + b + c$.

The primitive operators in computer algebra systems are given in Figure 1.6(a). Although Mathematica has an operator that constructs expressions, Maple and MuPAD do not. However, in both of these languages, the operation can be simulated with a procedure. (Implementation: Maple (txt), MuPAD (txt).)

MPL	Maple	Mathematica	MuPAD
Kind(u)	`whattype(u)` and `op(0,u)` for function names	`Head(u)`	`type(u)` and `op(u,0)` for undefined function names
Operand(u, i)	`op(i,u)`	`Part[u,i]` and `Numerator[u]` and `Denominator[u]` for fractions	`op(u,i)`
Number_of_operands(u)	`nops(u)`	`Length[u]`	`nops(u)`
Construct(f, L)	(simulated with a procedure)	`Apply[f,L]`	(simulated with a procedure)

(a) Primitive Structural Operators. (Implementation: Maple (mws), Mathematica (nb), MuPAD (mnb).)

MPL	Maple	Mathematica	MuPAD
Free_of(u, t)	`not(has(u,t))`	`FreeQ[u,x]`	`not(has(u,t))`
Substitute$(u, t=r)$	`subs(t=r,u)`	`ReplaceAll[u,t->r]` or `u/.t->r`	`subs(u,t=r)`

(b) Structure-based Operators. (Implementation: Maple (mws), Mathematica (nb), MuPAD (mnb).)

Figure 1.6. Operators in Maple, Mathematica, and MuPAD that are most similar to MPL's primitive structural operators and structure-based operators.

Structure-Based Operators

A *complete sub-expression* of an automatically simplified expression u is
either the expression u itself or an operand of some operator in u. In
terms of expression trees, the complete sub-expressions of u are either the
expression tree for u or one of its sub-trees. For example, for $a * (1+b+c)$,
the expression $1 + b + c$ is a complete sub-expression while $1 + b$ is not.

The next two MPL operators are based only on the structure of an
expression.

- *Free_of*(u, t). Let u and t (for target) be mathematical expressions.
  This operator returns **false** when t is identical to some complete sub-
  expression of u and otherwise returns **true**. For example,

$$Free\_of\,((a + b)\,c,\ a + b) \to \textbf{false}.$$

- *Substitute*$(u, t = r)$. Let u, t, and r be mathematical expressions.
  This operator forms a new expression with each occurrence of the
  target expression t in u replaced by the replacement expression r. The
  substitution occurs whenever t is structurally identical to a complete
  sub-expression of u. For example,

$$Substitute((a + b)\,c,\ a + b = x) \to x\,c.$$

The operators in computer algebra systems that are most similar to MPL's
structure-based operators are given in Figure 1.6(b).

The *Map* Operator

The *Map* operator provides another way to apply an operator to all operands
of the main operator of an expression. Let u be a mathematical expression with $n = $ *Number_of_operands*$(u) \geq 1$, and let $F(x)$ and $G(x, y, \ldots, z)$
be operators. The MPL *Map* operator has the two forms $Map(F, u)$ and
$Map(G, u, y, \ldots, z)$. The statement $Map(F, u)$ returns a new expression
with main operator *Kind*(u) and operands

$$F(\,Operand(u, 1)), \ldots, F(\,Operand(u, n)).$$

The statement $Map(G, u, y, \ldots, z)$ returns an expression with main operator *Kind*(u) and operands

$$G(\,Operand(u, 1), y, \ldots, z), \ldots, G(\,Operand(u, n), y, \ldots, z).$$

The *Map* operators in CAS languages are given in Figure 1.7.

MPL	Maple	Mathematica	MuPAD
$Map(F, u)$	`Map(F,u)`	`Map[F,u]`	`Map(u,F)`
$Map(G, u, y, z)$	`Map(G,u,y,z)`	`Map[G[#,y,z]&,u]`	`Map(u,G,y,z)`

Figure 1.7. The MPL *Map* operator and the corresponding operators in CAS languages. (Implementation: Maple (mws), Mathematica (nb), MuPAD (mnb).)

1.4 General Polynomial Expressions

In this section we describe the polynomial structure of an algebraic expression and MPL's primitive polynomial operators that obtain this structure.

Let $c_1, c_2, \ldots, c_r$ be algebraic expressions, and let $x_1, x_2, \ldots, x_m$ be algebraic expressions that are not integers or fractions. A *general monomial expression* (GME) in $\{x_1, x_2, \ldots, x_m\}$ has the form

$$c_1 c_2 \cdots c_r \, x_1^{n_1} x_2^{n_2} \cdots x_m^{n_m},$$

where the exponents n_j are non-negative integers and each c_i satisfies the independence property

$$Free\_of(c_i, x_j) \rightarrow \textbf{true}, \; j = 1, 2, \ldots, m. \tag{1.1}$$

The expressions x_j are called *generalized variables* because they mimic the role of variables, and the expressions c_i are called *generalized coefficients* because they mimic the role of coefficients. The expression $x_1^{n_1} \cdots x_m^{n_m}$ is called the *variable part* of the monomial, and if there are no generalized variables in the monomial, the variable part is 1. The expression $c_1 \cdots c_r$ is called the *coefficient part* of the monomial, and if there are no generalized coefficients in the monomial, the coefficient part is 1.

An expression u is a *general polynomial expression* (GPE) if it is either a GME or a sum of GMEs in $\{x_1, x_2, \ldots, x_m\}$. For example, $3\,a\,b\,x^2 \sin(x) - c\,x + d$ is a GPE with $x_1 = x$ and $x_2 = \sin(x)$. On the other hand, the expression $(\sin(x))\,x^2 + (\ln(x))\,x$ is not a GPE in x alone because the coefficients of powers of x ($\sin(x)$ and $\ln(x)$) do not satisfy the independence property in Equation 1.1.

Primitive Polynomial Operators

The following operators obtain the polynomial structure of an expression.

- *Polynomial_gpe(u, v)*.   Let u be an algebraic expression and let v be either a generalized variable x or a set S of generalized variables. The MPL operator *Polynomial_gpe(u, v)* returns **true** whenever u is a GPE in $\{x\}$ or in S, and otherwise returns **false**. For example, $Polynomial\_gpe(x^2 + y^2, \{x, y\}) \to$ **true**.

- $Degree\_gpe(u, x)$. Let $u = c_1 \cdots c_r \cdot x_1^{n_1} \cdots x_m^{n_m}$ be a monomial with non-zero coefficient part. The **degree** of u with respect to x_i is denoted by $\deg(u, x_i) = n_i$. By mathematical convention, the degree of the 0 monomial is $-\infty$. If u is a GPE in x_i that is a sum of monomials, then $\deg(u, x_i)$ is the maximum of the degrees of the monomials. If the generalized variable x_i is understood from context, we use the simpler notation $\deg(u)$. The MPL operator $Degree\_gpe(u, x)$ returns $\deg(u, x)$. For example, $Degree\_gpe(\sin^2(x) + b\sin(x) + c, \ \sin(x)) \to 2$.

- $Coefficient\_gpe(u, x, j)$. Let u be a GPE in x, and let j be a non-negative integer. The MPL operator $Coefficient\_gpe(u, x, j)$ returns the sum of the coefficient parts of all monomials of u with variable part x^j. For example, $Coefficient\_gpe(a\,x + b\,x + y, x, 1) \to a + b$.

- $Leading\_coefficient\_gpe(u, x)$. Let u be a GPE in x. The leading coefficient of a GPE $u \neq 0$ with respect to x is the sum of the coefficient parts of all monomials with variable part $x^{\deg(u, x)}$. The zero polynomial has, by definition, leading coefficient zero. The leading coefficient is represented by $\operatorname{lc}(u, x)$, and when x is understood from context by $\operatorname{lc}(u)$. The MPL operator $Leading\_coefficient\_gpe(u, x)$ returns $\operatorname{lc}(u, x)$. For example, $Leading\_coefficient\_gpe(a\,x + b\,x + y, x) \to a + b$.

- $Variables(u)$. The polynomial structure of an algebraic expression u depends on which expressions are chosen as the generalized variables. The MPL operator $Variables(u)$ selects a set of generalized variables so that the coefficients of all monomials in u are rational numbers. For example, $Variables(4x^3 + 3x^2 \sin(x)) \to \{x, \sin(x)\}$.

The operators in computer algebra languages that are most similar to MPL's polynomial operators are given in Figure 1.8.

1.5   Miscellaneous Operators

Some additional MPL operators that are used in our algorithms and exercises are given in Figure 1.9. Many more operators are defined in later chapters.

MPL	Maple	Mathematica	MuPAD
Polynomial_gpe(u,x)	type(u,polynom(anything,x))	PolynomialQ[u,x]	testtype(u, Type::PolyExpr(x))
Degree_gpe(u,x)	degree(u,x)	Exponent[u,x]	degree(u,x)
Coefficient_gpe(u,x,n)	coeff(u,x,n)	Coefficient[u,x,n]	coeff(u,x,n)
Leading_coefficient_gpe(u,x,n)	lcoeff(u,x)	Coefficient[u,x, Exponent[u,x]]	lcoeff(u,x)
Variables(u)	indets(u)	Variables[u]	indets(u,PolyExpr)

Figure 1.8. The polynomial operators in Maple, Mathematica, and MuPAD that are most similar to those in MPL. (Implementation: Maple (mws), Mathematica (nb), MuPAD (mnb).)

MPL	Maple	Mathematica	MuPAD
Return(u)	RETURN(u)	Return[u]	return(u)
Operand_list(u)	[op(u)]	Apply[List,u]	[op(u)]
Absolute_value(u) $\|u\|$	abs(u)	Abs[u]	abs(u)
Max({$n_1,\ldots,n_r$})	max($n_1,\ldots,n_r$)	Max[$n_1,\ldots,n_r$]	max($n_1,\ldots,n_r$)
Algebraic_expand(u)	expand(u)	Expand[u]	expand(u)
Numerator(u)	numer(u)	Numerator[u]	numer(u)
Denominator(u)	denom(u)	Denominator[u]	denom(u)
Derivative(u,x)	diff(u,x)	D[u,x]	diff(u,x)

Figure 1.9.   Miscellaneous   operators.   (Implementation: Maple   (mws),
Mathematica (nb), MuPAD (mnb).)

Further Reading

1.1 Computer Algebra Systems. Additional information on computer alge-
bra can be found in Akritas [2], Buchberger et al. [17], Davenport, Siret, and
Tournier [29],   Geddes, Czapor, and Labahn [39],Lipson [64], Mignotte [66],
Mignotte and Ştefănescu [67], Mishra [68], von zur Gathen and Gerhard [96],
Wester [100], Winkler [101], Yap [105], and Zippel [108].   Two older (but in-
teresting) discussions of computer algebra are found in Pavelle, Rothstein, and
Fitch [77] and Yun and Stoutemyer [107]. Simon ([89], and [90]), Wester [100]
(Chapter 3), and the web site

http://math.unm.edu/~wester/cas_review.html

give comparisons of commercial computer algebra software. Information about
computer algebra and computer algebra systems can be found at the web sites:

- SymbolicNet: http://www.SymbolicNet.org.
- Computer Algebra Information Network (CAIN):
  http://www.riaca.win.tue.nl/CAN/
- COMPUTER ALGEBRA, Algorithms, Systems and Applications:
  http://www-troja.fjfi.cvut.cz/~liska/ca/
- sci.math.symbolic:
  http://mathforum.org/discussions/about/sci.math.symbolic.html

The Association for Computing Machinery (ACM) has a *Special Interest Group
on Symbolic and Algebraic Manipulation (SIGSAM)*. This group publishes a quar-
terly journal the *SIGSAM Bulletin* which provides a forum for exchanging ideas
about computer algebra. In addition, SIGSAM sponsors an annual conference,
the *International Symposium on Symbolic and Algebraic Computation (ISSAC)*.
Information about SIGSAM is found at http://www.acm.org/sigsam. The main
research journal in computer algebra is the *Journal of Symbolic Computation*
(http://www.academicpress.com/jsc).

1.2 Mathematical Pseudo-language (MPL). The basic elements of MPL are described in Cohen [24], Chapter 2, and the basic concepts in computer algebra programming are described in Chapters 4 and 5.

1.3 Automatic Simplification and Expression Structure. The evaluation process and the structure of expressions is described in greater detail in Cohen [24], Chapter 3, and an algorithm for the *Free_of* operator is given in Chapter 5.

1.4 General Polynomial Expressions. Algorithms for the operators in this section are given in Cohen [24], Chapter 6.

1.5 Miscellaneous Operators. Cohen [24] has algorithms for *Algebraic_expand*, *Numerator*, and *Denominator* (Chapter 6), and *Derivative* (Chapter 5).

2

Integers, Rational Numbers, and Fields

The chapter is concerned with the numerical objects that arise in computer algebra including the integers, the rational numbers, and other classes of numerical expressions. In Section 2.1 we discuss the basic mathematical properties of the integers and describe some algorithms that are important for computer algebra. Section 2.2 is concerned with the manipulation of rational numbers. We define a standard form for a rational number and describe an algorithm that evaluates involved arithmetic expressions with integers and fractions to a rational number in standard form. In Section 2.3 we introduce the concept of a field, which is a mathematical system with axioms that describe in a general way the algebraic properties of the rational numbers and other classes of expressions that arise in computer algebra. We give a number of examples of fields and show that many transformations that are routinely used in the manipulation of mathematical expressions are logical consequences of the field axioms.

2.1 The Integers

In this section we describe some mathematical and computational properties of the integers

$$\mathbf{Z} = \{\ldots - 2, -1, 0, 1, 2, \ldots\}.$$

The following theorem gives the basic division property of the integers.[1]

Theorem 2.1. *For integers a and $b \neq 0$, there are unique integers q and r such that*

$$a = qb + r \tag{2.1}$$

and

$$0 \leq r \leq |b| - 1. \tag{2.2}$$

*The integer q is the **quotient** and is represented by the operator* $\mathrm{iquot}(a, b)$ *(for integer quotient). The integer r is the **remainder** and is represented by* $\mathrm{irem}(a, b)$.

Example 2.2.

$$
\begin{aligned}
8 &= q \cdot 3 + r = 2 \cdot 3 + 2, \\
8 &= q \cdot (-3) + r = (-2) \cdot (-3) + 2, \tag{2.3} \\
-8 &= q \cdot 3 + r = (-3) \cdot 3 + 1, \tag{2.4} \\
-8 &= q \cdot (-3) + r = 3 \cdot (-3) + 1. \tag{2.5}
\end{aligned}
$$

$\square$

In Theorem 2.1, the quotient and remainder are chosen so that $r \geq 0$. Another possibility is to choose the quotient and remainder so that

$$|r| \leq |b| - 1, \qquad r \cdot a \geq 0. \tag{2.6}$$

In this case, Equations (2.4) and (2.5) have the form

$$
\begin{aligned}
-8 &= q \cdot 3 + r = (-2) \cdot 3 - 2, \\
-8 &= q \cdot (-3) + r = 2 \cdot (-3) - 2.
\end{aligned}
$$

A third possibility is to choose the quotient and remainder so that

$$|r| \leq |b| - 1, \qquad r \cdot b \geq 0. \tag{2.7}$$

In this case, Equations (2.3) and (2.5) have the form

$$
\begin{aligned}
8 &= q \cdot (-3) + r = (-3) \cdot (-3) - 1, \\
-8 &= q \cdot (-3) + r = 2 \cdot (-3) - 2.
\end{aligned}
$$

Most computer algebra languages have operators similar to iquot and irem, although the remainder may satisfy either property (2.6) or (2.7) instead of (2.2) (see Figure 2.1).

---

[1] A proof of the theorem based on the formal axioms of the integers is given in Dean [31], pages 10–11.

MPL	Maple	Mathematica	MuPAD
$\text{iquot}(a, b)$, (2.2)	iquo(a, b), (2.6)	$\text{Quotient}[a, b]$, (2.7)	iquo(a, b), (2.6) a div b, (2.2)
$\text{irem}(a, b)$, (2.2)	irem(a, b), (2.6) a mod b, (2.2)	$\text{Mod}[a, b]$, (2.7)	irem(a, b), (2.6) a mod b, (2.2)

Figure 2.1. Integer operations in Maple, Mathematica, and MuPAD. For each operator, the remainder satisfies the property indicated by the equation number to the right of the expression. (Implementation: Maple (mws), Mathematica (nb), MuPAD (mnb).)

Definition 2.3.

1. *The integer $b \neq 0$ is a **divisor** of (or **divides**) the integer a if there is an integer q such that $a = q \cdot b$. We use the notation $b \mid a$ to indicate that b divides a and $b \nmid a$ if it does not.*

2. *A **common divisor** (or **common factor**) of the integers a and b is an integer c such that $c \mid a$ and $c \mid b$.*

In other words, $b \mid a$ if and only if $\text{irem}(a, b) = 0$. For example, $3 \mid 18$, and a common divisor of 18 and 45 is 3.

The properties of the divisor operator are given in the next theorem.

Theorem 2.4. *Let a, b, and c be integers.*

1. *Suppose $a \neq 0$, $b \neq 0$, $b \mid a$, and $a \mid b$. Then $a = \pm b$.*

2. *Suppose $c \neq 0$, $c \mid a$, and $c \mid b$. Then $c \mid (a + b)$.*

3. *Suppose $c \neq 0$ and $c \mid a$. Then $c \mid (a \cdot b)$.*

4. *Suppose $a \neq 0$, $b \neq 0$, $a \mid b$, and $b \mid c$. Then $a \mid c$.*

Proof: We prove (1) and leave the proofs of the other properties to the reader (Exercise 4). By definition, there are integers q_1 and q_2 such that $b = q_1 a$ and $a = q_2 b$. This implies $b (1 - q_1 q_2) = 0$, and since $b \neq 0$, $q_1 q_2 = 1$. Therefore, both q_1 and q_2 are either $+1$ or -1, and $a = \pm b$. $\square$

Definition 2.5. *Two integers a and b are **relatively prime** if their only common divisors are 1 and -1.*

For example, 16 and 21 are relatively prime.

Greatest Common Divisors

The greatest common divisor of two integers a and b is the largest (non-negative) common divisor of a and b. Although this description is intuitively appealing, a more formal definition is helpful for the development of an algorithm.

Definition 2.6. *Let a and b be integers. The greatest common divisor (gcd) of a and b (at least one of which is non-zero) is an integer d that satisfies the following three properties:*

1. *d is a common divisor of a and b.*

2. *If e is another common divisor of a and b, then $e \mid d$.*

3. *$d > 0$.*

The notation $\gcd(a, b)$ denotes the greatest common divisor.

If both $a = 0$ and $b = 0$, the above definition does not apply. In this case, by definition, $\gcd(0, 0) = 0$.

Property 2 is a roundabout way of saying that d is the largest common divisor of a and b.

Example 2.7.

$$
\begin{aligned}
\gcd(24, 18) &= 6, \\
\gcd(-24, 18) &= 6, \\
\gcd(34, 0) &= 34, \\
\gcd(17, 23) &= 1.
\end{aligned}
$$

Let's formally verify that $\gcd(24, 18) = d = 6$ using Definition 2.6. Since properties (a) and (c) are obviously true, we need only verify (b). Dividing 24 by 18, we have $24 = 1 \cdot 18 + 6$, and so, by Theorem 2.4(2), any common divisor e of 24 and 18 also divides d. $\qquad\Box$

The greatest common divisor is used to reduce a fraction to lowest terms. For example, to reduce $18/24$, we have $\gcd(18, 24) = 6$, and divide the numerator and denominator by the gcd to obtain $3/4$.

The next theorem gives three important properties of the greatest common divisor.

Theorem 2.8. *Let a and b be integers. Then,*

1. $\gcd(a, b)$ *exists;*

2. $\gcd(a, b)$ *is unique;*

3. $\gcd(b, 0) = |b|$.

Proof: While Part (1) may appear obvious from the intuitive idea of a greatest common divisor, it is included here to emphasize that we should not simply assume that there is an integer d that satisfies the second property in Definition 2.6. We omit the proof of this fact, but note that the algorithm that computes the gcd given later in this section implies that the gcd exists.[2]

To show Part (2), suppose either $a \neq 0$ or $b \neq 0$, and suppose d_1 and d_2 are both greatest common divisors. Property (2) in Definition 2.6 implies $d_1 \mid d_2$ and $d_2 \mid d_1$, and so by Theorem 2.4(1), $d_1 = \pm d_2$. Since $d_1 > 0$ and $d_2 > 0$, we have $d_1 = d_2$. If both $a = 0$ and $b = 0$, the gcd is unique by definition.

The proof of Part (3) is left to the reader (Exercise 12). $\square$

Euclid's Greatest Common Divisor Algorithm

A simple (but highly inefficient) approach for finding the gcd is to test all integers less than or equal to $\min(\{|a|, |b|\})$. A much more efficient algorithm, which uses the remainders in integer division, is based on the next theorem:

Theorem 2.9. *Let a and b $\neq 0$ be integers, and let $r = \text{irem}(a, b)$. Then,*

$$\gcd(a, b) = \gcd(b, r). \tag{2.8}$$

Proof: Let $d = \gcd(b, r)$. We show that d is the gcd of a and b by showing that it satisfies the three properties in Definition 2.6. The proof is based on the relationship

$$a = q\, b + r. \tag{2.9}$$

First, since $d \mid b$ and $d \mid r$, Theorem 2.4(2),(3) implies $d \mid a$, and so d satisfies Definition 2.6(1). Next, if e is any divisor of a and b, then Equation (2.9) implies $e \mid r$, and so e is a common divisor of b and r. Therefore, Definition 2.6(2) (applied to b and r) implies $e \mid d$, which means

---

[2]A non-algorithmic proof that the gcd exits, which is based on the formal axioms of the integers, is given in Akritas [2], page 36.

Definition 2.6(2) holds for a and b as well. Finally, since d is a gcd, it is positive, and Definition 2.6(3) is satisfied.   □

The gcd algorithm that is based on Equation (2.8) is known as Euclid's algorithm. It is one of the earliest known numerical algorithms (about 300 B.C.E.). When $b \neq 0$, define a sequence of integers $R_{-1}, R_0, R_1, \ldots$ with the scheme

$$
\begin{aligned}
R_{-1} &= a, \\
R_0 &= b, \\
R_1 &= \text{irem}(R_{-1}, R_0), \\
&\ \vdots \\
R_{i+1} &= \text{irem}(R_{i-1}, R_i), \\
&\ \vdots
\end{aligned}
\tag{2.10}
$$

The sequence (2.10) is called an *integer remainder sequence*. By Theorem 2.1,

$$0 \leq \cdots < R_{i+1} < R_i < \cdots < R_1 \leq |b| - 1, \tag{2.11}$$

and so some member of the sequence is zero. Let R_ρ be the first remainder that is zero. By repeatedly applying Equation (2.8), we have

$$
\begin{aligned}
\gcd(a, b) &= \gcd(R_{-1}, R_0) \\
&= \gcd(R_0, R_1) \\
&\ \vdots \\
&= \gcd(R_{\rho-1}, R_\rho) \\
&= \gcd(R_{\rho-1}, 0) \\
&= |R_{\rho-1}|.
\end{aligned}
\tag{2.12}
$$

Notice that we have included the absolute value operation in (2.12) even though the remainders in (2.11) are all non-negative. There are, however, two cases where the absolute value is needed. First, when $b \neq 0$ and $b \mid a$, we have $\text{irem}(a, b) = 0$, and so the remainder sequence terminates with $R_1 = 0$. Therefore,

$$\gcd(a, b) = \gcd(b, 0) = |b| = |R_0|.$$

The second case involves $R_0 = b = 0$ which means there are no iterations, and so

$$\gcd(a, b) = \gcd(a, 0) = |a| = |R_{-1}|.$$

```
     Procedure   Integer_gcd(a, b);
     Input
        a, b : integers;
     Output
        gcd(a, b);
     Local Variables
        A, B, R;
     Begin
1       A := a;   B := b;
2       while   B ≠ 0 do
3          R := Irem(A, B);
4          A := B;
5          B := R;
6       Return( Absolute_value(A))
     End
```

Figure 2.2. An MPL procedure for Euclid's greatest common divisor algorithm. (Implementation: Maple (txt), Mathematica (txt), MuPAD (txt).)

We summarize this discussion with the following theorem.

Theorem 2.10. *Let a and b be integers. Then*

$$\gcd(a, b) = |R_{\rho-1}|, \tag{2.13}$$

where the absolute value operation is not needed when $\rho \geq 2$.

Example 2.11. Using Euclid's algorithm,

$$\gcd(45, 18) = \gcd(18, 9) = \gcd(9, 0) = 9. \qquad \square$$

A procedure that obtains the greatest common divisor using Euclid's algorithm given in Figure 2.2.

The Extended Euclidean Algorithm

We obtain a useful relationship involving a, b, and $\gcd(a, b)$ by applying a back-substitution process to the remainder sequence. For example, if the remainder sequence terminates with $R_4 = 0$, we have

$$a = Q_1 b + R_1, \tag{2.14}$$
$$b = Q_2 R_1 + R_2, \tag{2.15}$$

$$R_1 = Q_3R_2 + R_3, \qquad\qquad (2.16)$$
$$R_2 = Q_4R_3,$$

where $Q_i = \text{iquot}(R_{i-2}, R_{i-1})$ and

$$\gcd(a, b) = R_3.$$

Using Equation (2.16) to substitute for R_3 in this expression, we have

$$\gcd(a, b) = R_1 - Q_3R_2. \qquad\qquad (2.17)$$

Next, using Equations (2.14) and (2.15) to substitute for R_1 and R_2 in Equation (2.17), we obtain

$$\gcd(a, b) = m\,a + n\,b, \qquad\qquad (2.18)$$

where

$$m = 1 + Q_2Q_3, \quad n = -Q_1 - Q_3 - Q_1Q_2Q_3. \qquad (2.19)$$

This discussion suggests the following theorem.

Theorem 2.12. *For integers a and b, there are integers m and n such that*

$$m\,a + n\,b = \gcd(a, b).$$

A constructive proof of the theorem is given by an algorithm that computes m and n. (See the discussion following Theorem 2.14 and Equation (2.27).)

Example 2.13. Let $a = 45$ and $b = 18$. The remainder sequence is

$$R_{-1} = 45, \quad R_0 = 18, \quad R_1 = 9, \quad R_2 = 0.$$

Therefore, since $Q_1 = 2$,

$$\gcd(45, 18) = |R_1| = 9 = 45 - Q_1 \cdot 18 = 1 \cdot 45 + (-2) \cdot 18,$$

and so $m = 1$ and $n = -2$. $\square$

An algorithm that obtains m and n along with $\gcd(a, b)$ is called the *extended Euclidean algorithm* and is simply a formalization of the back-substitution process. The algorithm is based on the following theorem.

Theorem 2.14. *For integers a and b, there are integers m_i and n_i such that*

$$m_i\,a + n_i\,b = R_i, \quad i = -1, 0, 1, ..., \rho.$$

Proof: First, for $i = -1$ and $i = 0$,

$$R_{-1} = a = 1 \cdot a + 0 \cdot b = m_{-1} a + n_{-1} b, \tag{2.20}$$
$$R_0 = b = 0 \cdot a + 1 \cdot b = m_0 a + n_0 b, \tag{2.21}$$

and so

$$m_{-1} = 1, \quad m_0 = 0, \tag{2.22}$$

and

$$n_{-1} = 0, \quad n_0 = 1. \tag{2.23}$$

Notice that Equations (2.20) and (2.21) include the case $b = 0$ because $\rho = 0$ when this occurs. So let's suppose $b \neq 0$ which implies $\rho \geq 1$. Consider the remainder sequence

$$a = R_{-1} = Q_1 R_0 + R_1,$$
$$b = R_0 = Q_2 R_1 + R_2,$$
$$\vdots$$
$$R_{i-2} = Q_i R_{i-1} + R_i, \tag{2.24}$$
$$\vdots$$

where $Q_i = \text{iquot}(R_{i-2}, R_{i-1})$, $R_i = \text{irem}(R_{i-2}, R_{i-1})$, and the sequence terminates when $R_\rho = 0$. Using Equation (2.24), we derive a recurrence relation for m_i and n_i:

$$
\begin{aligned}
R_i &= R_{i-2} - Q_i R_{i-1} \\
&= m_{i-2} a + n_{i-2} b - Q_i (m_{i-1} a + n_{i-1} b) \\
&= (m_{i-2} - Q_i m_{i-1}) a + (n_{i-2} - Q_i n_{i-1}) b. \tag{2.25}
\end{aligned}
$$

Therefore, since $R_i = m_i a + n_i b$, we obtain the two recurrence relations

$$m_i = m_{i-2} - Q_i m_{i-1}, \quad n_i = n_{i-2} - Q_i n_{i-1}. \tag{2.26}$$

These recurrence relations, together with the initial conditions (2.22) and (2.23), give a scheme for computing the values m_i and n_i. $\qquad \square$

Let's return now to the computation of m and n. Since

$$\gcd(a, b) = |R_{\rho-1}|,$$

and

$$R_{\rho-1} = m_{\rho-1} a + n_{\rho-1} b,$$

Procedure *Integer_ext_euc_alg(a, b)*;
Input
　　a, b : integers;
Output
　　the list $[\gcd(a, b), m, n]$;
Local Variables
　　$mpp, mp, npp, np, A, B, Q, R, m, n$;
Begin
1　　　$mpp := 1$; $mp := 0$; $npp := 0$; $np := 1$; $A := a$; $B := b$;
2　　　**while** $B \neq 0$ **do**
3　　　　$Q := Iquot(A, B)$;
4　　　　$R := Irem(A, B)$;
5　　　　$A := B$; $B := R$;
6　　　　$m = mpp - Q * mp$; $n := npp - Q * np$;
7　　　　$mpp := mp$; $mp := m$; $npp := np$; $np := n$;
8　　　**if** $A \geq 0$ **then**
9　　　　$Return([A, \ mpp, \ npp])$
10　　　**else**
11　　　　$Return([-A, \ -mpp, \ -npp])$
End

Figure 2.3. An MPL procedure for the extended Euclidean algorithm. (Implementation: Maple (txt), Mathematica (txt), MuPAD (txt).)

we have

$$m = \pm m_{\rho-1}, \qquad n = \pm n_{\rho-1}, \qquad (2.27)$$

where the plus signs apply when $R_{\rho-1} \geq 0$ and the minus signs apply otherwise.

Example 2.15. Let $a = 45$ and $b = 18$. We have

$$Q_1 = 2, \quad R_1 = 9, \quad m_1 = 1, \quad n_1 = -2,$$
$$Q_2 = 2, \quad R_2 = 0, \quad m_2 = -2, \quad n_2 = 5.$$

Therefore $\rho = 2$ and

$$\gcd(45, 18) = |R_1| = 9, \quad m = m_1 = 1, \quad n = n_1 = -2. \qquad \square$$

A procedure that computes $\gcd(a, b)$ and the integers m and n from the recurrence relations (2.26) with initial conditions (2.22) and (2.23) is given in Figure 2.3. The variables mp and np contain the previous values m_{i-1} and n_{i-1}, and mpp and npp contain the values for m_{i-2} and n_{i-2}.

Theorem 2.12 is important for computer algebra in both a theoretical sense and a computational sense. For example, the next theorem is based on this theorem. Recall that an integer $n > 1$ is *prime* if its only positive divisors are 1 and n.

Theorem 2.16. *Suppose that a, b, and c are integers.*

1. *If $c \mid (a b)$ and c and a are relatively prime, then $c \mid b$.*

2. *If $c \mid (a b)$ and c is prime, then $c \mid a$ or $c \mid b$.*

3. *If $a \mid c$, $b \mid c$, and $\gcd(a, b) = 1$, then $(a b) \mid c$.*

Proof: To prove (1), by Theorem 2.12 there are integers m and n such that $m c + n a = 1$. Therefore

$$m c b + n a b = b,$$

and since c divides each term in the sum on the left, $c \mid b$.

To prove (2), if $c \nmid a$, then, since c is prime, c and a are relatively prime and so by Part (1), $c \mid b$.

To prove (3), let $c = q_1 a$. Using Part (1), since $b \mid (q_1 a)$ and b and a are relatively prime, we have $b \mid q_1$. Therefore, $q_1 = q_2 b$, and $c = q_2 a b$. □

Prime Factorization of Positive Integers

The following theorem is known as the *Fundamental Theorem of Arithmetic*.[3]

Theorem 2.17. *An integer $n > 1$ can be factored uniquely as*

$$n = p_1^{n_1} \cdot p_2^{n_2} \cdots p_s^{n_s}, \tag{2.28}$$

where $p_1, p_2, \ldots, p_s$ are prime numbers with $p_i < p_{i+1}$ and $n_1, n_2, \ldots, n_s$ are positive integers.

For example, $60 = 2^2 \cdot 3 \cdot 5$.

For large integers the factorization problem is computationally much more difficult than the gcd problem. The references at the end of the chapter describe some approaches to this problem.

The prime factorization is obtained with the operator `ifactor` in both Maple and MuPAD and the operator `FactorInteger` in Mathematica. (Implementation: Maple (mws), Mathematica (nb), MuPAD (mnb).)

---

[3]A proof of the theorem is given in Dean [31] pages 23–24. A similar proof for the factorization of polynomials is given for Theorem 4.38 on page 138.

The Chinese Remainder Problem

We conclude this section with a discussion of the Chinese remainder problem, which is concerned with the solution of a system of integer remainder equations.

Definition 2.18. *Let $m_1, m_2, \ldots, m_r$ be distinct positive integers that satisfy*

$$\gcd(m_i, m_j) = 1, \quad \text{for } 1 \le i < j \le r. \tag{2.29}$$

*A collection of integers that satisfies the condition (2.29) is called **pairwise relatively prime**. The **remainder representation** for an integer x is the sequence of remainders $\text{irem}(x, m_1), \text{irem}(x, m_2), \ldots, \text{irem}(x, m_r)$.*

Example 2.19. Let $m_1 = 3$ and $m_2 = 4$. The remainder representations for $0 \le x \le 15$ in terms of m_1 and m_2 are given in the table:

x	$\text{irem}(x, 3)$	$\text{irem}(x, 4)$		x	$\text{irem}(x, 3)$	$\text{irem}(x, 4)$
0	0	0		8	2	0
1	1	1		9	0	1
2	2	2		10	1	2
3	0	3		11	2	3
4	1	0		12	0	0
5	2	1		13	1	1
6	0	2		14	2	2
7	1	3		15	0	3

Observe that for $0 \le x < m_1 m_2 = 12$, each x has a unique remainder representation, while for $x \ge 12$ the remainder representation is the same as for $x - 12$. □

Now let's consider the inverse problem where we are given a remainder representation and want to find an integer x that gives this representation. This problem is known as the *Chinese remainder problem*. Suppose we are given a sequence of pairwise relatively prime positive integers $m_1, m_2, \ldots, m_r$ and a sequence of integers $x_1, x_2, \ldots, x_r$, where $0 \le x_i < m_i$. The goal is to find an integer x that satisfies the remainder equations

$$\text{irem}(x, m_i) = x_i, \quad i = 1, 2, \ldots, r. \tag{2.30}$$

In the next example, we describe one approach to this problem.

Example 2.20. Consider the two remainder equations

$$\text{irem}(x, 5) = 4, \tag{2.31}$$

$$\text{irem}(x, 3) = 2. \tag{2.32}$$

We show that a solution is obtained with a sequence of divisions that involves the remainder sequence (2.10) for 5 and 3 that is given by Euclid's algorithm:

$$R_{-1} = 5, \quad R_0 = 3, \quad R_1 = 2, \quad R_2 = 1, \quad R_3 = 0.$$

First, observe that Equations (2.31) and (2.32) imply there are integers m and n such that

$$x = 5\,n + 4 = 3\,m + 2. \tag{2.33}$$

Once we find the integers m and n so that both of these sums give the same integer x, we will have a solution to the remainder equations. Solving Equation (2.33) for m, we obtain

$$
\begin{aligned}
m &= \frac{5\,n + 2}{3} \\
  &= n + \frac{2\,n + 2}{3},
\end{aligned} \tag{2.34}
$$

where the last expression is obtained by dividing the denominator 3 (which just happens to be R_0) into each of the coefficients of $5\,n + 2$. Since m is an integer, the fraction in Equation (2.34)

$$p = \frac{2\,n + 2}{3}$$

must also reduce to an integer. Solving this equation for n we have

$$
\begin{aligned}
n &= \frac{3\,p - 2}{2} \\
  &= p + \frac{p}{2} - 1,
\end{aligned} \tag{2.35}
$$

where the last expression is obtained by dividing the denominator $R_1 = 2$ into each of the coefficients of $3\,p - 2$. Since n is an integer, the fraction in Equation (2.35) $q = \frac{p}{2}$ must also reduce to an integer. Therefore,

$$p = 2\,q, \tag{2.36}$$

and the process terminates since the denominator of $2\,q$ is the remainder $R_2 = 1$. At this point, we obtain a solution to Equation (2.33) by assigning q an integer value and obtaining integer values for p, n, and m using Equations (2.36), (2.35), and (2.34). For example, if $q = 1$, we have $p = 2$, $n = 2$, $m = 4$ which gives solutions to Equation (2.33)

$$x = 5 \cdot 2 + 4 = 3 \cdot 4 + 2 = 14.$$

Therefore, $x = 14$ is a solution to the remainder equations (2.31) and (2.32). Notice that there are infinitely many solutions to the remainder equations because each integer q gives a distinct solution. □

The approach described in the last example gives an algorithm for the solution of two remainder equations. The process terminates since m_1 and m_2 are relatively prime, and so some member of the remainder sequence is $\gcd(m_1, m_2) = 1$. In Exercise 21 we describe a procedure that solves the Chinese remainder problem using this approach.

Another approach to the Chinese remainder problem is based on the extended Euclidean algorithm. Let m_1 and m_2 be two relatively prime positive integers, and consider the remainder equations

$$\text{irem}(x, m_1) \;=\; x_1, \tag{2.37}$$
$$\text{irem}(x, m_2) \;=\; x_2, \tag{2.38}$$

where $0 \le x_1 < m_1$ and $0 \le x_2 < m_2$. By the extended Euclidean algorithm, there are integers c and d such that

$$c\,m_1 + d\,m_2 = 1. \tag{2.39}$$

We use this relation to obtain a solution to the remainder equations. First, by multiplying both sides of this equation by x_1 and rearranging we have $d\,m_2\,x_1 = (-c\,x_1)\,m_1 + x_1$, which implies

$$\text{irem}(d\,m_2\,x_1, m_1) = x_1, \quad \text{irem}(d\,m_2\,x_1, m_2) = 0. \tag{2.40}$$

In other words, $d\,m_2\,x_1$ is a solution to Equation (2.37) but not to Equation (2.38) (unless $x_2 = 0$). In a similar way, by multiplying Equation (2.39) by x_2, we obtain

$$\text{irem}(c\,m_1 x_2, m_2) = x_2, \quad \text{irem}(c\,m_1 x_2, m_1) = 0, \tag{2.41}$$

which implies that $c\,m_1 x_2$ is a solution to Equation (2.38) but not necessarily to Equation (2.37). However, the relations (2.40) and (2.41) suggest that we can obtain a solution to both Equations (2.37) and (2.38) with the sum of the two partial solutions

$$w = c\,m_1 x_2 + d\,m_2 x_1. \tag{2.42}$$

Indeed, using the relation in Exercise 1(a), we have

$$\text{irem}(w, m_1) \;=\; \text{irem}(d\,m_2\,x_1, m_1) = x_1,$$
$$\text{irem}(w, m_2) \;=\; \text{irem}(c\,m_1\,x_2, m_2) = x_2.$$

Example 2.21. Let $m_1 = 5$, $m_2 = 3$, $x_1 = 4$, and $x_2 = 2$. Then, in Equation (2.39), $c = -1$ and $d = 2$, and from Equation (2.42), $w = 14$ is a solution to the remainder equations.   □

In the next theorem and its proof, we describe a general solution and algorithm for the Chinese remainder problem. The proof, which is similar to the above discussion, is based on the extended Euclidean algorithm.

Theorem 2.22. [Chinese Remainder Theorem] *Let $m_1, m_2, \ldots, m_r$ be positive integers that are pairwise relatively prime, and let $x_1, x_2, \ldots, x_r$ be integers with $0 \leq x_i < m_i$. Then, there is exactly one x in the interval*

$$0 \leq x < m_1 \cdot m_2 \cdots m_r \tag{2.43}$$

that satisfies the remainder equations

$$\mathrm{irem}(x, m_i) = x_i, \quad i = 1, 2, \ldots, r. \tag{2.44}$$

Proof: Observe that we obtain a unique solution by requiring that the solution be in the interval (2.43). We show first that there is some solution to the remainder equations (2.44) and then obtain a solution in this interval using integer division.

The proof is obtained with mathematical induction on the number of equations r. For the base case $r = 1$, integer division shows that $x = x_1$ is a solution to the first remainder equation. For the induction step, let's assume there is an integer s that satisfies the remainder equations

$$\mathrm{irem}(s, m_i) = x_i, \quad i = 1, \ldots, r - 1, \tag{2.45}$$

and show how to extend the process one step further to find an integer w that satisfies all of the remainder equations (2.44). Observe that Equation (2.45) implies

$$s = q_i \, m_i + x_i, \quad i = 1, \ldots, r - 1, \tag{2.46}$$

where $q_i = \mathrm{iquot}(s, m_i)$. In addition, for

$$n = m_1 \cdots m_{r-1}, \tag{2.47}$$

the condition (2.29) implies $\gcd(n, m_r) = 1$, and using the extended Euclidean algorithm, we obtain integers c and d such that

$$c\,n + d\,m_r = 1. \tag{2.48}$$

Let

$$\begin{aligned} w &= c\,n\,x_r + d\,m_r\,s, \tag{2.49}\\ m &= m_1 \cdots m_r. \end{aligned}$$

We show that w satisfies all of the remainder equations. First, for $1 \leq i \leq r - 1$, we use Equation (2.48) to eliminate $d\, m_r$ from w to obtain

$$w \; = \; c\,n\,x_r + (1 - c\,n)\,s = (c\,n\,x_r - c\,n\,s) + s. \qquad (2.50)$$

Using the equations (2.46) to eliminate the s on the far right, we obtain

$$w = (c\,n\,x_r - c\,n\,s + q_i\,m_i) + x_i.$$

Observe that by Equation (2.47), m_i divides each of the terms in parentheses, and therefore, the uniqueness property for integer division implies

$$\mathrm{irem}(w, m_i) = x_i, \qquad 1 \leq i \leq r - 1.$$

For $i = r$, we use Equation (2.48) to eliminate $c\,n$ from w to obtain

$$w = (d\,m_r\,s - d\,m_r\,x_r) + x_r.$$

Since m_r divides each term in parentheses, the uniqueness property for integer division implies $\mathrm{irem}(w, m_r) = x_r$, and, therefore, w satisfies all of the remainder equations.

Although w satisfies all of the remainder equations, it may lie outside the $0 \leq x < m$. To obtain a solution in the interval, divide w by m to obtain $w = q\,m + x$, where $0 \leq x < m$. To show that x satisfies all the remainder equations, we have (using the relation in Exercise 1(a))

$$
\begin{aligned}
\mathrm{irem}(x, m_i) &= \mathrm{irem}(-q \cdot m + w, m_i) \\
&= \mathrm{irem}((-q \cdot m_1 \cdots m_{i-1} \cdot m_{i+1} \cdots m_r)\,m_i + w, m_i) \\
&= \mathrm{irem}(w, m_i) \\
&= x_i.
\end{aligned}
$$

To show the uniqueness of the solution, suppose both x and x' satisfy the conditions in the theorem. Then, for $1 \leq i \leq r$ we can represent x and x' as

$$x = f_i\,m_i + x_i, \qquad x' = g_i\,m_i + x_i$$

which implies $(x - x') = (f_i - g_i)\,m_i$. Therefore, $m_i \mid (x - x')$, and since the integers $m_1, \ldots, m_r$ are relatively prime, Theorem 2.16(3) implies

$$m \mid (x - x'). \qquad (2.51)$$

However, since both x and x' are positive and lie within the interval (2.43), we have $-m < x - x' < m$. This inequality together with the condition (2.51) implies $x = x'$. $\square$

Example 2.23. Let

$$m_1 = 3, \quad m_2 = 4, \quad m_3 = 5,$$
$$x_1 = 1, \quad x_2 = 2, \quad x_3 = 4.$$

We find the solution to the three remainder equations

$$\mathrm{irem}(x, 3) = 1, \quad \mathrm{irem}(x, 4) = 2, \quad \mathrm{irem}(x, 5) = 4.$$

Following the proof of the theorem, we build up the solution in steps. If $r = 1$, then the solution so far is $s = x_1 = 1$ and $n = m_1 = 3$. By applying the extended Euclidean algorithm to $n = m_1$ and m_2, we obtain $a = -1$ and $b = 1$. Therefore, Equation (2.49) (with $r = 2$) gives $w = -2$. Notice that w satisfies the first two remainder equations, but is not in the proper range (≥ 0 and $< 3 \cdot 4$). Nevertheless, as in the proof of the theorem, we assume this is the solution so far and compute the remainder that gives a solution in the proper range at the end of the entire process.

At this point $r = 3$, $n = m_1 m_2 = 12$, and $s = -2$. Applying the extended Euclidean algorithm to n and m_3, we obtain $a = -2$ and $b = 5$. Therefore, Equation (2.49) (with $r = 3$) gives $w = -146$, and the solution to the problem is $x = \mathrm{irem}(-146, \ m_1 m_2 m_3) = 34$. ⊔

A procedure that finds a solution to the Chinese remainder problem is shown Figure 2.4. As in the proof of the theorem, the remainder operation that finds the solution in the proper range is done at the end of the procedure (line 11).

Exercises

1. Let a, b, c, $m \neq 0$, and $n > 0$ be integers. Show that

   (a) $\mathrm{irem}(a\,m + b, \ m) = \mathrm{irem}(b, \ m)$.

   (b) $\mathrm{irem}(a + b + c, \ m) = \mathrm{irem}(a + \mathrm{irem}(b + c, \ m), \ m)$.

   (c) $\mathrm{irem}(a + b, \ m) = \mathrm{irem}(\mathrm{irem}(a, \ m) + \mathrm{irem}(b, \ m), \ m)$.

   (d) $\mathrm{irem}(a\,b, \ m) = \mathrm{irem}(\mathrm{irem}(a, \ m) \cdot \mathrm{irem}(b, \ m), \ m)$.

   (e) $\mathrm{irem}(a^n, \ m) = \mathrm{irem}((\mathrm{irem}(a, \ m))^n, \ m)$.

2. Let b, $c > 0$, and $m \neq 0$ be integers.

   (a) Show that $\mathrm{irem}(c\,b, \ c\,m) = c \cdot \mathrm{irem}(b, \ m)$.

   (b) Show that the relationship in Part (a) may not hold if $c < 0$.

3. Let m be an integer. Show that $\mathrm{iquot}(m, 2) + \mathrm{iquot}(m - 1, 2) + 1 = m$.

4. Prove properties (2), (3), and (4) in Theorem 2.4.

Procedure   *Chinese_remainder*(M, X);
Input
 M : a list of distinct pairwise relatively prime positive integers;
 X : a list of non-negative integers with
  Number_of_operands(M) = *Number_of_operands*(X)
  and *Operand*(X, i) < *Operand*(M, i);
Output
 the solution described in Theorem 2.22;
Local Variables
 n, s, i, x, m, e, c, d;
Begin

```
1     n := Operand(M, 1);
2     s := Operand(X, 1);
3     for i from 2 to Number_of_operands(M) do
4         x := Operand(X, i);
5         m := Operand(M, i);
6         e := Integer_ext_euc_alg(n, m);
7         c := Operand(e, 2);
8         d := Operand(e, 3);
9         s := c * n * x + d * m * s;
10        n := n * m;
11    Return(Irem(s, n))
   End
```

Figure 2.4. An MPL procedure that obtains the solution to the system of remainder equations described in the Chinese remainder theorem. (Implementation: Maple (txt), Mathematica (txt), MuPAD (txt).)

5. Let u be a rational number. The *floor* of u (notation $\lfloor u \rfloor$) is the largest integer $\leq u$. The *ceiling* of u (notation $\lceil u \rceil$) is the smallest integer $\geq u$. Give procedures *Floor*(u) and *Ceiling*(u) that performs these operations. Assume that the input expression u is an algebraic expression. When u is not an integer or fraction, return the unevaluated form of the operators.

6. Give a procedure *Integer_divisors*(n) that finds the positive and negative divisors of an integer $n \neq 0$. The result should be returned as a set or a list. For example,

$$Integer\_divisors(15) \rightarrow \{1, -1, 2, -2, 3, -3, 5, -5, 15, -15\}.$$

This exercise is used in the procedure *Find_S_sets* described in Exercise 5, page 369.

7. Let n be an integer. Give a procedure *Number_of_digits*(n) that returns the number of digits in n.

8. Let $n \geq 0$ and $b \geq 2$ be integers. The integer n has a unique representation in the base b given by

$$n = b_0 + b_1 b + \cdots + b_k b^k \tag{2.52}$$

where $0 \leq b_i \leq b - 1$. For example, for $b = 2$,

$$5 = 1 + 0 \cdot 2 + 1 \cdot 2^2.$$

Give a procedure $Base\_rep(n, b)$ that obtains this representation. Since automatic simplification transforms the right side of Equation (2.52) to n, the procedure should return the representation as a list $[b_0, b_1, \ldots, b_k]$, where $b_k \neq 0$. For $n = 0$, return the empty list $[\ ]$.

9. Give a recursive procedure for Euclid's algorithm.

10. (a) Evaluate gcd(12768, 28424) using Euclid's algorithm.

    (b) Find integers m and n such that

$$m \cdot 12768 + n \cdot 28424 = \text{gcd}(12768, 28424).$$

11. The Fibonacci number sequence $f_0, f_1, f_2, \ldots$ is defined using the recursive definition:

$$f_j = \begin{cases} 1, & \text{when } j = 0 \text{ or } j = 1, \\ f_{j-1} + f_{j-2}, & \text{when } j > 1. \end{cases}$$

For $n \geq 2$, let $R_{-1} = f_n$ and $R_0 = f_{n-1}$ and consider the remainder sequence R_i in (2.10).

    (a) Show that $R_i = f_{n-(i+1)}$ for $i = -1, 0, \ldots, \rho - 1$ and that the remainder sequence terminates with $\rho = n - 1$ and $\text{gcd}(f_n, f_{n-1}) = 1$.

    (b) For this sequence, what is the relation between m_i and n_i in the extended Euclidean algorithm and the Fibonacci numbers?

12. Prove Theorem 2.8(3).

13. Suppose that a, b, u, and v are integers and $a = u \gcd(a, b)$, $b = v \gcd(a, b)$. Show that $\gcd(u, v) = 1$.

14. Suppose that $\gcd(a, b) = 1$ and t is a positive integer. Show that

$$\gcd(a^t, b^t) = 1.$$

15. Let a, b, and c be integers. Show that

$$\gcd(c\,a, c\,b) = |c| \gcd(a, b).$$

16. Let $a \neq 0$ and $b \neq 0$ be integers. Show that a and b are relatively prime if and only if there are integers m and n such that $m\,a + n\,b = 1$.

17. Give an example that shows that m and n in Theorem 2.12 are not unique.

18. Let $a \neq 0$ and $b \neq 0$ be integers, and define the *least common multiple* c of a and b as follows:

(a) $a \mid c$ and $b \mid c$.

(b) If $a \mid d$ and $b \mid d$, then $c \mid d$.

(c) $c > 0$.

We use the notation $\text{lcm}(a, b)$ to denote the least common multiple. For example, $\text{lcm}(4, 6) = 12$.

In this exercise, we derive the relation

$$\text{lcm}(a, b) = \frac{|a\,b|}{\gcd(a, b)} \qquad (2.53)$$

using the following steps:

(a) Suppose that $a \mid d$, $b \mid d$, $a = u \gcd(a, b)$, and $b = v \gcd(a, b)$. Show that there is an integer r such that $d = r\,u\,v \gcd(a, b)$.

(b) Show that both a and b divide $\dfrac{|a\,b|}{\gcd(a, b)}$.

(c) Derive Equation (2.53).

In addition:

(d) Show that there are integers m and n such that $\text{lcm}(a, b) = m\,a + n\,b$.

(e) Give a procedure *Integer_lcm*(a, b) that obtains $\text{lcm}(a, b)$.

19. Solve each of the following systems of remainder equations.

(a) $\text{irem}(x, 3) = 2$, $\text{irem}(x, 5) = 1$, $0 \le x < 15$.

(b) $\text{irem}(x, 3) = 2$, $\text{irem}(x, 4) = 1$, $\text{irem}(x, 5) = 2$, $0 \le x < 60$.

20. Suppose that $m_1 = 4$ and $m_2 = 6$, and consider the remainder equations

$$\text{irem}(x, m_1) = x_1, \qquad \text{irem}(x, m_2) = x_2, \qquad 0 \le x < m_1 m_2, \qquad (2.54)$$

where $0 \le x_1 < m_1$ and $0 \le x_2 < m_2$. (Notice that m_1 and m_2 are not relatively prime.)

(a) Find x_1 and x_2 so that there is more than one solution to the equations and inequality in (2.54).

(b) Find x_1 and x_2 so that the equations in (2.54) do not have a solution.

21. Let $m > 0$, $n > 0$, r, and s be integers with $\gcd(m, n) = 1$. Give a procedure *Chinese_remainder_2*(m, n, r, s) that obtains a solution to the two remainder equations $\text{irem}(x, m) = r$, $\text{irem}(x, n) = s$ that is based on the algorithm used in Example 2.20. The final solution should be in the range $0 \le x < m\,n$. Do not use the extended Euclidean algorithm.

2.2 Rational Number Arithmetic

In a mathematical sense, a rational number is a fraction a/b where a and $b \neq 0$ are integers. With this definition, each rational number can be represented in infinitely many ways (e.g., $1/2 = 2/4 = (-2)/(-4)$, etc.). The automatic simplification process removes this ambiguity by transforming a fraction to a standard form. Some typical examples[4] are

$$2/4 \to 1/2, \quad 2/(-4) \to (-1)/2 \quad (-2)/(-4) \to 1/2, \quad 4/1 \to 4.$$

These transformations suggest the following definition.

Definition 2.24. *Let a and b be integers. An expression a/b is a* **fraction in standard form** *if it satisfies the following two properties:*

1. $b > 1$.

2. $\gcd(a, b) = 1$.

The fractions $2/3$ and $(-2)/3$ are in standard form, while $4/6$, $2/(-3)$, $4/1$, and $0/2$ are not.

Each fraction can be transformed to either a fraction in standard form or an integer. This observation suggests the following definition.

Definition 2.25. *An expression is a* **rational number in standard form** *if it is either a fraction in standard form or an integer.*

The *Simplify_rational_number* Operator

Let u be an integer or a fraction with non-zero denominator. The operator *Simplify_rational_number(u)* transforms u to a rational number in standard form (Figure 2.5). In order to bypass the automatic simplification process in a CAS (which transforms fractions to standard form), we represent the fraction a/b with the function notation *FracOp(a, b)* instead of the usual infix notation.

Simplification of Rational Number Expressions

We consider next an algorithm that evaluates arithmetic expressions with integers and fractions. The input to the algorithm is described in the next definition.

---

[4]The displayed form of a fraction in a CAS may disguise the internal form.

Procedure *Simplify_rational_number*(u);
Input
    u : a fraction in function (*FracOp*) notation (with non-zero denominator)
      or an integer;
Output
    a fraction in standard form in function (*FracOp*) notation or an integer;
Local Variables
    n, d, g;
Begin
1   **if** *Kind*(u) = **integer then** *Return*(u)
2   **elseif** *Kind*(u) = *FracOp* **then**
3   n = *Operand*(u, 1);
4   d = *Operand*(u, 2);
5   **if** *Irem*(n, d) = 0 **then** *Return*(*Iquot*(n, d))
6   **else**
7   g := *Integer_gcd*(n, d);
8   **if** d > 0 **then**
9   *Return*(*FracOp*(*Iquot*(n, g), *Iquot*(d, g)))
10   **elseif** d < 0 **then**
11   *Return*(*FracOp*(*Iquot*(−n, g), *Iquot*(−d, g)))
    End

Figure 2.5. The MPL *Simplify_rational_number* procedure. (Implementation: Maple (txt), Mathematica (txt), MuPAD (txt).)

Definition 2.26. *An algebraic expression u is a* **rational number expression (RNE)** *if it satisfies one of the following rules:*

RNE-1. *u is an integer.*

RNE-2. *u is a fraction.*

At this point, we allow for the possibility of a fraction with zero denominator. The algorithm will recognize that such an expression is undefined.

RNE-3. *u is a unary or binary sum with operands that are RNEs.*

RNE-4. *u is a unary or binary difference with operands that are RNEs.*

RNE-5. *u is a binary product with operands that are RNEs.*

RNE-6. *u is a quotient with operands that are RNEs.*

RNE-7. *u is a power with a base that is an RNE and an exponent that is an integer.*

Infix notation	Example	Function name	Example
/ (fraction)	$2/3$ $(-2)/3$	$FracOp$	$FracOp(2,3)$ $FracOp(-2,3)$
$+$	$1 + 2 + 3/4$ $+3$	$SumOp$	$SumOp(1, 2, FracOp(3, 4))$ $SumOp(3)$
$-$	$4 - 5$ -2	$DiffOp$	$DiffOp(4,5)$ $DiffOp(2)$
$*$	$2 * 3 * (4 + 1)$	$ProdOp$	$ProdOp(2, 3, SumOp(4, 1))$
$\wedge$	$3 \wedge 2$	$PowOp$	$PowOp(3, 2)$
/ (quotient)	$(2 + 3)/4$	$QuotOp$	$QuotOp(SumOp(2, 3), 4)$
!	$5!$	$FactOp$	$FactOp(5)$

Figure 2.6. The MPL function names used for fractions and algebraic operators. Each name includes the suffix "Op" to avoid a conflict with a named function in a real CAS.

An MPL algorithm that evaluates RNEs is given in Figures 2.7, 2.8, and 2.9. To bypass the automatic simplification process, we represent fractions and algebraic operations[5] in the function notation given in Figure 2.6. In this notation, each infix operator is given a function name, and the operands associated with the operator are operands of the function. Observe that each name includes the suffix "Op" (which stands for "operator") to avoid a conflict with a named function in a real CAS.

In Figure 2.7, the main procedure $Simplify\_RNE(u)$ takes an RNE as input and returns either an integer, a fraction in standard form, or the global symbol **Undefined** when a division by zero is encountered. For example,

$$Simplify\_RNE(SumOp(FracOp(2, 3), FracOp(3, 4))) \rightarrow FracOp(17, 12),$$

$$Simplify\_RNE(PowOp(FracOp(4, 2), 3)) \rightarrow 8,$$

$$Simplify\_RNE(QuotOp(1, DiffOp(FracOp(2, 4), FracOp(1, 2)))) \rightarrow$$
$$\textbf{Undefined}.$$

The main procedure calls on the recursive procedure $Simplify\_RNE\_rec(u)$ (at line 1) which numerically evaluates u, and if the expression returned is not **Undefined**, it is transformed to a rational number in standard form (line 3). This is the only place in the algorithm that involves a gcd calculation.

---

[5]The factorial operator $FactOp$ is not used in this section. It is used, however, in the simplifier in Chapter 3.

Procedure *Simplify_RNE(u)*;
Input
   u : an **RNE** in function notation;
Output
   an integer, fraction (in function notation) in standard form, or
   the global symbol **Undefined**;
Local Variables v;
Begin
1    $v := Simplify\_RNE\_rec(u)$;
2   **if** $v = $ **Undefined then** *Return*(**Undefined**)
3   **else** *Return*(*Simplify_rational_number(v)*)
End

Figure 2.7. The MPL *Simplify_RNE* procedure. (Implementation: Maple (txt), Mathematica (txt), MuPAD (txt).)

In Figure 2.8, the procedure *Simplify_RNE_rec(u)* determines the type of u (**integer**, *FracOp*, *SumOp*, *DiffOp*, *ProdOp*, or *QuotOp*) and then performs the appropriate operation. The unary operations are handled in lines 5–9, while the binary operations are handled in lines 10–23. For each arithmetic operation, the operand or operands are recursively simplified (lines 6, 12, 13, and 21), and if some operand is **Undefined**, the procedure returns **Undefined** (lines 7, 14, 22).[6]

The actual calculations are obtained with the calls to procedures in lines 9, 16, 17, 18, 19, and 23. Two of the calculation procedures are shown in Figure 2.9, and the others are left to the reader (Exercise 3). The input to these procedures is either an integer or fraction, and the output is either a fraction (which is probably not in standard form) or (from *Evaluate_quotient* or *Evaluate_power*) the symbol **Undefined**.

In the procedure *Evaluate_quotient(u, v)* in Figure 2.9, the operators *Numerator_fun* and *Denominator_fun* obtain the numerator and denominator[7] of the v and w. (These operators can be applied to either integers or fractions (Exercise 3).) Since w can be an integer or a fraction (which is probably not in standard form), we determine if it represents 0 by ex-

---

[6]There are three places in the algorithm which lead to undefined expressions. They are line 3 in *Simplify_RNE_rec* when a fraction has zero denominator, line 1 in *Evaluate_quotient* when a quotient has a zero denominator, and line 12 in *Evaluate_power* when 0 is raised to a non-positive exponent.

[7]We cannot use the numerator and denominator operators in a CAS since these act on expressions in infix notation. The suffix "_fun" in the operators *Numerator_fun* and *Denominator_fun* indicate they apply to expressions in function notation.

Procedure *Simplify_RNE_rec(u)*;
Input
    u : an **RNE** in function notation;
Output
    an integer, fraction (in function notation), or the global symbol
    Undefined;
Local Variables *v, w*;
Begin

1   **if** *Kind(u)* = **integer then** *Return(u)*
2   **elseif** *Kind(u)* = *FracOp* **then**
3   **if** *Denominator_fun(u)* = 0 **then** *Return(***Undefined***)*
4   **else** *Return(u)*
5   **elseif** *Number_of_operands(u)* = 1 **then**
6   *v* := *Simplify_RNE_rec(Operand(u, 1))*;
7   **if** *v* = **Undefined then** *Return(***Undefined***)*;
8   **elseif** *Kind(u)* = *SumOp* **then** *Return(v)*
9   **elseif** *Kind(u)* = *DiffOp* **then** *Return(Evaluate_product(−1, v))*
10   **elseif** *Number_of_operands(u)* = 2 **then**
11   **if** *Kind(u)* ∈ {*SumOp, ProdOp, DiffOp, QuotOp*} **then**
12   *v* := *Simplify_RNE_rec(Operand(u, 1))*;
13   *w* := *Simplify_RNE_rec(Operand(u, 2))*;
14   **if** *v* = **Undefined or** *w* = **Undefined then**
           *Return(***Undefined***)*
15   **else**
16   **if** *Kind(u)* = *SumOp* **then**
           Return(Evaluate_sum(v, w))
17   **elseif** *Kind(u)* = *DiffOp* **then**
           Return(Evaluate_difference(v, w))
18   **elseif** *Kind(u)* = *ProdOp* **then**
           Return(Evaluate_product(v, w))
19   **elseif** *Kind(u)* = *QuotOp* **then**
           Return(Evaluate_quotient(v, w))
20   **elseif** *Kind(u)* = *PowOp* **then**
21   *v* := *Simplify_RNE_rec(Operand(u, 1))*;
22   **if** *v* = **Undefined then** *Return(***Undefined***)*
23   **else** *Return(Evaluate_power(v, Operand(u, 2)))*
    End

Figure 2.8. The MPL *Simplify_RNE_rec* procedure. (Implementation: Maple (txt), Mathematica (txt), MuPAD (txt).)

Procedure *Evaluate_quotient(v, w)*;
Input   *v, w* : an integer or a fraction in function notation with
            non-zero denominator;
Output
            a fraction in function notation or the global symbol **Undefined**;
Begin
1   **if** *Numerator_fun(w)* = 0 **then** *Return*(**Undefined**)
2   **else** *Return*(*FracOp*(*Numerator_fun(v)* ∗ *Denominator_fun(w)*,
                                    Numerator_fun(w) ∗ *Denominator_fun(v)*)))
End

Procedure *Evaluate_power(v, n)*;
Input
       v : an integer or a fraction in function notation with
            non-zero denominator;
       n : an integer;
Output
       a fraction in function notation or the global symbol **Undefined**;
Local Variables *s*;
Begin
1   **if** *Numerator_fun(v)* ≠ 0 **then**
2   **if** *n* > 0 **then**
3   *s* := *Evaluate_power(v, n − 1)*;
4   *Return*(*Evaluate_product(s, v)*)
5   **elseif** *n* = 0 **then** *Return*(1)
6   **elseif** *n* = −1 **then** *Return*(*FracOp*(*Evaluate_quotient(v)*,
                                                        Numerator_fun(v))))
7   **elseif** *n* < −1 **then**
8   *s* := *FracOp*(*Evaluate_quotient(v)*, *Numerator_fun(v)*);
9   *Return*(*Evaluate_power(s, −n)*)
10   **elseif** *Numerator_fun(v)* = 0 **then**
11   **if** *n* ≥ 1 **then** *Return*(0)
12   **elseif** *n* ≤ 0 **then** *Return*(**Undefined**)
End

Figure 2.9. The MPL *Evaluate_quotient* and *Evaluate_power* procedures. (Implementation: Maple (txt), Mathematica (txt), MuPAD (txt).)

amining *Numerator_fun(w)* (line 1). The procedure *Evaluate_power(v, n)*, which evaluates *v* to the exponent *n*, obtains the result by recursion on the power (lines 3 and 9).

When these procedures and the procedure *Simplify_rational_number* are implemented in a CAS, the only arithmetic operations used are the primitive integer operations addition, subtraction, multiplication, integer quotient, and integer remainder. In addition, the *Integer_gcd* operator called at line 7 of *Simplify_rational_number* only uses the integer remainder and the absolute value operations.

Some of the procedures in this section are used for numerical calculations in the automatic simplification algorithm described in Chapter 3.

Exercises

1. Let m be an integer and n/d a fraction in standard form. Show that $(m\,d + n)/d$ is a fraction in standard form.

2. (a) Let a, b, c, and d be integers, and suppose that

   $$\gcd(a,b) = 1, \quad \gcd(c,d) = 1, \quad \gcd(b,d) = 1.$$

   Show that $\gcd(a\,d + b\,c,\ b\,d) = 1$.

   (b) Use Part (a) to show that if the fractions a/b and c/d are in standard form, and if $\gcd(b,d) = 1$, then $(a\,d + b\,c)/(b\,d)$ is also a fraction in standard form.

   (c) Give an example that shows that Part(a) is not true if $\gcd(b,d) \neq 1$.

   (d) Suppose that a/b and c/d are fractions in standard form, where now b and d may not be relatively prime. Let $g = \gcd(b,d)$ and $l = \text{lcm}(b,d)$. Show that

   $$\frac{a \cdot \text{iquot}(d,g) + c \cdot \text{iquot}(b,g)}{l}$$

   is also a fraction in standard form. *Hint:* According to Exercise 18 on page 36, $l = \text{iquot}(b\,d,\ g)$.

3. Let u, v, and w be either integers or fractions (in function notation) with non-zero denominators. Give procedures for the following operators:

   (a) *Numerator_fun(u)*.

   (b) *Denominator_fun(u)*.

   (c) *Evaluate_sum(v, w)*.

   (d) *Evaluate_difference(v, w)*.

   (e) *Evaluate_product(v, w)*.

   The output of *Numerator_fun* and *Denominator_fun* is an integer, and the output for *Evaluate_sum*, *Evaluate_difference*, and *Evaluate_product* is a fraction in function notation.

2.3 Fields

A field is a general mathematical system with axioms that describe the basic
algebraic properties of number systems and other classes of expressions
that arise in computer algebra. In this section we give a formal definition
of a field, give a number of examples, and show that many transformations
routinely used in manipulations are logical consequences of the field axioms.

Definition 2.27. *Let* $\mathbf{F} = \{u, v, w, \ldots\}$ *be a set of expressions, and assume
that there is an addition operation* $u + v$ *and a multiplication operation* $u \cdot v$
that are defined for any u *and* v *in* $\mathbf{F}$. *The set* $\mathbf{F}$ *is a* **field** *if the operations
satisfy the following properties.*

Closure Properties

F-1. $u + v$ *is in* $\mathbf{F}$.

F-2. $u \cdot v$ *is in* $\mathbf{F}$.

Commutative Properties

F-3. $u + v = v + u$.

F-4. $u \cdot v = v \cdot u$.

Associative Properties

F-5. $u + (v + w) = (u + v) + w$.

F-6. $u \cdot (v \cdot w) = (u \cdot v) \cdot w$.

Distributive Properties

F-7. $u \cdot (v + w) = u \cdot v + u \cdot w$ *(left distributive property).*

F-8. $(u + v) \cdot w = u \cdot w + v \cdot w$ *(right distributive property).*

Identities

F-9. *There is an* **additive identity** 0 *in* $\mathbf{F}$ *such that* $u + 0 = u$ *for every*
u *in* $\mathbf{F}$.

F-10. *There is a* **multiplicative identity** 1 *in* $\mathbf{F}$ *such that* $u \cdot 1 = u$ *for
every* u *in* $\mathbf{F}$.

Inverses

F-11. *Each element u in* **F** *has an* **additive inverse** *in* **F** *denoted by $(-u)$ such that $u + (-u) = 0$.*

F-12. *Each element $u \neq 0$ in* **F** *has a* **multiplicative inverse** *in* **F** *denoted by u^{-1} such that $u \cdot u^{-1} = 1$.*

Although the field axioms are expressed in terms of a familiar notation that suggests the nature of the operations, we caution the reader not to assume more about the notation than is formally stated in the definition. For example, although $+$ is an addition-like operation and $\cdot$ is a multiplication-like operation, the operations can be quite different from what we ordinarily associate with addition and multiplication.[8] In a similar way, the notation for identities and inverses are only names for these expressions. For example, although the multiplicative inverse is expressed using the power notation u^{-1}, at this point it doesn't imply anything about a power operation.[9]

Example 2.28. A simple example of a field is the set **Q** of rational numbers, where the two field operations are ordinary addition and multiplication. In this case, the additive identity is the integer 0 and the multiplicative identity the integer 1. The additive inverse of a rational number u is $-u$, and the multiplicative inverse of $u = a/b \neq 0$ is $1/u = b/a$. In a similar way, the set **R** of (mathematical) real numbers is a field. □

Example 2.29. Consider the set of real numbers of the form $a + b\sqrt{2}$, where a and b are rational numbers. Let's show that this set with the operations of ordinary addition and multiplication of real numbers is a field. Since this set is included in **R**, we only verify the properties F-1, F-2, F-11, and F-12, and assume that the others are inherited from the properties of the real numbers. Since,

$$(a + b\sqrt{2}) + (c + d\sqrt{2}) = (a + c) + (b + d)\sqrt{2},$$

$$(a + b\sqrt{2}) \cdot (c + d\sqrt{2}) = (ac + 2bd) + (bc + ad)\sqrt{2},$$

the sum and product have the proper form, and so F-1 and F-2 are satisfied. The additive inverse of $a + b\sqrt{2}$ is $(-a) + (-b)\sqrt{2}$. To find the multiplicative inverse of

$$a + b\sqrt{2} \neq 0, \tag{2.55}$$

---

[8] For example, in Exercise 3 we describe a field with four expressions where $u + u = 0$ for all expressions in the field.

[9] For example, in the field $\mathbf{Z}_5$ (see page 52), there is an expression $u \neq 1$ such that $u = u^{-1}$.

we must find rational numbers x and y such that

$$(a + b\sqrt{2}) \cdot (x + y\sqrt{2}) = 1.$$

Expanding the left side of the equality, we obtain

$$(ax + 2by) + (bx + ay)\sqrt{2} = 1.$$

In this equation, $bx + ay = 0$ because, if this were not so, we could solve for $\sqrt{2}$ and conclude that it is a rational number. Therefore, we have two equations

$$bx + ay = 0, \qquad ax + 2by = 1,$$

and solving for x and y, we obtain

$$(a + b\sqrt{2})^{-1} = x + y\sqrt{2} = \frac{a}{a^2 - 2b^2} + \frac{-b}{a^2 - 2b^2}\sqrt{2}.$$

The multiplicative inverse can also be found by rationalizing the denominator:

$$\frac{1}{a + b\sqrt{2}} = \frac{1}{a + b\sqrt{2}} \frac{a - b\sqrt{2}}{a - b\sqrt{2}} = \frac{a}{a^2 - 2b^2} + \frac{-b}{a^2 - 2b^2}\sqrt{2}.$$

By the way, the denominator that appears in the expression on the right satisfies

$$a^2 - 2b^2 \neq 0. \tag{2.56}$$

To see why, first, if $b = 0$, then from the condition (2.55) we must have $a \neq 0$, and the condition (2.56) follows. Next, if $b \neq 0$, then if $a^2 - 2b^2 = 0$, we have $\sqrt{2} = \sqrt{a^2/b^2} = |a/b|$ which implies that $\sqrt{2}$ is a rational number. Since $\sqrt{2}$ is irrational, the condition (2.56) must be true.

The field described in this example is denoted by $\mathbf{Q}(\sqrt{2})$. The notation indicates that the field is constructed by adjoining the expression $\sqrt{2}$ to the base field $\mathbf{Q}$. $\square$

Example 2.30. Consider the set of expressions of the form $a + b\imath$ where $\imath = \sqrt{-1}$ and a and b are rational numbers. This set is a field with the two operations $+$ and $\cdot$ defined by

$$(a + b\imath) + (c + d\imath) = (a + c) + (b + d)\imath, \tag{2.57}$$

$$(a + b\imath) \cdot (c + d\imath) = (ac - bd) + (cb + ad)\imath. \tag{2.58}$$

(The addition and multiplication operations that are inside the parentheses on the right sides of the equations are addition and multiplication of rational numbers.) The additive inverse of $a + bi$ is $(-a) + (-b)\imath$, and, for $a + b\imath \neq 0$, the multiplicative inverse is

$$(a + b\imath)^{-1} = \frac{a}{a^2 + b^2} + \frac{(-b)}{a^2 + b^2}\imath$$

This field, which is denoted by $\mathbf{Q}(\imath)$, is known as the *Gaussian rational number* field.

In a similar way, the set $\mathbf{C}$ of complex numbers $a + b\imath$, where a and b are real numbers, is a field with the operations $+$ and $\cdot$ defined by Equations (2.57) and (2.58).   □

Example 2.31. Consider the set of expressions of the form

$$u = a + b\sqrt{2} + c\sqrt{3} + d\sqrt{2}\sqrt{3},$$

where a, b, c, and d are rational numbers together with the operations of ordinary addition and multiplication. We leave to the reader the verification that the sum and product of two of these expressions are also in this form and the derivation of expressions for the additive and multiplicative inverses (Exercise 2(a)). This field is denoted by $\mathbf{Q}(\sqrt{2}, \sqrt{3})$ where the notation indicates that the field is constructed from the base field $\mathbf{Q}$ by adjoining the two radicals $\sqrt{2}$ and $\sqrt{3}$. By grouping terms, we have

$$a + b\sqrt{2} + c\sqrt{3} + d\sqrt{2}\sqrt{3} = (a + b\sqrt{2}) + (c + d\sqrt{2})\sqrt{3},$$

which shows that we can also view this field as $(\mathbf{Q}(\sqrt{2}))(\sqrt{3})$, where the notation indicates that the field is constructed by first adjoining $\sqrt{2}$ to $\mathbf{Q}$ to obtain $\mathbf{Q}(\sqrt{2})$ and then adjoining $\sqrt{3}$ to $\mathbf{Q}(\sqrt{2})$.   □

Example 2.32. Consider the set of rational expressions of the form

$$\frac{u(x)}{v(x)}$$

where $u(x)$ and $v(x) \neq 0$ are polynomials in x with rational number coefficients. We interpret this definition in a broad sense to include simpler expressions such as polynomials, fractions, and integers. This set, together with the addition and multiplication operations defined by

$$\frac{u(x)}{v(x)} + \frac{r(x)}{s(x)} = \frac{Algebraic\_expand\,(u(x)\,s(x) + v(x)\,r(x))}{Algebraic\_expand(v(x)\,s(x))}$$

and

$$\frac{u(x)}{v(x)} \cdot \frac{r(x)}{s(x)} = \frac{Algebraic\_expand(u(x)\,r(x))}{Algebraic\_expand(v(x)\,s(x))},$$

is a field. In this case, the identities are the integers 0 and 1, the additive inverse of $u(x)/v(x)$ is the rational expression $((-1)\cdot u(x))/v(x)$ (with the numerator expanded), and if $u(x)/v(x) \neq 0$, the multiplicative inverse is $v(x)/u(x)$. The field of rational expressions is denoted by $\mathbf{Q}(x)$, where the notation indicates the field is obtained by adjoining the symbol x to the base field $\mathbf{Q}$.

Although the fields $\mathbf{Q}(x)$ and $\mathbf{Q}(\imath)$ use a similar notation to denote the field, the expressions in $\mathbf{Q}(x)$ are more complicated than those in $\mathbf{Q}(\imath)$. In the latter case, the symbol $\imath$ satisfies the simplification transformation $\imath^2 = -1$, and so involved rational expressions in the symbol $\imath$ simplify to the appropriate form. For example,

$$\frac{2+3\imath}{4+5\imath+6\imath^2} = 11/29 + (-16/29)\,\imath.$$

On the other hand, since the symbol x does not satisfy a simplification transformation,

$$\frac{2+3\,x}{4+5\,x+6\,x^2}$$

does not simplify. $\square$

Example 2.33. Not all of the familiar sets of numbers are fields. For example, the set $\mathbf{Z}$ of integers is not a field because it does not contain multiplicative inverses. However, $\mathbf{Z}$ does satisfy all of the field properties, except F-12. In addition, the integers satisfy the property:

$$\text{if } a \cdot b = 0, \text{ then either } a = 0 \text{ or } b = 0.$$

This property is described by saying that $\mathbf{Z}$ has no *zero divisors*. A mathematical system that satisfies field axioms F-1 through F-11 and has no zero divisors is called an *integral domain*. $\square$

Consequences of the Field Axioms

The field axioms embody the basic transformation rules that are needed to perform calculations with expressions in $\mathbf{F}$. Other properties and transformation rules that are used routinely in calculations are derived as logical consequences of the basic axioms. This process is illustrated in the proofs of the next two theorems.

Theorem 2.34. *Let* **F** *be a field.*

1. *The identities* 0 *and* 1 *in* **F** *are unique.*

2. *Every* u *in* **F** *has a unique additive inverse* $(-u)$, *and if* $u \neq 0$, *it has a unique multiplicative inverse* u^{-1}.

Proof: Notice that the field axioms do not say that identities and inverses are unique, only that they exist. Could there be a field that has, for example, two distinct additive identities? Of course, the answer is no, but this must be verified from the axioms.

To prove (1), suppose that there were two additive identities, say 0 and $0'$. Then, by F-9, with $0'$ acting as an additive identity,

$$0 = 0 + 0'.$$

Applying F-3 to the right side of the equality, we have

$$0 = 0' + 0.$$

Applying F-9 with 0 acting as an additive identity, we obtain

$$0 = 0'.$$

In a similar way, we can show that the multiplicative identity is unique.

Statement (2) follows in a similar manner (Exercise 4).   □

Theorem 2.35. *Let* u *and* v *be expressions in a field* **F**. *Then,*

1. $u \cdot 0 = 0$;

2. $(-u) = (-1) \cdot u$;

3. $(-0) = 0$;

4. $1^{-1} = 1$;

5. $(-1)^{-1} = -1$;

6. $(-u) \cdot (-v) = u \cdot v$;

7. *If* $u \neq 0$, *then* $(u^{-1})^{-1} = u$;

8. *If* $u \neq 0$ *and* $v \neq 0$, *then* $(u \cdot v)^{-1} = u^{-1} \cdot v^{-1}$.

Proof: To prove (1), by F-9,

$$u \cdot (u + 0) = u \cdot u,$$

and so by F-7,

$$u \cdot u + u \cdot 0 = u \cdot u.$$

Adding the additive inverse of $u \cdot u$ to both sides of this equality and using F-9, we have

$$(u \cdot u + u \cdot 0) + (-(u \cdot u)) = u \cdot u + (-(u \cdot u)) = 0. \qquad (2.59)$$

Next, using the field axioms, the left side of this equation can be transformed in the following way:

$$
\begin{aligned}
(u \cdot u + u \cdot 0) + (-(u \cdot u)) &= (-(u \cdot u)) + (u \cdot u + u \cdot 0) && \text{(by F-3)} \\
&= ((-(u \cdot u)) + u \cdot u) + u \cdot 0 && \text{(by F-5)} \\
&= (u \cdot u + (-(u \cdot u))) + u \cdot 0 && \text{(by F-3)} \\
&= 0 + u \cdot 0 && \text{(by F-11)} \\
&= u \cdot 0 + 0 && \text{(by F-3)} \\
&= u \cdot 0 && \text{(by F-9)}. \qquad (2.60)
\end{aligned}
$$

Therefore, by comparing Equations (2.59) and (2.60), we have $u \cdot 0 = 0$.

To prove (2), we show first that $(-1) \cdot u$ is the additive inverse of u. We have

$$
\begin{aligned}
u + (-1) \cdot u &= u \cdot 1 + u \cdot (-1) && \text{(by F-4 and F-10)} \\
&= u \cdot (1 + (-1)) && \text{(by F-7)} \\
&= u \cdot 0 && \text{(by F-11)} \\
&= 0 && \text{(by Part (1) of the theorem).}
\end{aligned}
$$

Therefore, by statement (2) of Theorem 2.34, $(-u) = (-1) \cdot u$.

The proofs of the remaining statements of the theorem are left to the reader (Exercise 5). $\square$

Theorem 2.36. *Suppose that u and v are in a field $\mathbf{F}$ with $u \cdot v = 0$. Then, either $u = 0$ or $v = 0$.*

This property is described by saying that a field has no *zero divisors*. The proof of the theorem is left to the reader (Exercise 6).

Subtraction, Division and Exponentiation in a Field

Although subtraction, division, and exponentiation are not directly part of the field definition, the operations are easily defined using the field operations.

Definition 2.37. *Let u and v be expressions in a field $\mathbf{F}$. The **subtraction** (or **difference**) operation is defined by*

$$u - v = u + (-v), \qquad (2.61)$$

where $(-v)$ is the additive inverse of v.

Using Theorem 2.35(2), subtraction can also be represented as

$$u - v = u + (-1) \cdot v,$$

which is the automatically simplified form for differences in most computer algebra systems.

Definition 2.38. *Let u and $v \neq 0$ be expressions in a field $\mathbf{F}$. The **division** or **quotient** operation is defined by*

$$u/v = u \cdot v^{-1}. \qquad (2.62)$$

The automatic simplification process in most computer algebra systems transforms a quotient to the form on the right in Equation (2.62).

Definition 2.39. *Let u be an expression in a field $\mathbf{F}$, and let n be an integer. The **exponentiation** or **power** operation u^n is defined with the recursive definition.*

1. *If $u \neq 0$ and $n = -1$, then u^{-1} is the multiplicative inverse of u.*

2. *If $u \neq 0$ and $n \neq -1$, then*

$$u^n = \begin{cases} u^{n-1} \cdot u, & \text{when } n \geq 1, \\ 1, & \text{when } n = 0, \\ \left(u^{-1}\right)^{(-n)}, & \text{when } n \leq -2. \end{cases}$$

3. *If $u = 0$, then*

$$0^n = \begin{cases} 0 & \text{when } n \geq 1, \\ \textbf{Undefined}, & \text{when } n \leq 0. \end{cases}$$

For example, a strict interpretation of the definition gives (for $u \neq 0$)

$$
\begin{aligned}
u^{-3} &= (u^{-1})^3 = (u^{-1})^2 \cdot u^{-1} = ((u^{-1})^1 \cdot u^{-1}) \cdot u^{-1} \\
&= (((u^{-1})^0 \cdot u^{-1}) \cdot u^{-1}) \cdot u^{-1} = ((1 \cdot u^{-1}) \cdot u^{-1}) \cdot u^{-1} \\
&= (u^{-1} \cdot u^{-1}) \cdot u^{-1}.
\end{aligned}
$$

The exponentiation operation satisfies the following properties.

Theorem 2.40. *Let $u \neq 0$ and $v \neq 0$ be expressions in a field $\mathbf{F}$, and let m and n be integers. Then,*

1. $(u \cdot v)^m = u^m \cdot v^m$;

2. $u^m \cdot u^n = u^{m+n}$;

3. $(u^m)^n = u^{m\,n}$.

When either $u = 0$ or $v = 0$, the properties hold as long as the expressions on both sides of the equal sign are defined.

Proof: To prove Part (1), first assume that $u \neq 0$ and $v \neq 0$. The proof for $m \geq 0$ uses mathematical induction. The base case $m = 0$ follows directly from Definition 2.39(2). For $m \geq 1$, assume that the relation is true for $m - 1$. Then, using Definition 2.39(2) and the induction hypothesis, we have

$$
(u \cdot v)^m = (u \cdot v)^{m-1} \cdot (u \cdot v) = u^{m-1} \cdot v^{m-1} \cdot u \cdot v = u^m \cdot v^m.
$$

For $m = -1$, the relation follows from Theorem 2.35(8). For $m \leq -2$, we have $-m \geq 2$ and so using the above case together with Definition 2.39(2) and Theorem 2.35(8), we have

$$
(u \cdot v)^m = ((u \cdot v)^{-1})^{-m} = (u^{-1} \cdot v^{-1})^{-m} = (u^{-1})^{-m} \cdot (v^{-1})^{-m} = u^m \cdot v^m.
$$

For the case when either $u = 0$ or $v = 0$, the relation is defined when $m \geq 1$ and follows from Definition 2.39(3).

The proofs for Parts (2) and (3) are left to the reader (Exercise 9). $\square$

Finite Fields

Some fields contain only a finite number of expressions. For example, let $p > 1$ be an integer, and consider the set $\mathbf{Z}_p = \{0, 1, \ldots, p-1\}$. Define an addition operation $\oplus_p$ and multiplication operation $\otimes_p$ by

$$
a \oplus_p b = \mathrm{irem}(a + b, \, p), \qquad a \otimes_p b = \mathrm{irem}(a \cdot b, \, p)
$$

where $+$ and $\cdot$ are the ordinary integer operations. For example, when $p = 5$, the operations are given in the following operation tables.

$\oplus_5$	0	1	2	3	4
0	0	1	2	3	4
1	1	2	3	4	0
2	2	3	4	0	1
3	3	4	0	1	2
4	4	0	1	2	3

$\otimes_5$	0	1	2	3	4
0	0	0	0	0	0
1	0	1	2	3	4
2	0	2	4	1	3
3	0	3	1	4	2
4	0	4	3	2	1

It is apparent from the tables that each element has an additive inverse, and each non-zero element has a multiplicative inverse. Since $\mathbf{Z}_5$ inherits the field properties F-3 through F-10 from the integers, it is a field.

However, $\mathbf{Z}_6$ is not a field. In this case, the $\otimes_6$ table is given in the following table.

$\otimes_6$	0	1	2	3	4	5
0	0	0	0	0	0	0
1	0	1	2	3	4	5
2	0	2	4	0	2	4
3	0	3	0	3	0	3
4	0	4	2	0	4	2
5	0	5	4	3	2	1

It is apparent that 2, 3, and 4 do not have multiplicative inverses.

The next theorem describes when $\mathbf{Z}_p$ is a field.

Theorem 2.41. *Let $p > 1$ be an integer. The set $\mathbf{Z}_p$ with the operations $\oplus_p$ and $\otimes_p$ is a field if and only if p is a prime number.*

Proof: Let's assume that p is prime and show that $\mathbf{Z}_p$ is a field. Since the closure properties F-1 and F-2 follow from the definitions of $\oplus_p$ and $\otimes_p$, and the properties F-3 through F-10 are inherited from the integers, we only need to verify the inverse properties. Let s be in $\mathbf{Z}_p$. The additive inverse of s is given by $p - s$.

We verify the multiplicative inverse property by giving an algorithm that obtains the inverse. Let $0 < s \leq p - 1$. Since p is prime, using the extended Euclidean algorithm, we obtain integers m and n such that

$$m \cdot s + n \cdot p = 1. \tag{2.63}$$

When m is in the range $0 < m \leq p - 1$, we can show that $s^{-1} = m$. Unfortunately, m may lie outside of this range although m cannot be a multiple of p. (If m were a multiple of p, then by Equation (2.63), $p \mid 1$ which is not true.) Therefore, to obtain the inverse, divide m by p to get

$$m = q \cdot p + r, \tag{2.64}$$

with $1 \leq r < p$. Using Equations (2.63) and (2.64), we obtain

$$
\begin{aligned}
s \otimes_p r &= \text{irem}(s \cdot r, \ p) \\
&= \text{irem}(s \cdot (m - q \cdot p), \ p) \\
&= \text{irem}((1 - p \cdot n) - s \cdot q \cdot p, \ p) \\
&= \text{irem}(1 - p \cdot (n + s \cdot q), \ p) \\
&= 1,
\end{aligned}
$$

which implies that $r = s^{-1}$. It is now a simple matter to give a procedure that finds s^{-1} (Exercise 11).

We have shown that whenever p is prime, $\mathbf{Z}_p$ is a field. The proof of the converse statement is left to the reader (Exercise 12). $\qquad\square$

Example 2.42. Let $p = 47$ and $s = 16$. The extended Euclidean algorithm gives $m = 3$, $n = -1$, and since m is in $\mathbf{Z}_{47}$, $16^{-1} = 3$.

Next let $p = 47$ and $s = 17$. In this case, $m = -11$ and $n = 4$, and since m is not in $\mathbf{Z}_{47}$, we obtain the inverse with $17^{-1} = \text{irem}(-11, 47) = 36$. $\quad\square$

The next theorem gives three simple arithmetic relationships in $\mathbf{Z}_p$.

Theorem 2.43. *Let p be a prime, and let u and v be in $\mathbf{Z}_p$. Then,*

1. $\underbrace{u \oplus_p \cdots \oplus_p u}_{p} = 0$, *where the notation indicates the sum has p copies of u;*

2. $(u \oplus_p v)^p = u^p \oplus_p v^p$;

3. $u^p = u$.

The relationship (3) is known as Fermat's little theorem.

Proof: Part (1) follows from

$$
\underbrace{u \oplus_p \cdots \oplus_p u}_{p} = \text{irem}(\underbrace{u + \cdots + u}_{p}, \ p) = \text{irem}(p \cdot u, \ p) = 0.
$$

To show Part (2), we have, using the binomial theorem,

$$
(u \oplus_p v)^p = \sum_{j=0}^{p} \left(\frac{p!}{(p-j)! \, j!} \right) \left(u^{p-j} \otimes_p v^j \right)
$$

$$
= u^p \oplus \left(\underbrace{\sum_{j=1}^{p-1} \left(u^{p-j} \otimes_p v^j \right) \oplus_p \cdots \oplus_p \left(u^{p-j} \otimes_p v^j \right)}_{\frac{p!}{(p-j)! \, j!}} \right) \oplus v^p. \quad (2.65)
$$

However, since p divides the integer $\dfrac{p!}{(p-j)!\,j!}$, for $1 \leq j \leq p-1$, we have by Part (1) of the theorem

$$\underbrace{\left(u^{p-j} \otimes_p v^j\right) \oplus_p \cdots \oplus_p \left(u^{p-j} \otimes_p v^j\right)}_{\frac{p!}{(p-j)!\,j!}} = 0$$

which eliminates the sum in parentheses in Equation (2.65). Therefore, $(u \oplus_p v)^p = u^p \oplus_p v^p$.

Part (3) is proved using mathematical induction on u. First, the relationship is true when $u = 0$ or $u = 1$. Suppose now that $2 \leq u \leq p-1$, and assume that the relationship holds for $u - 1$. Using Part (2) and the induction hypothesis, we have

$$u^p \;=\; ((u-1) \oplus_p 1)^p = (u-1)^p \oplus_p 1^p = (u-1) \oplus_p 1 = u.$$

Part (3) of the theorem has a number of useful consequences. First, it shows that Part (2) of the theorem is a simple relationship because both sides simplify to $u \oplus_p v$. Another application has to do with the computation of multiplicative inverses in $\mathbf{Z}_p$. For $u \neq 0$, (3) implies that $u^{p-2} \otimes_p u = 1$, and therefore $u^{-1} = u^{p-2}$. □

Other examples of finite fields are given in Exercise 3 and Exercise 7, page 164. Finite fields are used in some modern greatest common divisor and factorization algorithms for polynomials (see Chapter 9).

Algebraic Numbers

Definition 2.44. *A complex number c is called an* **algebraic number** *if it is the root of a polynomial equation $u(x) = 0$, where $u(x)$ is in $\mathbf{Q}[x]$ with positive degree.*[10] *The set of algebraic numbers is denoted by* **A**.

Example 2.45. The following are algebraic numbers because each one satisfies the polynomial equation on the right:

$$
\begin{aligned}
x &= 2/3, & 3x - 2 &= 0, \\
x &= \sqrt{2}, & x^2 - 2 &= 0, \\
x &= i = \sqrt{-1}, & x^2 + 1 &= 0, \\
x &= 7^{1/3}, & x^3 - 7 &= 0,
\end{aligned}
$$

---

[10]Although the terms *algebraic number* defined here and *algebraic expression* defined in Chapter 1 are similar, the mathematical concepts are quite different.

$$x = \frac{1 + \sqrt{5}}{2}, \qquad x^2 - x + 1 = 0, \qquad\qquad (2.66)$$

$$x = \sqrt{2 + \sqrt{2}}, \qquad x^4 - 4x^2 + 2 = 0,$$

$$x = \sqrt{2} + \sqrt{3}, \qquad x^4 - 10x^2 + 1 = 0. \qquad\qquad \square$$

Definition 2.46. *An expression u is an* **explicit algebraic number** *if it satisfies one of the following rules:*

EAN-1. *u is an integer or fraction.*

EAN-2. *$u = v^w$, with $v \neq 0$ an explicit algebraic number and w an integer or fraction.*

EAN-3. *u is a product or sum of explicit algebraic numbers.*

All of the expressions in Example 2.45 are explicit algebraic numbers. However, not all algebraic numbers are explicit. This is a consequence of Galois theory[11] which states that it is impossible to express the roots of all polynomial equations with degree ≥ 5 as explicit algebraic numbers. For example the polynomial equation

$$u = 4x^5 - 10x^2 + 5 = 0 \qquad\qquad (2.67)$$

has three distinct real roots and two distinct (non-real) complex roots, but none of them are explicit. In other words, some algebraic numbers can be expressed explicitly while others are only known implicitly as the root of an equation.

Definition 2.47. *An algebraic number that is not explicit is called an* **implicit algebraic number**.

The five roots of Equation (2.67) are implicit algebraic numbers.

Given an explicit algebraic number, it is often possible to find with some simple manipulations a polynomial that has the number as a root. For example, to find the polynomial associated with

$$x = \sqrt{2} + \sqrt{3}, \qquad\qquad (2.68)$$

first square both sides, and then move the integer part to the left side

$$x^2 - 5 = 2\sqrt{2}\sqrt{3}.$$

---

[11]The theory is named after the brilliant French mathematician Evariste Galois (1811-1832) who outlined its essential ideas. Galois was killed in a duel at the age of 20.

Squaring both sides again, we get

$$x^4 - 10\,x^2 + 1 = 0. \tag{2.69}$$

For a more complicated explicit algebraic number, the manipulations to find the polynomial are more involved. In Section 7.2 we give an algorithm that finds a polynomial that has a given explicit algebraic number as a root.

Now, let's reverse the process to find the roots of Equation (2.69). First, solving for x^2,

$$x^2 = 5 \pm 2\,\sqrt{2}\,\sqrt{3}. \tag{2.70}$$

Next, solving Equation (2.70) for x, we get the four roots

$$x = \pm\sqrt{5 \pm 2\,\sqrt{2}\,\sqrt{3}}. \tag{2.71}$$

Notice that the explicit form (2.71) is quite different from the form (2.68). Since Equation (2.71) gives all four roots of Equation (2.69), we have

$$\pm\sqrt{2} \pm \sqrt{3} = \pm\sqrt{5 \pm 2\,\sqrt{2}\,\sqrt{3}}. \tag{2.72}$$

It is easy to check this relationship by squaring both sides of the equation. However, this relationship points out a difficulty that arises with explicit algebraic numbers: namely, they may have more than one representation. This is not particularly surprising because by simply rationalizing a denominator, two representations can look different. For example,

$$(2 + 3\,\sqrt{2})^{-1} = -1/7 + 3/14\,\sqrt{2}.$$

In general, however, it is a difficult simplification problem to determine when two explicit algebraic expressions are the same number.[12] For example,

$$\sqrt{2} + \sqrt{3} + \sqrt{5} = \frac{\sqrt{2}\,\sqrt{5 \cdot 10^{1/4} + 10^{3/4} + \sqrt{3}\,\sqrt{7\sqrt{10} + 20}}}{10^{1/8}}. \tag{2.73}$$

Algebraic Number Fields

Definition 2.48. *An* **algebraic number field F** *is a field constructed from the base field* **Q** *by adjoining a finite number of algebraic numbers.*

Example 2.49. $\mathbf{Q}(\sqrt{2})$, $\mathbf{Q}(i)$, and $\mathbf{Q}(\sqrt{2}, \sqrt{3})$ are algebraic number fields. $\square$

---

[12]See Davenport, Siret, and Tournier [29] for a discussion of this problem.

We give algorithms for computations in some simple algebraic number fields in Section 4.3.

The set **A** of all (implicit or explicit) algebraic numbers is also a field[13] but not an algebraic number field in the sense of Definition 2.48 because it cannot be constructed from **Q** by adjoining a finite number of algebraic numbers.

Transcendental Numbers

Definition 2.50. *A complex number c that is not an algebraic number is called a* **transcendental number**.

The numbers π and e are real numbers that are transcendental.[14] In addition, although the number e^π is also transcendental, it is not known whether $\pi + e$, $\pi \cdot e$, or π^e are algebraic or transcendental.

Explicit Algebraic Numbers: Theory versus Practice

In Exercise 15, we describe an operator *Explicit_algebraic_number(u)* that returns **true** when u is an explicit algebraic number and otherwise returns **false**. Since we assume that the simplification context for the operator is automatic simplification, there are algebraic expressions that the operator determines are explicit algebraic numbers even though the expressions are undefined in a mathematical sense. For example, the expression

$$u = \frac{1}{2^{1/2} + 3^{1/2} - \left(5 + 2 \cdot 2^{1/2} \cdot 3^{1/2}\right)^{1/2}}$$

is undefined mathematically because the denominator simplifies to 0 (see Equation (2.72)), but this is not obtained by automatic simplification in Maple, Mathematica or MuPAD. For this reason, implementations of the operator in these systems obtain *Explicit_algebraic_number(u)* → **true**. In fact, it is a difficult problem to determine when an explicit algebraic number simplifies to 0. We return to this point in Section 7.2.

Exercises

1. Consider the set of real numbers that have the form $a + b\, 2^{1/3} + c\, 2^{2/3}$ where a, b, and c are rational numbers. Show that this set with the operations of ordinary addition and multiplication is a field.

---

[13] This is proved in Theorem 7.16 on page 289.
[14] For proofs of the transcendence of π and e, see Hardy and Wright [43].

2. (a) Verify that $\mathbf{Q}(\sqrt{2}, \sqrt{3})$ described in Example 2.31 is a field.

   (b) Let u be in $\mathbf{Q}(\sqrt{2}, \sqrt{3})$. Give a procedure *Mult_inv_23(u)* that finds the multiplicative inverse in this field.

3. Let x and y be symbols, and consider the set of expressions $\mathbf{F} = \{0, 1, x, y\}$ with the operations defined by

$\oplus$	0	1	x	y
0	0	1	x	y
1	1	0	y	x
x	x	y	0	1
y	y	x	1	0

$\otimes$	0	1	x	y
0	0	0	0	0
1	0	1	x	y
x	0	x	y	1
y	0	y	1	x

   Show that $\mathbf{F}$ is a field.

4. Prove Theorem 2.34(2). *Hint:* For the uniqueness of the additive inverse, suppose that there are two expressions x and y such that $u + x = 0$ and $u + y = 0$. Show that $x = y$. In a similar way, show that the multiplicative inverse is unique.

5. Prove properties (3) through (8) of Theorem 2.35.

6. Prove Theorem 2.36.

7. Let u and v be expressions in a field $\mathbf{F}$. Show that the division operation satisfies the following.

   (a) $u/1 = u$.

   (b) $u/(-1) = (-u)$.

   (c) If $u \neq 0$ and $v \neq 0$, $(u/v)^{-1} = v/u$.

8. Let $u \neq 0$ be an expression in a field $\mathbf{F}$. Prove that the exponentiation operation satisfies the following.

   (a) $\left(u^{-1}\right)^n = u^{-n}$.

   (b) $\left(u^{-1}\right)^n = \left(u^n\right)^{-1}$.

   Do not use Theorem 2.40 in this exercise.

9. (a) Prove Theorem 2.40(2). *Hint:* Use mathematical induction on $n \geq 0$, and consider the three cases $m + n - 1 \geq 0$, $m + n - 1 = 0$, and $m + n - 1 \leq -1$. The relations in Exercise 8 are useful here.

   (b) Prove Theorem 2.40(3). *Hint:* The relations in Exercise 8 are useful here.

10. Are Theorem 2.43(1) and Theorem 2.43(2) true when p is not a prime? Either prove statements (1) and (2) in this setting or give examples that show that the statements are not always true.

11. Let p be prime, and let s, $t \neq 0$ be in $\mathbf{Z}_p$.

(a) Give a procedure *Multiplicative_inverse_p*(t, p) that obtains the multiplicative inverse t^{-1}.

(b) Give a procedure *Division_p*(s, t, p) that obtains s/t in $\mathbf{Z}_p$.

12. Let $p > 1$ be an integer. Show that if $\mathbf{Z}_p$ is a field, then p is prime.

13. Show that each of the following is an algebraic number by finding a (non-zero) polynomial with rational number coefficients that each one satisfies.

(a) $\sqrt{7} + \sqrt{11}$.

(b) $\sqrt{\sqrt{7} + \sqrt{11}}$.

(c) $\dfrac{1}{1 + \sqrt{2}}$.

(d) $1 + \sqrt{2}\sqrt{3}$.

(e) $\sqrt{2} + \sqrt{3} + \sqrt{5}$.

14. Using the fact that π is transcendental, show that each of the following is also transcendental.

(a) π^2.

(b) $1/\pi$.

15. Give a procedure *Explicit_algebraic_number*(u) that returns **true** if an algebraic expression is an explicit algebraic number and **false** otherwise.

16. Suppose that u is an explicit algebraic number. Give a procedure

$$Nest\_depth(u)$$

that gives the **nesting depth** of u which is defined using the following rules.

(a) If u is an integer or fraction, then $Nest\_depth(u) \to 0$.

(b) If $u = a^r$, where r is a fraction, then

$$Nest\_depth(u) \to 1 + Nest\_depth(a).$$

(c) If u is a sum or a product with n operands then

$$Nest\_depth(u) \quad \to \quad Max(\{Nest\_depth(Operand(u, 1)),$$
$$\ldots, Nest\_depth(Operand(u, n))\})\,.$$

For example,

$$Nest\_depth\left(5 \cdot \left(1 + 2^{1/2}\right)^{1/3}\right) \to 2.$$

17. Let u be an algebraic expression. Give a procedure $Variables\_2(u)$ that returns the set of generalized variables of u that are not explicit algebraic numbers. For example,

$$Variables\_2(2^{1/2}\, x + 3 \cdot 5^{2/3}\, y + z) \to \{x, y, z\}.$$

Note: The *Variables* operator described on page 12 is one of our primitive polynomial operators.

18. Give a procedure $Alg\_collect(u)$ that collects explicit algebraic number coefficients of like terms in a sum u. If u is not a sum, then return u. For example,

$$Alg\_collect \left(\sqrt{2}\, x\, y + 3\, x\, y + \sqrt{3}\, x + 4x \right) \to \left(\sqrt{2} + 3 \right) x\, y + \left(\sqrt{3} + 4 \right) x,$$

$$Alg\_collect \left(\sqrt{2}\, \sqrt{3} + \sqrt{2}\, z + \sqrt{3}\, z \right) \to \sqrt{2}\, \sqrt{3} + \left(\sqrt{2} + \sqrt{3} \right) z,$$

$$Alg\_collect \left(\sqrt{2}\, x\, y + 2\, x\, y + x \right) \to \left(\sqrt{2} + 2 \right) x\, y + x,$$

$$Alg\_collect \left(a \left(\sqrt{2}\, x + x \right) \right) \to a \left(\sqrt{2}\, x + x \right).$$

In the last example, collection is not done because the input expression is not a sum.

19. A Gaussian rational number c is in standard form if it has one of the following forms:

   (a) c is a rational number in standard form.

   (b) $c = \imath$.

   (c) $c = b \cdot \imath$, where $b \neq 0, 1$ is a rational number in standard form.

   (d) $c = a + b \cdot \imath$, where $a \neq 0$ and $b \neq 0, 1$ are rational numbers in standard form.

   (e) $c = a + \imath$, where $a \neq 0$ is a rational number in standard form.

In forms (c), (d), and (e) the order of the operands in a sum or product is a significant property of the standard form. (For example, the expression $\imath \cdot b$ is out of order, and so is not in standard form.) In this exercise, we ask you to give an algorithm $Simplify\_Gaussian(u)$ that transforms a Gaussian rational number expression (defined below) to either a Gaussian rational number in standard form or the symbol **Undefined** when a division by zero is encountered. The *Gaussian rational number expressions (GRNE)* are the ones that satisfy one of the following rules:

GRNE-1. u is an integer.

GRNE-2. u is a fraction. At this point, we allow for the possibility of a fraction with zero denominator. The algorithm will recognize that such an expression is undefined.

GRNE-3. u is the symbol $\imath$.

GRNE-4. u is a unary or binary sum with operands that are GRNEs.

GRNE-5. u is a unary difference or a binary difference with operands that are GRNEs.

GRNE-6. u is a binary product with operands that are GRNEs.

GRNE-7. u is a quotient with operands that are GRNEs.

GRNE-8. u is a power with a base that is a GRNE and an exponent that is an integer.

To bypass the automatic simplification rules in a CAS, represent the arithmetic operations and fractions with the function forms given in Figure 2.6 on page 39. For example,

$$Simplify\_Gaussian(SumOp($$
$$FracOp(1,2), QuotOp(1, SumOp(3, \imath))))$$
$$\rightarrow \quad SumOp(FracOp(4,5), ProdOp(FracOp(-1, 10), \imath)).$$

Further Reading

2.1 The Integers. See Birkhoff and Mac Lane [10], Dean [31], or Pinter [79] for derivations of properties of the integers. Euclid's algorithm appears in Book Seven of Euclid's *Elements*, Propositions 1 and 2. See Calinger [19], pages 119-120 for Euclid's original approach. An explicit formula for the gcd is given in Polézzi [80]. See Cormen, Leiserson and Rivest [28] and von zur Gathen and Gerhard [96] for algorithms for the prime factorization of positive integers. Libbrecht [61] has an historical essay on the Chinese remainder theorem. Davis and Hersh [30] has an interesting discussion of the Chinese remainder theorem and its generalizations.

2.3 Fields. See Birkhoff and Mac Lane [10], Dean [31], or Pinter [79] for a discussion of fields and algebraic numbers. A history of the field concept is given in Kleiner [52], [53]. The theory and application of finite fields is described in Lidl and Niederreiter [62]. See Jeffrey and Rich [49] for a discussion of some simplification problems with square roots. An interesting discussion of algebraic and transcendental numbers is given in Stevenson [93].

3

Automatic Simplification

The automatic simplification process is defined as a collection of algebraic and trigonometric simplification transformations that is applied to an expression as part of the evaluation process. In this chapter we take an in-depth look at the algebraic component of this process.

In Section 3.1 we describe the automatic simplification process in an informal way, focusing on which transformations should be included in the process and which ones are best handled by other operators. Next, we give a precise definition of an automatically simplified algebraic expression which includes an order relation that describes actions of the additive and multiplicative commutative transformations.

In Section 3.2 we describe the basic automatic simplification algorithm. Although the algorithm is based only on the rules of elementary algebra, it is quite involved. To handle a large variety of expressions, the automatic simplification process includes over 30 rules (or rule groups), some of which are recursive. The problem addressed in this section is how to organize this involved process into a coherent algorithm.

3.1 The Goal of Automatic Simplification

The transformation rules in automatic simplification are motivated by the field axioms (which are assumed to hold for expressions as well as for number fields) and by the transformations that are logical consequences of these axioms. We begin by examining the role of each of the axioms in the automatic simplification process.

The Basic Distributive Transformations

The (right) distributive property (field axiom F-8) has the form

$$(a + b) \cdot c = a \cdot c + b \cdot c. \qquad (3.1)$$

This transformation is applied during automatic simplification in a right to left manner to combine constant (integer or fraction) coefficients of like terms in a sum. A typical transformation is

$$2\,x + y + (3/2)\,x \;=\; (7/2)\,x + y.$$

There are, however, some similar manipulations that are omitted from automatic simplification. One natural generalization of this manipulation involves the collection of non-constant coefficients of like terms as well as the constant ones as is done in

$$3\,x + a\,x + y = (3 + a)\,x + y. \qquad (3.2)$$

There are two reasons why this transformation is omitted from automatic simplification. First, there is uncertainty about which parts of an expression are the like terms and which parts are the coefficients. This happens with the expression

$$a\,x + b\,x + b\,y,$$

for if we collect coefficients of x, we obtain $(a + b)\,x + b\,y$, while if we collect coefficients of b, we obtain $a\,x + b\,(x+y)$. In other words, since the structure of an expression by itself does not identify semantic roles such as coefficient and like term, the operation is ambiguous.

A second reason to avoid collecting non-constant coefficients has to do with its relationship to the *Algebraic_expand* operator. Since this operator applies Equation (3.1) in the opposite (left to right) direction, the collection of non-constant coefficients cancels out some of its actions. For these reasons, it is common practice in computer algebra systems to collect only rational number coefficients with automatic simplification. Unfortunately, this means that coefficients that are not rational numbers but play the semantic role of constants (e.g., π, $\sqrt{2}$) are also not collected. For example, in

$$2\,x + 3\,x + \sqrt{2}\,x = 5\,x + \sqrt{2}\,x,$$

we combine only the integer coefficients of x.

Since automatic simplification uses the tree structure to recognize like terms, it is only able to combine coefficients of like terms with a similar structure. This means that the manipulation

$$\sin(x) + 2\,\sin(x) = 3\,\sin(x)$$

is obtained with automatic simplification, while

$$1 + x + 2(1 + x) = 3(1 + x) \tag{3.3}$$

is not. In the latter case, the $1 + x$ at the far left is not a complete sub-expression of the sum and so is not structurally the same as the $1 + x$ in the product $2(1 + x)$.

There is another way to use distributive properties to simplify the expression (3.3). If we apply the left distributive property (field axiom F-7)

$$a \cdot (b + c) = a \cdot b + a \cdot c$$

in a left to right fashion to distribute the constant 2 over $1 + x$, and then use the distributive property again to combine coefficients of x, we obtain

$$1 + x + 2(1 + x) = 3 + 3x. \tag{3.4}$$

This manipulation is more involved because it involves the application of the distributive property both to distribute and collect constants. Because of this, there is some question about whether it belongs in automatic simplification or should be obtained with the *Algebraic_expand* operator. In addition, if constants are distributed over sums, then what should be done with $2(3 + x)(4 + x)$? Should the 2 be distributed over one of the two sums in the product and if so which one? This uncertainty is reflected in the different approaches taken by computer algebra software. Maple and MuPAD achieve manipulations like Equation (3.4) with automatic simplification,[1] while Mathematica does not. However, all three systems achieve the manipulation

$$x + 1 + (-1)(x + 1) = 0, \tag{3.5}$$

even though it also involves both the distribution and collection of constants.

With these considerations in mind, we include the following transformation in our automatic simplification algorithm.

Definition 3.1. *The* **basic distributive transformation** *in automatic simplification refers to the collection of integer and fraction coefficients of like terms in a sum.*

Observe that the basic distributive transformation does not distribute constants over sums[2] and so does not obtain the simplifications in Equations (3.4) and (3.5).

---

[1] Automatic simplification in Maple and MuPAD distributes a constant over a single sum but does not change $2(3 + x)(4 + x)$. (Implementation: Maple (mws), MuPAD (mnb).)

[2] A modification of the automatic simplification algorithm that does distribute constants over sums is described in Exercise 8, page 108.

As a result of this transformation, a simplified sum cannot have two operands with the same like term.[3]

The Basic Associative Transformations

The associative transformations are based on the additive and multiplicative associative field axioms (F-5 and F-6):

$$a + (b + c) = (a + b) + c, \tag{3.6}$$

$$(a \cdot b) \cdot c = a \cdot (b \cdot c). \tag{3.7}$$

The associative properties suggest that the parentheses in Equations (3.6) and (3.7) are superfluous, and so a natural way to incorporate associativity into automatic simplification is simply to eliminate the parentheses. This means automatic simplification should perform the following manipulations:

$$(w + x) + (y + z) = w + x + y + z,$$

$$((w \cdot x) \cdot y) \cdot z = w \cdot x \cdot y \cdot z,$$

$$(((u + v) \cdot w) \cdot x + y) + z = (u + v) \cdot w \cdot x + y + z.$$

The tree structures that correspond to these simplifications are shown in Figures 3.1 and 3.2. In a structural sense, the associative transformations tend to reduce the number of levels in an expression tree by removing superfluous addition or multiplication operators and reassigning their operands to appropriate instances of the same operator in the tree.

When the automatic simplification process is applied to a particular sub-expression of an expression, the additive associative transformation is always applied before the distributive transformations. The next two examples show how this order of application affects the simplification process.

Example 3.2. Consider the manipulation

$$x + (x + y) = x + x + y = 2x + y.$$

In this case, we first apply an associative transformation that modifies the structure of the expression and then apply a distributive transformation that collects coefficients of like terms. On the other hand, if the distributive transformation were applied before the associative transformation, it would not combine the two unlike terms x and $x + y$, and so after applying the associative transformation, the simplified form would be $x + x + y$.   □

---

[3]This statement also requires that the basic associative transformation for sums be applied before the basic distributive transformation (see Definition 3.5 and Example 3.2).

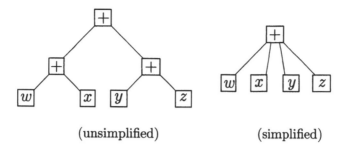

(a) Tree structures for $(w + x) + (y + z) = w + x + y + z$.

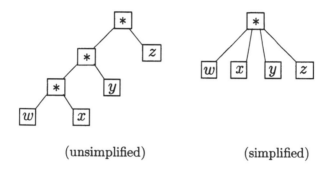

(b) Tree structures for $((w \cdot x) \cdot y) \cdot z = w \cdot x \cdot y \cdot z$.

Figure 3.1. Tree transformations with the basic associative transformations.

Example 3.3. Sometimes the additive associative transformation inhibits an application of the distributive transformation. This happens with the expression

$$(1 + x) + 2(1 + x). \tag{3.8}$$

(We considered a similar expression in Equation (3.3) where parentheses were omitted from the $1 + x$ at the left.) In this unsimplified form, both instances of $1 + x$ are syntactically identical, which suggests that a distributive transformation might collect the coefficients of $1 + x$. However, since the associative transformation is applied first, the parentheses about the $1 + x$ on the left are removed and the operands 1 and x become operands of the main $+$ operator. As we pointed out in the discussion following

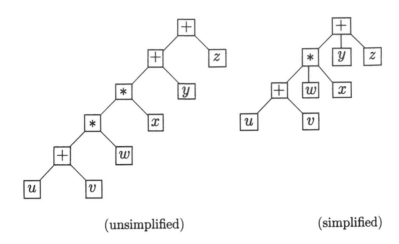

(unsimplified)   (simplified)

(c) Tree structures for $(((u + v) \cdot w) \cdot x + y) + z = (u + v) \cdot w \cdot x + y + z.$

Figure 3.2. Tree transformations with the basic associative transformations (continued).

Equation (3.3), when the expression is in this form, the distributive transformation does not recognize the two instances of $1 + x$ as like terms, and so the simplified form of the expression (3.8) is $1 + x + 2(1 + x)$. □

The next example shows how the multiplicative associative transformation paves the way for the distributive transformation by transforming the operands of a sum to a common form that makes it easy to recognize like terms.

Example 3.4. Consider the manipulation

$$
\begin{aligned}
2(x\,y\,z) + 3x\,(y\,z) + 4(x\,y)\,z &= 2\,x\,y\,z + 3\,x\,y\,z + 4\,x\,y\,z \\
&= 9\,x\,y\,z.
\end{aligned}
$$

Since the sum on the left has three terms with different structures, the recognition of like terms is more complicated than it needs to be. However, when the simplification process is applied recursively to the operands of the sum, the multiplicative associative transformation puts each term in the standard form shown in the sum on the right. Since each term in this new sum has a similar form, it is now easy to recognize the like terms. ☐

With these considerations in mind, we include the following transformations in our automatic simplification algorithm.

Definition 3.5. *The* **basic associative transformations** *refer to the transformations that modify the structure of an expression u in the following ways.*

1. *Suppose that u is a sum. If s is an operand of u that is also a sum, then the operator of s is removed from the expression tree, and the operands of s are affixed to the main operator of u. At a given level in an expression, this transformation is applied before the basic distributive transformation (Definition 3.1, page 65).*

2. *Suppose that u is a product. If p is an operand of u that is also a product, then the operator of p is removed from the expression tree, and the operands of p are affixed to the main operator of u. At a given level in an expression, this transformation is applied before the basic power transformation (3.14) in Definition 3.8, page 74.*

In Examples 3.9 and 3.10 below, we show how the product associative transformation affects the action of the power transformation (3.14).

The basic associative transformations imply two more properties of automatically simplified expressions. As a result of these transformations, a simplified sum cannot have an operand that is also a sum, and a simplified product cannot have an operand that is a product.

The Basic Commutative Transformations

The commutative transformations are based on the field commutative properties (axioms F-3 and F-4)

$$a + b = b + a, \tag{3.9}$$

$$a \cdot b = b \cdot a. \tag{3.10}$$

The next example shows how these transformations prepare an expression for an application of a distributive transformation.

Example 3.6. Consider the manipulation

$$
\begin{aligned}
2\,z\,x\,y + 3\,y\,x\,z + 4\,x\,z\,y &= 2\,x\,y\,z + 3\,x\,y\,z + 4\,x\,y\,z \\
&= 9\,x\,y\,z.
\end{aligned}
$$

In the first line, the three operands of the main operator $+$ are transformed recursively by commutative transformations to a standard form. In this form, it is easy to recognize the like terms and obtain the simplified form. $\square$

In order to include the commutative transformations in a simplification algorithm, we need to describe the standard forms for a sum and a product that are produced by these transformations. There are two ways to define these standard forms. The approach that is most often used is to define an order relation for expressions and to reorder the operands in a sum or product using this relation. The Mathematica and MuPAD systems use this approach as will our simplification algorithm. Another approach, used by the Maple system, is described later in this section.

Let's take a brief look at the order relation approach. Although the general definition of an order relation is quite involved, a few of its rules are easy to state. For example, for symbols the order is based on the lexicographical (i.e., alphabetical) order of the symbol names. For example, in

$$b + c + a = a + b + c,$$

the operands of the sum on the left are out of order and are reordered by commutative transformations to the standard form on the right. Another order rule[4] places constants in a product before symbols as is done in

$$x \, 3 \, a = 3 \, a \, x.$$

A third rule orders powers in a sum that have a common base according to their exponents as is done in

$$x^3 + x^2 + x^4 = x^2 + x^3 + x^4.$$

The rules that are used to determine the order of more complicated expressions are quite involved although they are ultimately based on rules similar to those given above. (A complete description of the reordering process is given in Definition 3.26 starting on page 84.) For example, the order relation used by our algorithm produces the standard form shown on the right:

$$3 \, (x^2 + 2) \, y \, x^3 \, (x^2 + 1) = 3 \, x^3 \, (1 + x^2) \, (2 + x^2) \, y.$$

One advantage of basing commutative transformations on an order relation is that expressions are often displayed in a form that resembles the conventions used in mathematical notation. There are, however, other

---

[4]In MuPAD, an integer or fraction operand in a product is represented internally as the last operand even though the displayed form indicates it is the first operand.

$In[1] := \mathbf{u} = \mathbf{Expand}[(\mathbf{x}^2 + \mathbf{b} * \mathbf{x}) * (\mathbf{c} + \mathbf{x})]$

$Out[1] = \mathrm{b\,c\,x} + \mathrm{b\,x}^2 + \mathrm{c\,x}^2 + \mathrm{x}^3$

$In[2] := \mathbf{ReplaceAll}[\mathbf{u}, \mathbf{x} - > \mathbf{a}]$

$Out[2] = \mathrm{a}^3 + \mathrm{a}^2\,\mathrm{b} + \mathrm{a}^2\,\mathrm{c} + \mathrm{a\,b\,c}$

$In[3] := \mathbf{D}[\mathbf{Cos}[\mathbf{x}] + \mathbf{x}^2 * \mathbf{Sin}[\mathbf{x}], \mathbf{x}]$

$Out[3] = \mathrm{x}^2\,\mathrm{Cos}[\mathrm{x}] - \mathrm{Sin}[\mathrm{x}] + 2\,\mathrm{x}\,\mathrm{Sin}[\mathrm{x}]$

Figure 3.3. An interactive dialogue with the Mathematica system that shows the effect of commutative transformations in automatic simplification. (Implementation: Mathematica (nb), MuPAD (mnb).)

instances where the reordering of an expression can confuse the presentation. Both of these possibilities arise in the dialogue with the Mathematica system shown in Figure 3.3. At *In[1]*, the **Expand** operator creates a polynomial that is ordered by increasing powers of x and assigned to u. At *In[2]*, the symbol x in u is replaced by the symbol a which significantly changes the order in the expression. Since the ordering scheme is oblivious to the semantics of an operation, it has transformed the expression to a form that makes the operation appear awkward. At *In[3]*, the result of the differentiation has been reordered so that the displayed form obscures the application of the differentiation rules. Results similar to these are inevitable when an order relation determines the actions of the commutative transformations.

The Maple system uses another approach to determine the actions of the commutative transformations. In this system, the simplified form of a sum or product is determined by the simplified form of the expression when it is first encountered in a session. For example, consider the Maple dialogue shown in Figure 3.4. At the first prompt, the expression $a * (z + x + y)$ is assigned to u. At this point, Maple stores the sub-expression $z + x + y$ and the expression $a * (z + x + y)$ in a table that defines these forms as the simplified forms of these expressions. At the second prompt, the statement includes two sub-expressions that are equivalent by commutative transformations to $z + x + y$. At this point the Maple system searches the table to see if a simplified form of this expression or any of its sub-

---

```
> u := a*(z + x + y);
```

$$u := a\,(z + x + y)$$

```
> v := b*(x + y + z) + c*(y + z + x);
```

$$v := b\,(z + x + y) + c\,(z + x + y)$$

```
> w := (3*z + 2*x + y - x - 2*z)*(x*x + a - x^2);
```

$$w := a\,(z + x + y)$$

Figure 3.4. An interactive dialogue with the Maple system that illustrates the actions of the commutative transformations. (Implementation: Maple (mws).)

expressions can be transformed by commutative transformations to one already in the table. Since $z + x + y$ is already in the table, this form is used in the final simplified displayed expression. In a similar way, at the third prompt an involved expression is simplified to a form that is already in the table.

Each of the two approaches that determines the actions of the commutative transformations has its strengths and weaknesses. Let's contrast the two approaches by comparing the Mathematica dialogue in Figure 3.3 to a similar dialogue with the Maple system (Figure 3.5). The first statement in the Maple dialogue assigns an expression in powers of x to a variable u. However, in this case the simplified form is harder to comprehend than the more systematic form produced by the Mathematica system. On the other hand, the next two Maple manipulations are much easier to follow than the Mathematica ones because the expressions are not reordered. The last statement in the Maple session shows the **sort** operator which redefines the internal structure of a polynomial so that the powers now appear in decreasing order.

From a user's perspective, it is difficult to make a general statement about which of the two approaches produces more readable results. If expressions are entered in the Maple system in a systematic way, there is a good chance the results will be displayed in a reasonable form. We didn't do this in the first statement in Figure 3.5 because $x^2 + b*x$ has decreasing powers of x while $c + x$ has increasing powers of x. If we had reversed the

```
> u := expand( (x^2+b*x)*(c+x) );
```

$$u := x^2\,c + x^3 + b\,x\,c + b\,x^2$$

```
> v := subs(x=a,u);
```

$$v := a^2\,c + a^3 + b\,a\,c + b\,a^2$$

```
> diff(cos(x) + x^2*sin(x), x);
```

$$-\sin(x) + 2\,x\,\sin(x) + x^2\,\cos(x)$$

```
> sort(u,x);
```

$$x^3 + c\,x^2 + b\,x^2 + b\,c\,x$$

Figure 3.5. An interactive dialogue with the Maple system that is similar to the Mathematica dialogue in Figure 3.3. (Implementation: Maple (mws).)

order of one of these, the result would be in a more systematic form. On the other hand, in some cases a system that uses an order relation has a better chance of reordering an awkward expression into a more understandable form. With either approach, however, it is inevitable that some expressions will be displayed in an awkward way.

The following transformations are included in our automatic simplification algorithm.

Definition 3.7. *The* **basic commutative transformations** *refer to transformations that are based on the additive and multiplicative commutative properties that reorder the operands in a sum or a product to a standard form. The standard form is based upon the order relation that is defined in Definition 3.26 starting on page 84.*

The Basic Power Transformations

The power transformations are based on the three relations in Theorem 2.40. For m and n integers,

$$a^m \cdot a^n = a^{m+n}, \tag{3.11}$$

$$(a^m)^n = a^{m \cdot n}, \tag{3.12}$$

$$(a \cdot b)^n = a^n \cdot b^n. \tag{3.13}$$

In both the real field $\mathbf{R}$ and the complex field $\mathbf{C}$, the power operation can be generalized to allow exponents that are also members of the field. In these more general settings, the transformation (3.11) is true, while the transformations (3.12) and (3.13) are only true when n is an integer (see Example 3.11). With these considerations in mind, we assume that similar transformations are applied to mathematical expressions during automatic simplification.

Definition 3.8. *Let n be an integer. The following* **basic power transformations** *are applied during automatic simplification:*

$$u^v \cdot u^w \to u^{v+w}, \tag{3.14}$$

$$(u^v)^n \to u^{v \cdot n}, \tag{3.15}$$

$$(u \cdot v)^n \to u^n \cdot v^n. \tag{3.16}$$

As a result of these transformations, the simplified form of a power with an integer exponent cannot have a base that is also a power or a product. In addition, the simplified form of a product cannot have two operands that have a common base.

The power transformations in MPL are similar to those in Maple, Mathematica, and MuPAD (Figure 3.6).

Example 3.9. Consider the simplification:

$$
\begin{aligned}
(x \cdot y) \cdot (x \cdot y)^2 &= (x \cdot y) \cdot (x^2 \cdot y^2) \\
&= x \cdot y \cdot x^2 \cdot y^2 \\
&= x^3 y^3.
\end{aligned}
$$

In the first step, the second operand $(x \cdot y)^2$ is simplified (recursively) using the transformation (3.16). In the second step, a basic associative transformation is applied to each operand of the main product. In the third step, the transformation (3.14) is applied to give the final form.[5] Notice how the associative transformation prepares the expression for the final power transformation.   □

---

[5]Although the actual path taken by our algorithm for the last two steps is more involved than what is shown in this example, the final result is the same.

Input	MPL	Maple	Mathematica	MuPAD
$x^2 * x^3$	x^5	x^5	x^5	x^5
$x^{1/2} * x^{1/3}$	$x^{5/6}$	$x^{5/6}$	$x^{5/6}$	$x^{5/6}$
$x^a * x^b$	x^{a+b}	$x^a x^b$	x^{a+b}	$x^a x^b$
$(x^2)^3$	x^6	x^6	x^6	x^6
$(x^a)^2$	x^{2a}	$(x^a)^2$	x^{2a}	x^{2a}
$(x^2)^{1/2}$	$(x^2)^{1/2}$	$(x^2)^{1/2}$	$(x^2)^{1/2}$	$(x^2)^{1/2}$
$(x^{1/2})^2$	x	x	x	x
$(x^2)^a$	$(x^2)^a$	$(x^2)^a$	$(x^2)^a$	$(x^2)^a$
$(x * y)^2$	$x^2 y^2$	$x^2 y^2$	$x^2 y^2$	$x^2 y^2$
$(x * y)^{1/3}$	$(x\ y)^{1/3}$	$(x\ y)^{1/3}$	$(x\ y)^{1/3}$	$(x\ y)^{1/3}$
$(x * y)^a$	$(x\ y)^a$	$(x\ y)^a$	$(x\ y)^a$	$(x\ y)^a$

Figure 3.6. Power transformations in automatic simplification in MPL, Maple, Mathematica, and MuPAD. (Implementation: Maple (mws), Mathematica (nb), MuPAD (mnb).)

Example 3.10. Consider the simplification

$$(x \cdot y) \cdot (x \cdot y)^{1/2} = x \cdot y \cdot (x \cdot y)^{1/2}.$$

In this case, the second operand $(x \cdot y)^{1/2}$ is in simplified form, and the final result is obtained by applying a basic associative transformation. Notice that automatic simplification does not obtain $(x \cdot y)^{3/2}$ because the associative transformation converts the expression to a form where the power transformation (3.16) does not apply. □

Example 3.11. The following examples show that the transformations (3.15) and (3.16) do not hold in a complex setting when n is not an integer.[6] Indeed, if the transformations were true, we would obtain the following:

$$1 = 1^{1/2} = (\imath^4)^{1/2} \neq \imath^2 = -1,$$

$$1 = 1^{1/2} = ((-1) \cdot (-1))^{1/2} \neq (-1)^{1/2}(-1)^{1/2} = \imath^2 = -1. \qquad □$$

The Basic Difference Transformations

Definition 3.12. *The following* **basic difference transformations** *are applied by automatic simplification.*

1. Each unary difference is replaced by the product

$$-u \rightarrow (-1) \cdot u. \qquad (3.17)$$

---

[6]See Pennisi [78], pp. 112–113 for exponent relationships in a complex setting.

2. *Each binary difference is replaced by the sum*

$$u - v \rightarrow u + (-1) \cdot v. \tag{3.18}$$

As a result of these transformations, automatically simplified expressions do not contain the difference operator. Most computer algebra systems apply these transformations during automatic simplification.

The Basic Quotient Transformation

Definition 3.13. *During automatic simplification the following* **basic quotient transformation** *is applied:*

$$\frac{u}{v} \rightarrow u \cdot v^{-1}. \tag{3.19}$$

As a result of this transformation, automatically simplified expressions do not contain the quotient operator. Most computer algebra systems apply the quotient transformation during automatic simplification although this may not be apparent from the displayed form of an expression which may still contain a quotient operator.

Example 3.14. The quotient transformation prepares an expression for a power transformation as is done in

$$\frac{a \cdot x^3}{x} = (a \cdot x^3) \cdot x^{-1} = a \cdot x^3 \cdot x^{-1} = a \cdot x^2. \qquad \square$$

The Basic Identity Transformations

Definition 3.15. *The following* **basic identity transformations** *are applied by automatic simplification:*

$$
\begin{align}
u + 0 \quad &\rightarrow \quad u, \tag{3.20}\\
u \cdot 0 \quad &\rightarrow \quad 0, \tag{3.21}\\
u \cdot 1 \quad &\rightarrow \quad u, \tag{3.22}\\
0^w \quad &\rightarrow \quad \begin{cases} 0, & \text{if } w \text{ is a positive integer or fraction,} \\ \textbf{Undefined}, & \text{otherwise,} \end{cases} \tag{3.23}\\
1^w \quad &\rightarrow \quad 1, \tag{3.24}
\end{align}
$$

$$v^0 \rightarrow \begin{cases} 1, & \text{if } v \neq 0, \\ \textbf{Undefined}, & \text{if } v = 0, \end{cases} \tag{3.25}$$

$$v^1 \rightarrow v. \tag{3.26}$$

In light of Definition 3.15, automatically simplified expressions satisfy the following properties.

1. A sum cannot have 0 as an operand.

2. A product cannot have 0 or 1 as an operand.

3. A power cannot have 0 or 1 as a base or an exponent.

The transformations (3.20)–(3.26) are considered basic because each one appears in some form in our algorithm. Other manipulations involving identities are consequences of these basic identity transformations, numerical calculations with identity elements, and other transformations in this section (Exercise 1, page 107). These include the following transformations with quotients and differences:

$$\frac{u}{0} \rightarrow \textbf{Undefined}, \tag{3.27}$$

$$\frac{0}{u} \rightarrow \begin{cases} 0, & \text{if } u \neq 0, \\ \textbf{Undefined}, & \text{if } u = 0, \end{cases} \tag{3.28}$$

$$u/1 \rightarrow u, \tag{3.29}$$

$$u - 0 \rightarrow u, \tag{3.30}$$

$$0 - u \rightarrow (-1) \cdot u. \tag{3.31}$$

Although transformations similar to (3.20)–(3.31) are achieved by automatic simplification in most computer algebra systems, some systems transform powers with base 0 to other forms (see Figure 3.7). For example, although 0^0 is usually considered indeterminate, some mathematicians define 0^0 as 1, and this transformation is obtained by both Maple and Mu-PAD.[7] In addition, in MPL the expression 0^w is considered undefined when w is not a positive integer or fraction. Certainly, this is appropriate when w is a negative integer or fraction, and when w is some other algebraic expression, we take the conservative (safe) approach and simply assume that the expression is undefined. On the other hand, when w is not a non-zero integer or fraction, Maple transforms $0^w \rightarrow 0$, and both Mathematica and MuPAD consider 0^w as an unevaluated power.

Unfortunately, since the transformations (3.23), (3.25), and (3.28) include test conditions, they may give results that are not mathematically

---

[7]Knuth, Graham, and Patashnik [56] (page 162) assume that $0^0 = 1$. Their motivation is based on some limiting cases of the binomial theorem.

	MPL	Maple	Mathematica	MuPAD
0^0	**Undefined**	1	Indeterminate	1
0^{-2}	**Undefined**	error message	ComplexInfinity	error message
0^w	**Undefined**	0	0^w	0^w

Figure 3.7. Transformations of powers with base 0 in MPL, Maple, Mathematica, and MuPAD. (Implementation: Maple (mws), Mathematica (nb), MuPAD (mnb).)

legitimate in a larger simplification context. For example, for the transformation

$$\left(\sin^2(x) + \cos^2(x) - 1\right)^0 \to 1,$$

the base is not transformed to 0 by automatic simplification. Therefore, automatic simplification does not recognize this as the undefined expression 0^0 and transforms it instead to 1.

The Basic Unary Transformations

Definition 3.16. *The following* **basic unary transformations** *are applied during automatic simplification:*

$$\cdot x \quad \to \quad x, \tag{3.32}$$

$$+x \quad \to \quad x. \tag{3.33}$$

These transformations apply to unary operations that appear in an expression and to those that are created during the course of the simplification. As a result of these transformations, a simplified expression cannot contain a unary product or sum.

The Basic Numerical Transformations

Definition 3.17. *The following* **basic numerical transformations** *are applied during automatic simplification:*

1. *the multiplication of constant operands in a product,*

2. *the addition of constant operands in a sum,*

3. *the evaluation of powers with an integer or fraction base and an integer exponent, and*

Input	MPL	Maple	Mathematica	MuPAD
$4^{1/2}$	$4^{1/2}$	$4^{1/2}$	2	2
$54^{1/3}$	$54^{1/3}$	$54^{1/3}$	$3 \cdot 2^{1/3}$	$54^{1/3}$

Figure 3.8. Numerical power transformations with radicals in automatic simplification in MPL, Maple, Mathematica, and MuPAD. (Implementation: Maple (mws), Mathematica (nb), MuPAD (mnb).)

4. the evaluation of factorials with a non-negative integer operand.

All arithmetic operations result in rational numbers in the standard form described in Definition 2.25 on page 37.

As a result of these transformations, automatically simplified expressions have the following properties.

1. A sum or product can have at most one operand that is constant.

2. A power with an integer exponent cannot have a constant base.

3. A factorial cannot have a non-negative integer operand.

Although some computer algebra systems evaluate radicals during automatic simplification (e.g., $4^{1/2} = 2$, $54^{1/3} = 3 \cdot 2^{1/3}$ (Figure 3.8)), these transformations are not included in our algorithm.

The "Undefined" Transformation

We take the view that if some operand of an expression is undefined, then the entire expression is considered undefined. For example, $x + 0^{-1/2}$ is undefined because the second operand of the sum $0^{-1/2}$ is undefined. With these considerations in mind, we include the following in automatic simplification.

Definition 3.18. *The following "Undefined" transformation is applied during automatic simplification:*

> *if u is a compound expression with an operand that is the symbol* **Undefined***, then automatic simplification transforms u to* **Undefined***.*

Basic Algebraic Expressions

In order to describe the automatic simplification algorithm, we need precise
definitions for the expressions that are the input and output of the algorithm. The following definition describes the class of input expressions.

Definition 3.19. *A* **basic algebraic expression (BAE)** *is an expression
that satisfies one of the following rules.*

BAE-1. *u is an integer.*

BAE-2. *u is a fraction.*

BAE-3. *u is a symbol.*

BAE-4. *u is a product with one or more operands that are BAEs.*

BAE-5. *u is a sum with one or more operands that are BAEs.*

BAE-6. *u is a quotient with two operands that are BAEs.*

BAE-7. *u is a unary or binary difference where each operand is a BAE.*

BAE-8. *u is a power where both operands are BAEs.*

BAE-9. *u is a factorial with an operand that is a BAE.*

BAE-10. *u is a function form with one or more operands that are BAEs.*

Example 3.20. The following are BAEs:

$$2/4, \quad a \cdot (x + x), \quad a + b^3/b, \quad b - 3 \cdot b, \quad a + (b + c) + d,$$

$$2 \cdot 3 \cdot x \cdot x^2, \quad f(x)^1, \quad + x^2 - x, \quad 0^{-3}, \quad \cdot x \quad 2/(a - a), \quad 3! \ \square$$

The BAEs are similar to conventional algebraic expressions, except now
products and sums can have one or more operands. By including these
expressions in the definition, the automatically simplified expressions (see
Definition 3.21) are included in the BAEs. Notice that we have included
unary products as well as unary sums because they can be created in the
course of a simplification. All unary products and sums are eliminated
during the simplification process.

At this point, we allow expressions that are transformed by automatic
simplification to a mathematically undefined form because the algorithm
recognizes when these situations arise and returns the symbol **Undefined**.
For example, in the last example, the expressions 0^{-3} and $2/(a - a)$ are
mathematically undefined.

Automatically Simplified Algebraic Expressions

In Chapter 1 we described some of the structural properties of automatically simplified algebraic expressions. The next definition, which gives a complete description of these expressions, defines the goal of automatic simplification. The properties in the definition are the result of the basic transformations given in the beginning of this section.

Definition 3.21. *An expression u is an* **automatically simplified algebraic expression** **(ASAE***) if it satisfies one of the following rules.*

ASAE-1. *u is an integer.*

ASAE-2. *u is a fraction in standard form.*

ASAE-3. *u is a symbol except the symbol* **Undefined***.*

The next rule describes the simplified form of a product. An important simplification transformation for products is the basic power transformation $u^v \cdot u^w = u^{v+w}$. Since the relation holds when an exponent is understood to be 1 (e.g., $b \cdot b^2 = b^3$), it is useful to view each non-constant operand of a product as a power. With these considerations in mind, we define the *base* and *exponent* of an ASAE u as

$$\text{base}(u) = \begin{cases} u, & \text{when } u \text{ is a symbol, product, sum,} \\ & \text{factorial, or function} \\ Operand(u,1), & \text{when } u \text{ is a power,} \\ \textbf{Undefined}, & \text{when } u \text{ is an integer or fraction,} \end{cases}$$

$$\text{exponent}(u) = \begin{cases} 1, & \text{when } u \text{ is a symbol, product, sum,} \\ & \text{factorial, or function,} \\ Operand(u,2), & \text{when } u \text{ is a power,} \\ \textbf{Undefined}, & \text{when } u \text{ is an integer or fraction.} \end{cases}$$

For example,

$$\text{base}(x^2) = x, \quad \text{exponent}(x^2) = 2,$$

$$\text{base}(x) = x, \quad \text{exponent}(x) = 1.$$

The implementation of procedures for these operators is left to the reader (Exercise 1).

The base operator is used to define the simplified form of a product.

ASAE-4. *u is a product with two or more operands* $u_1, u_2, \ldots, u_n$ *that satisfy the following properties.*

1. *Each operand* u_i *is an ASAE which can be either an integer ($\neq$ 0, 1), fraction, symbol ($\neq$ **Undefined**), sum, power, factorial, or function. (An operand of a product cannot be a product.) An expression that satisfies this property is called an **admissible factor** of a product.*

2. *At most one operand* u_i *is a constant (integer or fraction).*

3. *If* $i \neq j$, *then* $\mathrm{base}(u_i) \neq \mathrm{base}(u_j)$.

4. *If* $i < j$, *then* $u_i \lhd u_j$.

The operator $\lhd$ in ASAE-4-4 refers to the order relation on expressions that is used to define the actions of the commutative properties. The definition of $\lhd$ is given later in the section (see Definition 3.26 starting on page 84).

Example 3.22. The expression $2 \cdot x \cdot y \cdot z^2$ is a simplified product. It can be easily checked that this expression satisfies ASAE-4-1, ASAE-4-2, and ASAE-4-3, and once $\lhd$ is defined, we shall see that it also satisfies ASAE-4-4. On the other hand, the following are not simplified:

$$2 \cdot (x \cdot y) \cdot z^2, \quad \text{(ASAE-4-1 is not satisfied)},$$
$$1 \cdot x \cdot y \cdot z^2, \quad \text{(ASAE-4-1 is not satisfied)},$$
$$2 \cdot x \cdot y \cdot z \cdot z^2, \quad \text{(ASAE-4-3 is not satisfied)}. \qquad \square$$

The next rule describes the simplified form of a sum. An important simplification transformation for a sum is the basic distributive transformation that collects constant coefficients of like terms. In order to describe the result of this transformation, we need to precisely define "like term." With this consideration in mind, we define the *like-term part* and *constant part* of an ASAE u with the operators

$$\text{term}(u) = \begin{cases} \cdot u, & \text{when } u \text{ is a symbol, sum, power, factorial,} \\ & \text{or function,} \\ u_2 \cdots u_n, & \text{when } u = u_1 \cdots u_n \text{ is a product} \\ & \text{and } u_1 \text{ is constant,} \\ u, & \text{when } u = u_1 \cdots u_n \text{ is a product} \\ & \text{and } u_1 \text{ is not constant,} \\ \textbf{Undefined}, & \text{when } u \text{ is an integer or fraction,} \end{cases}$$

$$\text{const}(u) = \begin{cases} 1, & \text{when } u \text{ is a symbol, sum, power, factorial,} \\ & \text{or function,} \\ u_1, & \text{when } u = u_1 \cdots u_n \text{ is a product} \\ & \text{and } u_1 \text{ is constant,} \\ 1, & \text{when } u = u_1 \cdots u_n \text{ is a product} \\ & \text{and } u_1 \text{ is not constant,} \\ \textbf{Undefined}, & \text{when } u \text{ is an integer or fraction.} \end{cases}$$

The definitions assume that a simplified product has at least two operands, and when a constant appears as an operand, it is the first operand.[8] Observe that when u is not an integer or fraction, the term operator returns a product with one or more operands. When the term operator is applied to an operand of a sum, it returns the part used to determine the like terms, while the const operator returns the corresponding coefficient part. For example,

$$\text{term}(x) = \cdot x, \quad \text{term}(2 \cdot y) = \cdot y, \quad \text{term}(x \cdot y) = x \cdot y,$$

$$\text{const}(x) = 1, \quad \text{const}(2 \cdot y) = 2, \quad \text{const}(x \cdot y) = 1.$$

Notice that $\text{term}(u)$ is either a product or **Undefined**. This means expressions such as x and $2 \cdot x$ have like-term parts that are structurally the same. The implementation of procedures for these operators is left to the reader (Exercise 1).

The term operator is used to define the simplified form of a sum.

ASAE-5. *u is a sum with two or more operands $u_1, u_2, \ldots, u_n$ that satisfy the following properties.*

1. *Each operand u_i is an ASAE that can be either an integer ($\neq 0$), fraction, symbol ($\neq$ **Undefined**), product, power, factorial, or function. (An operand of a sum cannot a sum.) An expression that satisfies this property is called an **admissible term** of a sum.*

2. *At most one operand of u is a constant (integer or fraction).*

---

[8]See rule O-7 on page 87.

3. *If $i \neq j$, then* $\text{term}(u_i) \neq \text{term}(u_j)$.

4. *If $i < j$, then $u_i \lhd u_j$.*

Example 3.23. The sum $2 \cdot x + 3 \cdot y + 4 \cdot z$ is an ASAE. On the other hand, the following expressions are not simplified:

$$
\begin{array}{ll}
1 + (x + y) + z, & \text{(ASAE-5-1 is not satisfied)}, \\
1 + 2 + x, & \text{(ASAE-5-2 is not satisfied)}, \\
1 + x + 2 \cdot x, & \text{(ASAE-5-3 is not satisfied)}, \\
z + y + x, & \text{(ASAE-5-4 is not satisfied)}.
\end{array}
$$

$\square$

ASAE-6. *u is a power v^w that satisfies the following properties.*

1. *The expressions v and w are ASAEs.*

2. *The exponent w is not 0 or 1.*

3. *If w is an integer, then the base v is an ASAE which is a symbol($\neq$ **Undefined**), sum, factorial, or function.*

4. *If w is not an integer, then the base v is any ASAE except 0 or 1.*

Example 3.24. The expressions x^2, $(1 + x)^3$, 2^m, $(x \cdot y)^{1/2}$, and $(x^{1/2})^{1/2}$ are ASAEs, while 2^3, $(x^2)^3$, $(x\,y)^2$, $(1 + x)^1$, and 1^m are not. $\square$

ASAE-7. *u is a factorial with an operand that is any ASAE except a non-negative integer.*

Example 3.25. The expressions $n!$ and $(-3)!$ are ASAEs, while $3!$ is not. $\square$

ASAE-8. *u is a function with one or more operands that are ASAEs.*

The $\lhd$ Order Relation

The $\lhd$ order relation defines the actions of the basic commutative transformations, and, in a simplified sum or product, the operands are ordered according to this relation. Since the operands of these expressions are simplified recursively, it is sufficient to define the order relation for expressions that are automatically simplified.

Definition 3.26. *Let u and v be distinct ASAEs. The $\lhd$ order relation is defined by the following rules.*

O-1. *Suppose that u and v are both constants (integers or fractions). Then,*

$$u \lhd v \to u < v.$$

This notation indicates the value of $u \lhd v$ is obtained by evaluating the order relation $<$ for integers or fractions. For example,

$$2 \lhd 5/2 \to 2 < 5/2 \to \textbf{true}.$$

O-2. *Suppose that u and v are both symbols. Then, $u \lhd v$ is defined by the lexicographical order of the symbols. The lexicographical order is similar to an alphabetical order except now the individual characters are constructed from upper and lower case letters and digits which are ordered as*

$$0, 1, \ldots, 9, A, B, \ldots, Z, a, b, \ldots, z.$$

For example, $a \lhd b$, $v1 \lhd v2$, and $x1 \lhd xa$.

O-3. *Suppose that u and v are either both products or both sums with operands*

$$u_1, u_2, \ldots, u_m \quad \text{and} \quad v_1, v_2, \ldots, v_n.$$

1. *If $u_m \neq v_n$, then $u \lhd v \to u_m \lhd v_n$.*

2. *If there is an integer k with $1 \leq k \leq \min(\{m, n\}) - 1$ such that*

   $$u_{m-j} = v_{n-j}, \quad j = 0, 1, \ldots, k-1 \quad \text{and} \quad u_{m-k} \neq v_{n-k},$$

   then $u \lhd v \to u_{m-k} \lhd v_{n-k}$.

3. *If*

   $$u_{m-k} = v_{n-k}, \quad k = 0, 1, \ldots, \min(\{m, n\}) - 1,$$

   then $u \lhd v \to m < n$.

In O-3, since u and v are ASAE*s*, their operands satisfy

$$u_i \lhd u_j, \quad v_i \lhd v_j, \quad \text{for } i < j.$$

Therefore, the most significant operand in each expression (with respect to $\lhd$) is the last operand. The value of $u \lhd v$ is determined by comparing corresponding operands in each expression starting with the most significant operands u_m and v_n. Rule O-3-1 says that if $u_m \neq v_n$, then the value $u \lhd v$ is determined by evaluating $u_m \lhd v_n$ with a recursive application of the rules. On the other hand, according to Rule O-3-2, if $u_m = v_n$ the focus shifts one operand to the left to u_{m-1} and v_{n-1}, and if $u_{m-1} \neq v_{n-1}$, then $u \lhd v = u_{m-1} \lhd v_{n-1}$. Again, if $u_{m-1} = v_{n-1}$, the focus shifts one more step to the left to u_{m-2} and v_{n-2}. This process continues until either two unequal operands are found (say $u_{m-k} \neq v_{n-k}$) in which case $u \lhd v \rightarrow u_{m-k} \lhd v_{n-k}$, or until either u or v runs out of operands in which case condition O-3-3 applies. For example,

$$a + b \lhd a + c, \quad a + c + d \lhd b + c + d, \quad c + d \lhd b + c + d.$$

O-4. *Suppose that u and v are both powers.*

    1. If $\text{base}(u) \neq \text{base}(v)$, then

$$u \lhd v \rightarrow \text{base}(u) \lhd \text{base}(v).$$

    2. If $\text{base}(u) = \text{base}(v)$, then

$$u \lhd v \rightarrow \text{exponent}(u) \lhd \text{exponent}(v).$$

In other words, if the bases are different, the order is determined by the order of bases, and if the bases are the same, the order is determined by the order of the exponents.

Example 3.27. Using O-4, O-1, and O-3-1, we have

$$(1 + x)^2 \lhd (1 + x)^3 \lhd (1 + y)^2. \tag{3.34}$$

$\square$

O-5. *If u and v are both factorials, then*

$$u \lhd v \rightarrow \text{Operand}(u, 1) \lhd \text{Operand}(v, 1).$$

The order of two factorials is obtained by comparing their operands.

O-6. *Suppose that u and v are functions.*

    1. If $\text{Kind}(u) \neq \text{Kind}(v)$, then $u \lhd v \rightarrow \text{Kind}(u) \lhd \text{Kind}(v)$.
    2. Suppose that $\text{Kind}(u) = \text{Kind}(v)$, and the operands of the functions are given by $u_1, u_2, \ldots, u_m$ and $v_1, v_2, \ldots, v_n$.
        (a) If $u_1 \neq v_1$, then $u \lhd v \rightarrow u_1 \lhd v_1$.
        (b) If there is an integer k with $1 \leq k \leq \min(\{m, n\})$ such that

$$u_j = v_j, \quad j = 1, \ldots, k - 1 \quad \text{and} \quad u_k \neq v_k,$$

    then $u \lhd v \rightarrow u_k \lhd v_k$.

(c) If

$$u_k = v_k, \qquad k = 1, \ldots, \min(\{m, n\}),$$

then $u \lhd v \to m < n$.

When two functions have different names, the order is determined by the function names. For example, $f(x) \lhd g(x)$. The comparison of functions with the same name is similar to the comparison of two products or two sums, except now the first operand is the most significant operand. For example, $f(x) \lhd f(y)$ and $g(x) \lhd g(x, y)$.

The rules given so far apply when either u and v are both constants or $Kind(u) = Kind(v)$. The remaining rules apply when u and v are different types.

O-7. *If u is an integer or fraction and v is any other type, then $u \lhd v$.*

As a consequence of this rule, whenever a constant is an operand in a simplified sum or product, it must be the first operand.

O-8. *Suppose that u is a product. If v is a power, sum, factorial, function, or symbol, then[9]*

$$u \lhd v \to u \lhd \cdot v. \tag{3.35}$$

In other words, in this rule the order is determined by viewing both expressions as products and recursively applying O-3. Notice that we streamline the rule list by not mentioning explicitly the case where u is a product and v a constant because this is handled through rule O-13, where u and v are interchanged together with a recursive application of O-7. For the remaining rules, we omit cases that are handled by earlier rules through rule O-13.

Example 3.28. This rule is a natural way to compare a product and a power because it implies that an expression like $a \cdot x^2 + x^3$ with increasing

---

[9]Strictly speaking, $\lhd$ is only defined for automatically simplified expressions. Although the expression $\cdot v$ in this rule is not automatically simplified, the order $u \lhd \cdot v$ is still well defined because it simply provides a way to compare the last operand of u to v. In fact, the rule can be defined without the unary product. If $u = u_1 \cdots u_n$, then the transformation (3.35) is equivalent to

$$u \lhd v \to \begin{cases} u_n \lhd v, & \text{if } u_n \neq v, \\ \textbf{true}, & \text{if } u_n = v. \end{cases}$$

We use the unary product form of the rule, however, because the rules are somewhat easier to describe in this format. These comments also apply to rules O-9, O-10, and O-11.

powers of x is an ASAE. Indeed, using O-8 followed by O-3-1, O-4-2, and O-1, we have

$$a \cdot x^2 \lhd x^3 \to a \cdot x^2 \lhd \cdot x^3 \to x^2 \lhd x^3 \to 2 < 3 \to \textbf{true}. \qquad \square$$

O-9. *Suppose that u is a power. If v is a sum, factorial, function, or symbol, then*

$$u \lhd v \to u \lhd v^1.$$

In other words, in this rule the order is determined by viewing both expressions as powers and recursively applying rule O-4.

Example 3.29. Using O-9 followed by O-4-1, O-3-2, and O-2, we have

$$(1+x)^3 \lhd (1+y) \to (1+x)^3 \lhd (1+y)^1 \to (1+x) \lhd (1+y) \to x \lhd y \to \textbf{true}.$$

$$\square$$

O-10. *Suppose that u is a sum. If v is a factorial, function, or symbol, then*

$$u \lhd v \to u \lhd +v.$$

In other words, in this rule the order is determined by viewing both expressions as sums and recursively applying rule O-3.

Example 3.30. Using O-10, O-3-1, and O-2, we have $1 + x \lhd y$. $\qquad \square$

O-11. *Suppose that u is a factorial. If v is a function or symbol, then*

    1. *If $Operand(u, 1) = v$ then $u \lhd v \to \textbf{false}$.*

    2. *If $Operand(u, 1) \neq v$ then $u \lhd v \to u \lhd v!$.*

In other words when $Operand(u, 1) \neq v$, the order is determined by viewing both expressions as factorials and then recursively applying rule O-5. For example, this rule implies $m! \lhd n$.

O-12. *Suppose that u is a function, and v is a symbol.*

    1. *If $Kind(u) = v$, then $u \lhd v \to \textbf{false}$.*

    2. *If $Kind(u) \neq v$, then $u \lhd v \to Kind(u) \lhd v$.*

$v \uparrow$ $u \rightarrow$	constant	·	<	+	!	function	symbol
constant	O-1 ($\vee$)	true	true	true	true	true	true
·	false	O-3	$u \lhd \cdot v$	$u \lhd \cdot v$	$u \lhd \cdot v$	$u \lhd \cdot v$	$u \lhd \cdot v$
<	false	$\cdot u \lhd v$	O-4	$u \lhd v^1$	$u \lhd v^1$	$u \lhd v^1$	$u \lhd v^1$
+	false	$\cdot u \lhd v$	$u^1 \lhd v$	O-3	$u \lhd +v$	$u \lhd +v$	$u \lhd +v$
!	false	$\cdot u \lhd v$	$u^1 \lhd v$	$+u \lhd v$	O-5	if $Operand(u,1) = v$ then false else $u \lhd v!$	if $Operand(u,1) = v$ then false else $u \lhd v!$
function	false	$\cdot u \lhd v$	$u^1 \lhd v$	$+u \lhd v$	if $u = Operand(v,1)$ then true else $u! \lhd v$	O-6	if $Kind(u) = v$ then false else $Kind(u) \lhd v$
symbol	false	$\cdot u \lhd v$	$u^1 \lhd v$	$+u \lhd v$	if $u = Operand(v,1)$ then true else $u! \lhd v$	if $u = Kind(v)$ then true else $u \lhd Kind(v)$	O-2 (lexicographical)

Figure 3.9. A summary of the rules for $u \lhd v$. The entries along the diagonal correspond to rules O1 through O-6. The entries above the diagonal correspond to rules O-7 through O-12, while those below the diagonal correspond to the complementary rules implied by rule O-13.

Example 3.31. Rule O-12-1 implies $x \lhd x(t)$, and Rules O-12-2 and O-2 imply $x \lhd y(t)$. □

O-13. *If u and v do not satisfy the conditions in any of the above rules, then*

$$u \lhd v \to \mathbf{not}(v \lhd u).$$

Example 3.32. If $u = x$ and $v = x^2$, then the first 12 rules do not apply. However, using rules O-13, O-9, O-4-2, and O-1, we have

$$x \lhd x^2 \to \mathbf{not}(x^2 \lhd x) \to \mathbf{not}(x^2 \lhd x^1) \to \mathbf{not}(2 < 1) \to \mathbf{not}(\mathbf{false}) \to \mathbf{true}.$$
□

A summary of the $\lhd$ rules, which includes the complementary rules implied by O-13, is given in Figure 3.9.

The $\lhd$ transformation rules can be readily expressed in a procedure that determines the order of two expressions. The details of the procedure are left to the reader (Exercise 2).

Exercises

1. Let u be an ASAE in function notation. Give a procedure for each of the following operators.

    (a) An operator $Base(u)$ for base(u).

    (b) An operator $Exponent(u)$ for exponent(u).

    (c) An operator $Term(u)$ for term(u).

    (d) An operator $Const(u)$ for const(u).

    These operators are used in the automatic simplification algorithm described in Section 3.2.

2. Let u and v be ASAEs in function notation. Give a procedure

    $$Compare(u, v)$$

    that evaluates the order relation $u \lhd v$. The operator returns **true** when $u \lhd v$ and **false** otherwise. *Note*: To obtain the order of two symbols the following commands can be used: in Maple, the `lexorder` command; in Mathematica, the `Order` command; and in MuPAD, the `sysorder` command. (Implementation: Maple (mws), Mathematica (nb), MuPAD (mnb).)

    This operator is used in the automatic simplification algorithm described in Section 3.2.

3. Let u be a BAE in function notation. Give a procedure $ASAE(u)$ that returns **true** if u is an ASAE and **false** otherwise.

4. Determine $u \lhd v$ for each of the following pairs of expressions. In each case, give the sequence of rules used to determine $u \lhd v$.

   (a) $u = a + b$, $v = c$.

   (b) $u = a \cdot b$, $v = a + b^2$.

   (c) $u = c^2$, $v = a + b^4$.

   (d) $u = (2 + 4 \cdot x)^3$, $v = y^2$.

   (e) $u = c \cdot (x + 4 \cdot y^2)$, $v = (a + 3 \cdot b^2) \cdot (x + 4 \cdot y^2)$.

5. Give the automatically simplified form of each of the following expressions. Indicate which simplification rules are applied to determine the simplified form.

   (a) $m \cdot c^2$.

   (b) $\dfrac{m \cdot v^2}{2}$.

   (c) $a \cdot x^2 + b \cdot x + c$.

   (d) $((x^{1/2})^{1/2})^8$.

   (e) $((x \cdot y)^{1/2} \cdot z^2)^2$.

6. Show that each of the following expressions is an ASAE.

   (a) $1 + x + 2 \cdot (1 + x)$.

   (b) $x^2 \cdot (x^2)^y$.

   (c) $3 \cdot x^3 \cdot (1 + x^2) \cdot (2 + x^2) \cdot y$.

3.2 An Automatic Simplification Algorithm

In this section we describe an automatic simplification algorithm that is based on the basic algebraic transformations described in Section 3.1. The algorithm does the following:

1. For u a BAE, the algorithm returns either an ASAE or the symbol **Undefined**.

2. For u an ASAE, the algorithm returns u.

The Main Simplification Procedure

The main procedure of the algorithm is given in Figure 3.10. In order to bypass a CAS's automatic simplification process, we assume that the input expression is in function notation (see Figure 2.6, page 39) instead of infix notation. For clarity, however, we use infix notation in our discussion and examples.

Procedure *Automatic_simplify(u)*;
Input
      u : a **BAE** *in function notation;*
Output
      An ASAE in function notation or the symbol **Undefined***;*
Local Variables
      v;
Begin
1   **if** *Kind(u)* ∈ {**integer**, **symbol**} **then**
2   *Return(u)*;
3   **elseif** *Kind(u)* = *FracOp* **then**
4   *Return(Simplify_rational_number(u))*
5   **else**
6   *v := Map(Automatic_simplify, u)*;
7   **if** *Kind(v)* = *PowOp* **then**
8   *Return(Simplify_power(v))*
9   **elseif** *Kind(v)* = *ProdOp* **then**
10   *Return(Simplify_product(v))*
11   **elseif** *Kind(v)* = *SumOp* **then**
12   *Return(Simplify_sum(v))*
13   **elseif** *Kind(v)* = *QuotOp* **then**
14   *Return(Simplify_quotient(v))*
15   **elseif** *Kind(v)* = *DiffOp* **then**
16   *Return(Simplify_difference(v))*
17   **elseif** *Kind(v)* = *FactOp* **then**
18   *Return(Simplify_factorial(v))*
19   **else**
20   *Return(Simplify_function(v))*
      End

Figure 3.10. The main MPL procedure in the *Automatic_simplify* algorithm. (Implementation: Maple (txt), Mathematica (txt), MuPAD (txt).)

To begin, since integers and symbols are in simplified form, the procedure simply returns the input expression (lines 1–2). For fractions, the simplified form is obtained (at line 4) with the *Simplify_rational_number* procedure described in Section 2.2. For other compound expressions, we first recursively simplify the operands (line 6) and then apply the appropriate simplification operator (lines 7–20).

Because of the large number of basic transformations described in Section 3.1 and the complex interactions between them, the automatic simplification algorithm is quite involved. In Figure 3.11, we list the operators

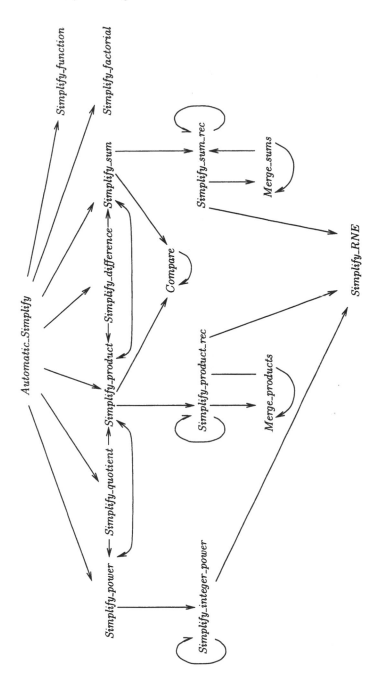

Figure 3.11. The relationship between the operators in the automatic simplification algorithm.

called by *Automatic_simplify* along with other operators that make up the
algorithm. An arrow from one operator to another operator indicates that
the first operator calls on the second operator, and a circular reference
indicates the operator is recursive. Each of the operators is defined by a
collection of transformation rules, and in order to guide the reader through
the maze of rules, we give a large number of examples to illustrate how the
rules are applied.

Simplification of Powers

The *Simplify_power* operator, which transforms a power v^w to either an
ASAE or the symbol **Undefined**, is based on the following transformations:

1. the basic power transformations (3.15) and (3.16),

2. the basic identity transformations (3.23), (3.24), (3.25), and (3.26),

3. the basic numerical transformation in Definition 3.17(3), and

4. the "Undefined" transformation in Definition 3.18.

The operator is defined in terms of a main sequence of non-recursive transformation rules (Definition 3.33) which calls on another sequence of recursive rules (Definition 3.34).

Definition 3.33. *Let $u = v^w$ where the base $v = Operand(u, 1)$ and the
exponent $w = Operand(u, 2)$ are either **ASAEs** or the symbol **Undefined**.
The operator Simplify_power(u) is defined by the following transformation
rules.*

SPOW-1. *If $v = $ **Undefined** or $w = $ **Undefined** then*

$$Simplify\_power(u) \rightarrow \textbf{Undefined}.$$

SPOW-2. *Suppose that $v = 0$. If w is a positive integer or fraction, then*

$$Simplify\_power(u) \rightarrow 0,$$

   otherwise

$$Simplify\_power(u) \rightarrow \textbf{Undefined}.$$

SPOW-3. *If $v = 1$, then*

$$Simplify\_power(u) \rightarrow 1.$$

SPOW-4. *If w is an integer, then*

$$Simplify\_power(u) \rightarrow Simplify\_integer\_power(v, w).$$

(Simplify_integer_power is defined in Definition 3.34 below.)

SPOW-5. *If the first four rules do not apply, then*

$$Simplify\_power(u) \rightarrow u.$$

Rule SPOW-1 states that a power with an **Undefined** operand inherits this **Undefined** value. This happens, for example, with $(1/0)^2$ because the simplified form of the base is **Undefined**.

Rules SPOW-2 and SPOW-3 are the basic identity transformations (3.23) and (3.24). Rule SPOW-4, which simplifies powers with integer exponents, is defined by the *Simplify_integer_power* operator (see Definition 3.34 below). The last rule SPOW-5, which handles all other cases, simply returns u without change.

Definition 3.34. *Consider the expression v^n where $v \neq 0$ is an ASAE and n is an integer. The operator Simplify_integer_power(v, n) is defined by the following transformation rules.*

SINTPOW-1. *If $Kind(v) \in \{\textbf{integer}, FracOp\}$, then*

$$Simplify\_integer\_power(v, n) \rightarrow Simplify\_RNE(PowOp(v, n)).$$

Rule SINTPOW-1, performs the numerical computation of powers using the *Simplify_RNE* operator described in Section 2.2.

SINTPOW-2. *If $n = 0$, then Simplify_integer_power$(v, n) \rightarrow 1$.*

SINTPOW-3. *If $n = 1$, then Simplify_integer_power$(v, n) \rightarrow v$.*

Rules SINTPOW-2, and SINTPOW-3, achieve the two basic identity transformations (3.25) and (3.26). In SINTPOW-2, since SPOW-2 above is applied before this rule, we assume that $v \neq 0$.

SINTPOW-4. *Suppose that $Kind(v) = PowOp$ with base*

$$r = Operand(v, 1)$$

and exponent

$$s = Operand(v, 2).$$

In addition, let

$$p = Simplify\_product(ProdOp(s, n)). \tag{3.36}$$

(Simplify_product is given in Definition 3.37.)

1. *If p is an integer, then*

$$Simplify\_integer\_power(v, n) \rightarrow Simplify\_integer\_power(r, p).$$
$$(3.37)$$

2. *If p is not an integer, then*

$$Simplify\_integer\_power(v, n) \rightarrow PowOp(r, p). \qquad (3.38)$$

This rule, which applies when the base v is also a power, is based on the power transformation $(r^s)^n = r^{s \cdot n}$ in Equation (3.15). The *Simplify_product* operator in Equation (3.36), which simplifies the new product $s \cdot n$, is described in Definition 3.37. Observe that the transformation (3.37) is recursive because the new power r^p may not be simplified (see Example 3.35 below). On the other hand, if p is not an integer, then r^p is an ASAE because r is the base of the simplified power r^s, and so r^p is returned.

Example 3.35. Consider the simplification of the expression

$$((x^{1/2})^{1/2})^8,$$

where $v = (x^{1/2})^{1/2}$ and $n = 8$. Observe that ASAE-6 implies that v is an ASAE. To apply rule SINTPOW-4, let $r = x^{1/2}$ and $s = 1/2$. We have $p = s \cdot n = 4$, and so $r^p = (x^{1/2})^4$ which is not in simplified form. However, a recursive application of *Simplify_integer_power* in the transformation (3.38) obtains the simplified form x^2. $\qquad \square$

SINTPOW-5. *Suppose that $Kind(v) = ProdOp$, and let*

$$r = Map(Simplify\_integer\_power, v, n). \qquad (3.39)$$

Then,

$$Simplify\_integer\_power(v, n) \rightarrow Simplify\_product(r). \qquad (3.40)$$

This rule, which applies when v is a product, is based on the power transformation

$$v^n = (v_1 \cdots v_m)^n = v_1^n \cdots v_m^n$$

in Equation (3.16). The expression r that is defined with the *Map* operator is a new product $r = r_1 \cdots r_m$, where

$$r_i = Simplify\_integer\_power(v_i, n)$$

is the simplified form of v_i^n. Notice that Equation (3.39) involves a recursive application of *Simplify_integer_power*, because the new power v_i^n may not be in simplified form. In addition, since the new product r may not be in simplified form, it is simplified in transformation (3.40) with the *Simplify_product* operator. These points are illustrated in Example 3.36.

Example 3.36. Consider the simplification of

$$\left((x \cdot y)^{1/2} \cdot z^2 \right)^2,$$

where

$$v = v_1 \, v_2 = (x \cdot y)^{1/2} \cdot z^2, \quad n = 2.$$

Then,

$$
\begin{aligned}
r \; &= \; Simplify\_integer\_power\left((x \cdot y)^{1/2}, 2 \right) \\
&\quad \cdot \, Simplify\_integer\_power\left(z^2, 2 \right) \\
&\rightarrow \; (x \cdot y) \cdot z^4.
\end{aligned} \tag{3.41}
$$

Notice that the expression (3.41) is not a simplified product because of the parentheses about $x \cdot y$. Using the transformation (3.40), we obtain the simplified form

$$Simplify\_product(r) \rightarrow x \cdot y \cdot z^4. \qquad \square$$

SINTPOW-6. *If none of the rules apply, then*

$$Simplify\_integer\_power(v, n) \rightarrow PowOp(v, n).$$

The last rule applies when v is a symbol, sum, factorial, or function. In this rule, v^n is already simplified, and so this expression is returned in function notation.

Simplification of Products

The *Simplify_product* operator, which transforms a product to either an ASAE or the symbol **Undefined**, is based on the following transformations:

1. the basic associative transformation (Definition 3.5(2)),

2. the basic commutative transformation for products (Definition 3.7),

3. the basic power transformation (3.14),

4. the basic identity transformations (3.21) and (3.22),

5. the basic unary transformation (3.32),

6. the basic numerical transformation in Definition 3.17(1), and

7. the basic "Undefined" transformation (Definition 3.18).

This operator is defined in terms of a main sequence of non-recursive transformation rules (Definition 3.37) together with two sequences of recursive rules (Definition 3.38 and Definition 3.42).

Definition 3.37. *Let u be a product with one or more operands that are ASAEs, and let $L = [u_1, \ldots, u_n]$ be the list of the operands of u. The Simplify_product(u) operator is defined by the following transformation rules.*

SPRD-1. *If* **Undefined** $\in L$, *then*

$$Simplify\_product(u) \to \textbf{Undefined}.$$

SPRD-2. *If $0 \in L$, then $Simplify\_product(u) \to 0$.*

Rule SPRD-1 states that a product with an operand that is undefined inherits the simplified form **Undefined**. Rule SPRD-2 is based on the identity transformation (3.21). Both rules are included at this point because the expressions **Undefined** and 0 are not created at any other point during the simplification of a product (which means the rules never have to be recursively reapplied), and the presence of these expressions has such a profound effect on the output.

SPRD-3. *If $L = [u_1]$, then $Simplify\_product(u) \to u_1$.*

Rule SPRD-3 states that a unary product simplifies to its operand.

SPRD-4. *Suppose that the first three rules do not apply, and let*

$$v = Simplify\_product\_rec(L). \qquad (3.42)$$

The Simplify_product_rec operator, which is given in Definition 3.38, returns a list with zero or more operands, and if v has two or more operands, they satisfy the conditions of ASAE-4. There are three cases.

1. *If $v = [v_1]$, then $Simplify\_product(u) \to v_1$.*
2. *If $v = [v_1, v_2, \ldots, v_s]$ with $s \geq 2$, then*

$$Simplify\_product(u) \to Construct(ProdOp, v).$$

3. *If $v = [\ ]$, then $Simplify\_product(u) \to 1$.*

Rule SPRD-4-1 is invoked when the operands of u simplify to a product with single operand v_1, in which case the simplified form is just the operand v_1. For example, using the rules in Definition 3.38 and Definition 3.42, we have

$$v = Simplify\_product\_rec([a^{-1}, b, a]) \rightarrow [b],$$

and therefore by SPRD-4-1,

$$Simplify\_product(a^{-1} \cdot b \cdot a) \rightarrow b.$$

Rule SPRD-4-2 corresponds to the case where the simplified form of a product is also a product. For example, since

$$v = Simplify\_product\_rec([c, 2, b, c, a]) \rightarrow [2, a, b, c^2],$$

we have

$$Simplify\_product(c \cdot 2 \cdot b \cdot c \cdot a) \rightarrow 2 \cdot a \cdot b \cdot c^2.$$

Rule SPRD 4-3 is included since an integer 1 that is created during the simplification of a product is not included in the operand list (see rules SPRDREC-1-1 and SPRDREC-1-3). (It is convenient to do this because ASAE-4-1 states that an operand of a simplified product cannot be the integer 1.) For this reason, $Simplify\_product\_rec(L)$ may return the empty list, and when this happens, the product simplifies to 1 (see Example 3.39).

Definition 3.38. *Let $L = [u_1, u_2, \ldots, u_n]$ be a non-empty list with $n \geq 2$ non-zero ASAEs. The operator $Simplify\_product\_rec(L)$ (for "Simplify product recursive") returns a list with zero or more operands. The operator is defined by the following transformation rules.*

SPRDREC-1. *Suppose that $L = [u_1, u_2]$ and neither operand is a product.*

1. *Suppose that both u_1 and u_2 are constants, and let*

$$P = Simplify\_RNE(ProdOp(u_1, u_2)).$$

If $P = 1$, then

$$Simplify\_product\_rec(L) \rightarrow [\,],$$

and if $P \neq 1$, then

$$Simplify\_product\_rec(L) \rightarrow [P].$$

2. *(a) If $u_1 = 1$, then $Simplify\_product\_rec(L) \rightarrow [u_2]$.*
   (b) If $u_2 = 1$, then $Simplify\_product\_rec(L) \rightarrow [u_1]$.

3. *Suppose that* $\text{base}(u_1) = \text{base}(u_2)$, *and let*

$$S = Simplify\_sum(SumOp(\text{exponent}(u_1), \text{exponent}(u_2)))$$

and

$$P = Simplify\_power(PowOp(\text{base}(u_1), S)).$$

If $P = 1$, *then*

$$Simplify\_product\_rec(L) \rightarrow [\ ],$$

and if $P \neq 1$, *then*

$$Simplify\_product\_rec(L) \rightarrow [P].$$

4. *If* $u_2 \lhd u_1$, *then*

$$Simplify\_product\_rec(L) \rightarrow [u_2, u_1].$$

5. *If the first four rules do not apply, then*

$$Simplify\_product\_rec(L) \rightarrow L.$$

Rule SPRDREC-1 applies when L has two operands and neither one is a product. (The case when one of the operands is a product is handled in rule SPRDREC-2.) Rule SPRDREC-1-1 is the basic numerical transformation in Definition 3.17(1). Observe that when the product $P = 1$, the empty list is returned. This is done so that the integer 1 created during the simplification process is not included as an operand of a product (see also rules SPRD-4-3 and SPRDREC-1-3).

Rule SPRDREC-1-2 is based on the basic identity transformation (3.22).

Rule SPRDREC-1-3 performs the basic power transformation (3.14). The *Simplify_sum* operator (Exercise 7) simplifies the sum of the exponents, and since the new power may not be an ASAE, the rule includes an application of *Simplify_power*. Again, when the product $P = 1$, the empty list is returned.

Example 3.39. If $L = [a, a^{-1}]$, then $S = 0$ and $a^S = a^0$, which is not in simplified form. Applying the *Simplify_power* operator, we have $P = 1$ and so $Simplify\_product\_rec([a, a^{-1}]) \rightarrow [\ ]$.   □

Rule SPRDREC-1-4 is invoked when the two operands are out of order with respect to the $\lhd$ order relation. When the last rule SPRDREC-1-5 is invoked, the earlier conditions that lead to a modification of the input do not apply, and so the input is returned without change.

SPRDREC-2. *Suppose that $L = [u_1, u_2]$, and suppose that at least one of its two operands is a product.*

1. *If u_1 is a product with operands $p_1, p_2, \ldots, p_s$ and u_2 is a product with operands $q_1, q_2, \ldots, q_t$, then*

   $$Simplify\_product\_rec(L) \rightarrow$$
   $$Merge\_products([p_1, p_2, \ldots, p_s], [q_1, q_2, \ldots, q_t]).$$

   (Merge_products is given in Definition 3.42.)

2. *If u_1 is a product with operands*

   $$p_1, p_2, \ldots, p_s$$

   and u_2 is not a product, then

   $$Simplify\_product\_rec(L) \rightarrow Merge\_products([p_1, p_2, \ldots, p_s], [u_2]).$$

3. *If u_1 is not a product and u_2 is a product with operands*

   $$q_1, q_2, \ldots, q_t,$$

   then

   $$Simplify\_product\_rec(L) \rightarrow Merge\_products([u_1], [q_1, q_2, \ldots, q_t]).$$

Rule SPRDREC-2, which is based on the basic associative transformation in Definition 3.5(2), applies when L has two operands and at least one operand is a product. This rule (together with SPRDREC-3) guarantees that a simplified product will not have an operand that is also a product. The *Merge_products* operator, which is given in Definition 3.42, applies the associative property and combines and re-orders the operands of the two products to create a new operand list. This rule is recursive because *Merge_products* invokes one of the earlier SPRDREC rules.

Example 3.40. To simplify the expression $u = (2 \cdot a \cdot c \cdot e) \cdot (3 \cdot b \cdot d \cdot e)$, we have by SPRDREC-2-(a)

$$Simplify\_product\_rec([2 \cdot a \cdot c \cdot e, \quad 3 \cdot b \cdot d \cdot e])$$
$$\rightarrow Merge\_products([2, a, c, e], \quad [3, b, d, e])$$
$$\rightarrow [6, a, b, c, d, e^2].$$

Therefore, the original expression u simplifies to $6 \cdot a \cdot b \cdot c \cdot d \cdot e^2$. □

In SPRDREC-2-2 and SPRDREC-2-3, only one of the operands is a product, and so the remaining operand (u_2 or u_1) is merged with the operands of the product ($p_1, \ldots, p_s$ or $q_1, \ldots, q_t$).

SPRDREC-3. *Suppose that* $L = [u_1, u_2, \ldots, u_n]$ *with* $n > 2$, *and let*

$$w = Simplify\_product\_rec(Rest(L)). \tag{3.43}$$

   1. If u_1 *is a product with operands* $p_1, p_2, \ldots, p_s$, *then*

$$Simplify\_product\_rec(L) \rightarrow Merge\_products([p_1, p_2, \ldots, p_s], w).$$

   2. If u_1 *is not a product, then*

$$Simplify\_product\_rec(L) \rightarrow Merge\_products([u_1], w).$$

Rule SPRDREC-3 is invoked when the input list L has more than two operands. The first rule SPRDREC-3-1 is similar to the associative transformation SPRDREC-2-1. It merges the operands of the product u_1 with the list of operands that make up the simplified form of $Rest(L)$. In a similar way, rule SPRDREC-3-2 merges the single operand u_1 with the operands of w.

Example 3.41. Suppose that $L = [a \cdot b,\ c,\ b]$. Then, applying the rules SPRDREC-3-1 and SPRDREC-1-4, we have

$$w = Simplify\_product\_rec([c, b]) \rightarrow [b, c].$$

Therefore,

$$
\begin{aligned}
Simplify\_product\_rec([a \cdot b,\ b,\ c]) \quad &= \quad Merge\_products([a, b],\ [b, c]) \\
&\rightarrow \quad [a,\ b^2,\ c].
\end{aligned}
$$
$\square$

The transformation rules for $Merge\_products(p, q)$ are given in the next definition. When

$$p = [p_1, \ldots, p_m], \qquad q = [q_1, \ldots, q_n]$$

represent the operands of two simplified products, the operator applies the basic product associative transformation and, using $Simplify\_product\_rec$, combines and reorders the operands of p and q into a new operand list.

Definition 3.42. *Let* p *and* q *be lists of zero or more admissible factors in* $\lhd$ *order. In addition, if* p *or* q *has two or more operands, the operands satisfy the properties of ASAE-4. The operator* $Merge\_products(p, q)$ *is defined using the following rule sequence:*

MPRD-1. *If* $q = [\ ]$, *then* $Merge\_products(p, q) \rightarrow p$.

MPRD-2. *If $p = [\ ]$, then $Merge\_products(p, q) \rightarrow q$.*

MPRD-3. *Suppose that p and q are non-empty lists, and let $p_1 = First(p)$ and $q_1 = First(q)$. Define*

$$h = Simplify\_product\_rec([p_1, q_1]). \qquad (3.44)$$

There are four possibilities for h.

1. *If $h = [\ ]$, then*

   $$Merge\_products(p, q) \rightarrow Merge\_products(Rest(p), Rest(q)).$$

2. *If $h = [h_1]$ (a one operand list), then*

   $$Merge\_products(p, q) \rightarrow$$

   $$Adjoin(h_1, Merge\_products(Rest(p), Rest(q))).$$

3. *If $h = [p_1, q_1]$, then*

   $$Merge\_products(p, q) \rightarrow Adjoin(p_1, Merge\_products(Rest(p), q)).$$

4. *If $h = [q_1, p_1]$, then*

   $$Merge\_products(p, q) \rightarrow Adjoin(q_1, Merge\_products(p, Rest(q))).$$

Rules MPRD-1 and MPRD-2 are invoked when at least one of the input lists is empty. In this case, there is no merging to be done, and so the other list is returned. These rules serve as termination conditions for the recursion.

Rule MPRD-3 includes four cases. If MPRD-3-1 applies, $p_1 \cdot q_1$ simplifies to 1 which is not included in the operand list. In this case, the operator returns the result of a recursive application of the *Merge_products* rules to the rest of the two lists. If MPRD-3-2 applies, p_1 and q_1 are combined (using SPRDREC-1-1 or SPRDREC-1-3) to give a new operand $h_1 \neq 1$ which is placed at the beginning of the output list. If MPRD-3-3 applies, then $p_1 \lhd q_1$, which means that p_1 is placed at the beginning of the output list while the rest of p is merged with q. Finally, if MPRD-3-4 applies, then $q_1 \lhd p_1$ and so q_1 is placed at the beginning of the output list while p is merged with the rest of q.

Example 3.43. Let's consider the role of the *Merge_products* operator in the simplification of the product $u = (a \cdot c \cdot e) \cdot (a \cdot c^{-1} \cdot d \cdot f)$. Observe that the two operands of the main product operator are both products in simplified form. To obtain the simplification, we apply rule SPRD-4 with $L = [a \cdot c \cdot e, \quad a \cdot c^{-1} \cdot d \cdot f]$ and SPRDREC-2-1 with $p = [a, c, e]$ and $q = [a, c^{-1}, d, f]$. To merge the two lists, we apply MPRD-3 followed by SPRDREC-1-3 to obtain

$$h = Simplify\_product\_rec([a, a]) = [a^2].$$

Therefore, using MPRD-3-2,

$$Merge\_products\left([a, c, e], [a, c^{-1}, d, f]\right)$$
$$= Adjoin\left(a^2, Merge\_products\left([c, e], [c^{-1}, d, f]\right)\right). \qquad (3.45)$$

Applying the *Merge_products* recursively, we have, using SPRDREC-1-3,

$$h = Simplify\_product\_rec([c, c^{-1}]) = [\ ],$$

and therefore, using MPRD-3-1,

$$Merge\_products([c, e], [c^{-1}, d, f])$$
$$= Merge\_products([e], [d, f]). \qquad (3.46)$$

Again, applying the rules recursively, we have, using SPRDREC-1-4,

$$h = Simplify\_product\_rec([e, d]) = [d, e],$$

and therefore, using MPRD-3-4,

$$Merge\_products([e], [d, f])$$
$$= Adjoin(d, Merge\_products([e], [f])). \qquad (3.47)$$

Again, applying the rules recursively, using SPRDREC-1-5,

$$h = Simplify\_product\_rec([e, f]) = [e, f],$$

and therefore, using MPRD-3-3,

$$Merge\_products([e], [f])$$
$$= Adjoin(e, Merge\_products([\ ], [f])). \qquad (3.48)$$

Finally, applying the rules one last time, using MPRD-1, we have

$$Merge\_products([\ ], [f]) = [f]. \qquad (3.49)$$

Now, using Equations (3.49), (3.48), (3.47), (3.46), and (3.45), the recursion unwinds, and

$$Merge\_products([a, c, e], [a, c^{-1}, d, f]) = [a^2, d, e, f].$$

Therefore,

$$Simplify\_product\_rec([a \cdot c \cdot e, \quad a \cdot c^{-1} \cdot d \cdot f]) \rightarrow [a^2, d, e, f],$$

and

$$Simplify\_product((a \cdot c \cdot e) \cdot (a \cdot c^{-1} \cdot d \cdot f)) \rightarrow a^2 \cdot d \cdot e \cdot f. \qquad \square$$

Simplification of Sums

The *Simplify_sum* operator, which simplifies a sum to either an ASAE or the symbol **Undefined**, is based on the following transformations:

1. the basic distributive transformation in Definition 3.1,

2. the basic associative transformation in Definition 3.5(1),

3. the basic commutative transformation for sums in Definition 3.7,

4. the basic identity transformation (3.20),

5. the basic unary transformation (3.33),

6. the basic numerical transformation in Definition 3.17(2), and

7. the basic "Undefined" transformation in Definition 3.18.

The operator is defined in terms of the following procedures.

1. The main procedure *Simplify_sum* is defined by a sequence of non-recursive transformation rules. The rules are similar to those for *Simplify_product* in Definition 3.37 although there is no analogue for rule SPRD-2.

2. The procedure *Simplify_sum_rec* is defined by recursive transformation rules that are similar to the ones for *Simplify_product_rec* in Definition 3.38. In this case the analogue of SPRDREC-1-3 is based on the basic distributive transformation in Definition 3.1.

3. The procedure *Merge_sums* is defined by a sequence of recursive transformation rules that are similar to the ones for *Merge_products* in Definition 3.42.

We leave the details of the transformation rules and operators to the reader (Exercise 7).

Simplification of Quotients, Differences, Factorials, and Functions

The operator *Simplify_quotient*, which simplifies quotients, is based on the basic quotient transformation $u/v = u \cdot v^{-1}$. Observe that this transformation creates a new power and a new product which are simplified using the *Simplify_power* and *Simplify_product* operators.

The operator *Simplify_difference(u)* is based on the basic difference transformations $-u = (-1) \cdot u$ and $u - v = u + (-1) \cdot v$. Observe that these transformations create either a new product or a new sum which is simplified using either the *Simplify_product* or *Simplify_sum* operators.

The operator *Simplify_factorial(u)* is based on the basic numerical transformation Definition 3.17(4) which evaluates a factorial with a non-negative integer operand. If the operand of a factorial is **Undefined**, then the procedure returns **Undefined**. In other cases, the input expression u is returned without change.

The operator *Simplify_function(u)* applies to function forms. Although there are no function transformations in our algorithm, some function argument may be **Undefined**, in which case the operator returns **Undefined**. In other cases, return the input argument u without change.

The details of procedures for these operators are left to the reader (Exercise 6).

Algebraic Properties of ASAEs

Consider the set of all ASAEs, and, for u and v in this set, define the sum of u and v as

$$Automatic\_simplify(u + v)$$

and the product of u and v as

$$Automatic\_simplify(u \cdot v).$$

This set together with these operations satisfies all the field axioms except the distributive axioms F-7 and F-8 and the additive inverse axiom F-11. Axioms F-1, F-2, F-3, F-4, F-5, and F-6 follow directly from the transformation rules in this section. Axiom F-10 follows from SPRDREC-1-1 and SPRDREC-1-2, and Axiom F-9 follows from similar rules for sums (Exercise 7). Axiom F-12 follows from SINTPOW-4 and SPRDREC-1-3.

On the other hand, the distributive axioms F-7 and F-8 are not always true because expansion is not included in automatic simplification. For example, for a, b, and c symbols, the expressions $a \cdot (b+c)$ and $a \cdot b + a \cdot c$ are both already simplified but these expressions are not the same as expression trees. In addition, F-11 is not true because $(-1) \cdot (a + b)$ does not have an additive inverse. Indeed,

$$Automatic\_simplify((a + b) + (-1) \cdot (a + b)) \to a + b + (-1) \cdot (a + b)$$

rather than the 0 expression.

Exercises

1. Let u be a symbol. Indicate which of the simplification transformations listed in this section are used to perform the following transformations.

    (a) $u/0 \to$ **Undefined**.

    (b) If $u \neq 0$, $0/u \to 0$.

    (c) $u/1 \to u$.

    (d) $u - 0 \to 0$.

    (e) $0 - u \to (-1) \cdot u$.

2. Which of the simplification transformations in this section are used to transform the quotient (in function notation) $QuotOp(1, 2)$ to the fraction $FracOp(1, 2)$?

3. Let x and y be symbols. Indicate which simplification transformations in this section are used to obtain each of the following transformations.

    (a) $x/x \to 1$.

    (b) $(x/y) \cdot (y/x) \to 1$

4. Trace the flow of the $Simplify\_product\_rec$ operator for each of the following input lists u.

    (a) $u = [c, b, a]$.

    (b) $u = [a, a^{-1}, 1]$.

    (c) $u = [b, a, a^{-1}, 1]$.

    (d) $u = [a, c, b, d, e]$.

5. Give procedures for the following operators.

    (a) $Simplify\_power(u)$ (Definition 3.33, page 94).

    (b) $Simplify\_integer\_power(v, n)$ (Definition 3.34, page 95).

    (c) $Simplify\_product(u)$ (Definition 3.37, page 98).

    (d) $Simplify\_product\_rec(L)$ (Definition 3.38, page 99).

    (e) $Merge\_products(p, q)$ (Definition 3.42, page 102).

6. Give procedures for the following operators.

    (a) $Simplify\_quotient(u)$.

(b) *Simplify_difference(u)*.

(c) *Simplify_factorial(u)*.

(d) *Simplify_function(u)*.

(See the discussion starting on page 106.)

7. (a) Give transformation rules for the operators.

    i. *Simplify_sum(u)*.

    ii. *Simplify_sum_rec(L)*.

    iii. *Merge_sums(p, q)*.

    (See the discussion starting on page 105.)

    (b) Give procedures for these operators.

8. Modify the automatic simplification algorithm so that for a product with two operands, one a constant and the other a sum, the constant is distributed over the sum. For example, the modification should obtain

$$Automatic\_simplify(2 \cdot (x + y)) \rightarrow 2 \cdot x + 2 \cdot y.$$

If a product has more than two operands, the transformation does not occur. For example, $2 \cdot (3 + x) \cdot (4 + x)$ is not changed. *Hint:* The best place to apply this transformation is in SPRD-4.

9. For $\imath = \sqrt{-1}$, modify the automatic simplification algorithm so that all field operations in $\mathbf{Q}(\imath)$ (including integer powers and quotients) are performed by automatic simplification. (An algorithm for simplifying expressions with Gaussian rational numbers is described in Exercise 19, page 61.)

10. Let N be a set of variables that are designated as integer variables. Define an *integer expression* as one constructed using the symbols in N, the integers, and the addition, multiplication, and power (with non-negative integer exponents) operations.

    (a) Let u be an ASAE. Give a procedure *Integer_expression(u)* that returns **true** if u is an integer expression and **false** otherwise. Use function notation for u.

    (b) Modify the automatic simplification algorithm so that the transformations (3.15) and (3.16) are applied when n is an integer expression.

11. Let u be an ASAE in function notation and let x be a symbol. Give a procedure *Derivative_fun(u, x)* that obtains the derivative of u with respect to x. Be sure to return the derivative as an ASAE.

Further Reading

3.1 The Goal of Automatic Simplification. Discussions of automatic simplification are given in Fateman [36], Korsvold [57], Nievergelt, Farrar, and Reingold

[75], Tobey, Bobrow, and Zilles [95], Weissman [99], Wooldridge [103], Wulf et al. [104], and Yun and Stoutemyer [107]. Stoutemyer [94] gives an interesting discussion of the misuse of some automatic simplification transformation rules. Moses [71] has an interesting discussion of algebraic simplification.

4

Single Variable Polynomials

In this chapter we are concerned with the mathematical and computational properties of polynomials with one variable with coefficients from a field $\mathbf{F}$. All the algorithms considered here are ultimately based on polynomial division.

In Section 4.1 we introduce the basic definitions, describe an algorithm for polynomial division, and then give an algorithm for polynomial expansion. In Section 4.2 we examine the gcd problem for polynomials and give versions of Euclid's algorithm and the extended Euclidean algorithm in this setting. In Section 4.3 we use the procedures in the first two sections to perform arithmetic operations for expressions in simple algebraic number fields and to extend the polynomial operations to this setting. In Section 4.4 we describe an algorithm for partial fraction expansion that is based on polynomial division, polynomial expansion, and the extended Euclidean algorithm.

Since the mathematical definitions and theorems apply for any coefficient field $\mathbf{F}$, we present the ideas in this general setting. In a computational setting, however, the algorithms make most sense for coefficient fields where computation is effective and efficient.

4.1 Elementary Concepts and Polynomial Division

Let u be a polynomial of the form

$$u = u_n x^n + u_{n-1} x^{n-1} + \cdots + u_1 x + u_0,$$

where the coefficients u_j are from a field $\mathbf{F}$. The notation $\mathbf{F}[x]$ represents the set of all such polynomials.

Recall that $\mathrm{lc}(u, x)$ represents the leading coefficient of a polynomial and $\deg(u, x)$ its degree.[1] A polynomial with $\mathrm{lc}(u, x) = 1$ is called a *monic* polynomial.

The operators lc and deg satisfy the properties in the following theorem.

Theorem 4.1.

Let u and v be polynomials in $\mathbf{F}[x]$.

1. $\mathrm{lc}(u\,v) = \mathrm{lc}(u)\,\mathrm{lc}(v)$.

2. *If $u \neq 0$ and $v \neq 0$, then $\deg(u\,v) = \deg(u) + \deg(v)$.*

3. $\deg(u \pm v) \leq \max(\{\deg(u), \deg(v)\})$.

Proof: To prove (1), first, when either $u = 0$ or $v = 0$, both sides of the equality are 0. Next, if both $u \neq 0$ and $v \neq 0$, the highest order monomial in $u\,v$ is $\mathrm{lc}(u)\,\mathrm{lc}(v)x^{\deg(u)+\deg(v)}$. Since both $\mathrm{lc}(u)$ and $\mathrm{lc}(v)$ are not zero, and since there are no zero divisors in a field (Theorem 2.36, page 50), the product $\mathrm{lc}(u)\,\mathrm{lc}(v) \neq 0$, and so (1) follows.

The proof for Part (2) is similar to the proof for Part (1).

Property (3) follows from the definition of degree. Notice that this relationship is an inequality because addition (or subtraction) may eliminate the leading terms of the polynomials.   □

Theorem 4.2. *Let u and v be polynomials in $\mathbf{F}[x]$. If $u\,v = 0$, then either $u = 0$ or $v = 0$.*

This theorem, which is a consequence of Theorem 4.1(1), states that there are no zero divisors in $\mathbf{F}[x]$. The proof of the theorem is left to the reader (Exercise 3).

Polynomial Division

For single variable polynomials, the familiar process of long division of u by v is defined formally using recurrence relations.

---

[1]When the variable x is evident from context, we use the simpler notations $\mathrm{lc}(u)$ and $\deg(u)$.

Definition 4.3. *Let u and $v \neq 0$ be two polynomials in $\mathbf{F}[x]$, and consider the sequence of quotients q_i and remainders r_i:*

$$q_0 = 0, \quad r_0 = u,$$

$$q_i = q_{i-1} + \frac{\mathrm{lc}(r_{i-1})}{\mathrm{lc}(v)} \, x^{\deg(r_{i-1}) - \deg(v)}, \tag{4.1}$$

$$r_i = r_{i-1} - \frac{\mathrm{lc}(r_{i-1})}{\mathrm{lc}(v)} \, x^{\deg(r_{i-1}) - \deg(v)} \, v. \tag{4.2}$$

The iteration terminates when

$$\deg(r_i) < \deg(v). \tag{4.3}$$

*If the process terminates with $i = \sigma$, then q_σ is the **quotient** of u divided by v, and r_σ is the **remainder**. We also represent the quotient and the remainder by the operators $\mathrm{quot}(u, v, x)$ and $\mathrm{rem}(u, v, x)$.*

Example 4.4. Let $u = 5\,x^2 + 4\,x + 1$ and $v = 2\,x + 3$. Then,

$$q_0 = 0, \quad r_0 = 5\,x^2 + 4\,x + 1,$$
$$q_1 = 5/2\,x, \quad r_1 = -7/2\,x + 1,$$
$$q_2 = 5/2\,x - 7/4, \quad r_2 = 25/4.$$

Since $\deg(r_2) < \deg(v)$, the process stops after two iterations ($\sigma = 2$), and

$$u = q_\sigma\,v + r_\sigma = (5/2\,x - 7/4)\,(2\,x + 3) + 25/4. \qquad \square$$

Polynomial division satisfies the properties in the following theorem.

Theorem 4.5. *Let u and $v \neq 0$ be polynomials in $\mathbf{F}[x]$.*

1. *The polynomial division process terminates.*

2. *Let q_σ and r_σ be defined by the iteration in Definition 4.3. Then,*

$$u = q_\sigma\,v + r_\sigma \tag{4.4}$$

and

$$\deg(r_\sigma) < \deg(v). \tag{4.5}$$

3. *The quotient q_σ and the remainder r_σ are unique in the sense that they are the only polynomials that satisfy both conditions (4.4) and (4.5).*

The property of the remainder in the inequality (4.5) is called the *Euclidean property* of polynomial division.

Proof: To show (1), let's assume that we have obtained r_{i-1} (which does not satisfy the termination condition (4.3)), and show that

$$\deg(r_i) < \deg(r_{i-1}). \tag{4.6}$$

This condition implies that the inequality (4.3) is eventually satisfied for some r_i.

Since the process does not terminate with r_{i-1}, we have

$$\deg(r_{i-1}) \geq \deg(v). \tag{4.7}$$

It is convenient to express the recurrence relation for r_i in the form

$$r_i = \left(r_{i-1} - \mathrm{lc}(r_{i-1})\, x^{\deg(r_{i-1})} \right) \tag{4.8}$$
$$- \left(v - \mathrm{lc}(v)\, x^{\deg(v)} \right) \left(\frac{\mathrm{lc}(r_{i-1})}{\mathrm{lc}(v)} \right) x^{\deg(r_{i-1}) - \deg(v)}$$

which is equivalent to the form (4.2). There are two cases. First, if v is the monomial $\mathrm{lc}(v) x^{\deg(v)}$, then Equation (4.8) implies

$$\deg(r_i) = \deg \left(r_{i-1} - \mathrm{lc}(r_{i-1})\, x^{\deg(r_{i-1})} \right) < \deg(r_{i-1}). \tag{4.9}$$

On the other hand, when $v - \mathrm{lc}(v) x^{\deg(v)} \neq 0$, Equation (4.8) and Theorem 4.1(2),(3) imply that

$$
\begin{aligned}
\deg(r_i) \;\leq\; & \max \left(\left\{ \deg \left(r_{i-1} - \mathrm{lc}(r_{i-1})\, x^{\deg(r_{i-1})} \right), \right. \right. \\
& \qquad \left. \left. \deg \left(x^{\deg(r_{i-1}) - \deg(v)} \left(v - \mathrm{lc}(v)\, x^{\deg(v)} \right) \right) \right\} \right) \\
\leq\; & \max \left(\left\{ \deg \left(r_{i-1} - \mathrm{lc}(r_{i-1})\, x^{\deg(r_{i-1})} \right), \right. \right. \\
& \qquad \left. \left. (\deg(r_{i-1}) - \deg(v)) + \deg \left(v - \mathrm{lc}(v)\, x^{\deg(v)} \right) \right\} \right) \\
<\; & \deg(r_{i-1}).
\end{aligned}
$$

Therefore, the inequality (4.6) is satisfied, and in this case, $\deg(r_\sigma) < \deg(v)$.

To show (2), first observe that if $\deg(v) > \deg(u)$, there are no iterations ($\sigma = 0$), and Equation (4.4) is satisfied with $q_\sigma = 0$ and $r_\sigma = u$. If $\deg(v) \leq \deg(u)$, there is at least one iteration. From Equations (4.1) and (4.2), we have

$$q_i\, v + r_i = q_{i-1}\, v + r_{i-1}, \qquad i = 1, 2, \ldots, \sigma, \tag{4.10}$$

and therefore

$$u = q_0 v + r_0 = q_1 v + r_1 = \cdots = q_\sigma v + r_\sigma.$$

The inequality (4.5) is satisfied because it is the stopping criterion for the iteration.

To verify the uniqueness property (3), suppose that there are two pairs of polynomials $q_{\sigma_1}, r_{\sigma_1}$ and $q_{\sigma_2}, r_{\sigma_2}$ that satisfy the conditions (4.4) and (4.5). Then,

$$u = q_{\sigma_1} v + r_{\sigma_1} = q_{\sigma_2} v + r_{\sigma_2},$$

and so

$$(q_{\sigma_1} - q_{\sigma_2}) v = r_{\sigma_2} - r_{\sigma_1}. \tag{4.11}$$

First, if $\deg(v) = 0$, then the inequality (4.5) implies $r_{\sigma_1} = r_{\sigma_2} = 0$, and so $q_{\sigma_1} = q_{\sigma_2}$ as well. Suppose next that $\deg(v) > 0$. If $q_{\sigma_1} - q_{\sigma_2} \neq 0$, then

$$\deg((q_{\sigma_1} - q_{\sigma_2}) v) \geq \deg(v). \tag{4.12}$$

However, (4.5) implies

$$\deg(r_{\sigma_2} - r_{\sigma_1}) < \deg(v), \tag{4.13}$$

and since the conditions (4.11), (4.12), and (4.13) cannot all be true, it must be that $q_{\sigma_1} - q_{\sigma_2} = 0$, and so $q_{\sigma_1} = q_{\sigma_2}$ and $r_{\sigma_1} = r_{\sigma_2}$. $\square$

The approach that we used to prove the uniqueness property is called a *degree argument* because it is based primarily on the properties of degree. Degree arguments are frequently used to obtain properties of polynomials.

Theorem 4.6. *Let u be a polynomial in $\mathbf{F}[x]$ with positive degree, and suppose that c in $\mathbf{F}$ is a root of $u = 0$. Then, u can be factored as $u = (x - c)\, q$, where q is in $\mathbf{F}[x]$.*

The proof of the theorem, which follows directly from Theorem 4.5, is left to the reader (Exercise 13).

In Figure 4.1, we give a procedure for polynomial division and functions that extract the quotient and remainder.[2] Observe that in line 10 we have

---

[2]When $\deg(v) = 0$, the termination of *Polynomial_division* at line 6 depends on

$$\deg(0) \to -\infty. \tag{4.14}$$

This operation is obtained by the MPL operator *Degree_gpe* used at lines 3, 4, and 11. If a CAS's degree operator is used instead, it must also obtain the operation (4.14). Both Maple's **degree** operator and Mathematica's **Exponent** operator obtain (4.14), while MuPAD's **degree** operator obtains $\deg(0) \to 0$. If the MuPAD implementation uses its **degree** operator, line 6 must be replaced by

while $m \geq n$ **and** $r \neq 0$ **do** .

Procedure *Polynomial_division(u, v, x)*;
Input
    u, v : GPEs in x with $v \neq 0$;
    x : a symbol;
Output
    a list $[q, r]$ with quotient q and remainder r;
Local Variables
    q, r, m, n, lcv, s, lcr;
Begin
1    $q := 0$;
2    $r := u$;
3    $m := Degree\_gpe(r, x)$
4    $n := Degree\_gpe(v, x)$
5    $lcv := Leading\_Coefficient\_gpe(v, x)$;
6   **while** $m \geq n$ **do**
7       $lcr := Leading\_Coefficient\_gpe(r, x)$;
8       $s := lcr/lcv$;
9       $q := q + s * x^{m-n}$;
10       $r := Algebraic\_expand((r - lcr * x^m) - (v - lcv * x^n) * s * x^{m-n})$;
11       $m := Degree\_gpe(r, x)$;
12    $Return([q, r])$
End

$$Quotient(u, v, x) \overset{\text{function}}{:=} Operand(Polynomial\_division(u, v, x), 1);$$

$$Remainder(u, v, x) \overset{\text{function}}{:=} Operand(Polynomial\_division(u, v, x), 2);$$

Figure 4.1. An MPL procedure for polynomial division of u by v in $\mathbf{F}[x]$. (Implementation: Maple (txt), Mathematica (txt), MuPAD (txt).)

used the iteration formula for r_i in Equation (4.8) rather than the formula in Equation (4.2). In fact, as long as all the field operations in $\mathbf{F}$ are obtained by automatic simplification, we can use Equation (4.2). However, when some field operations are not included in automatic simplification, Equation (4.2) may not eliminate the leading term from r_{i-1}, and the algorithm may not terminate. In Exercise 7, we give an example where the procedure with line 10 replaced by the formula in Equation (4.2) does not terminate.

Although the *Polynomial_division* procedure terminates whenever u and v are GPEs in x, for some coefficient fields it may return q_σ and r_σ in an inappropriate form. In Maple, Mathematica, and MuPAD, two fields where it does give appropriate results are $\mathbf{Q}$ and $\mathbf{Q}(\imath)$, because all numerical simplifications (in lines 8 and 10) are handled by automatic simplification. For example, for the polynomials in $\mathbf{Q}(\imath)[x]$:

$$u = (2 + 4\imath)\, x^2 + (-1 - 8\imath)\, x + (-3 + 3\imath),$$

$$v = (1 + 2\imath)\, x + (1 - \imath),$$

all three systems obtain with two iterations $q_\sigma = 2\, x - 3$ and $r_\sigma = 0$.

On the other hand, consider the polynomials in $\mathbf{Q}(\sqrt{2})[x]$:

$$u = (2 - 4\sqrt{2})\, x^2 + (-1 + 4\sqrt{2})\, x + (-3 + 3\sqrt{2}),$$

$$v = (1 - 2\sqrt{2})\, x + (1 - \sqrt{2}),$$

where $u = (2\, x - 3)\, v$. In this case, all three systems obtain the involved expressions

$$q_\sigma = \frac{2 - 4\sqrt{2}}{1 - 2\sqrt{2}}\, x + \frac{-1 + 4\sqrt{2} - \frac{10}{1 - 2\sqrt{2}} + 6\frac{\sqrt{2}}{1 - 2\sqrt{2}}}{1 - 2\sqrt{2}},$$

$$r_\sigma = -3 + 3\sqrt{2} + \frac{9}{1 - 2\sqrt{2}} - 5\frac{\sqrt{2}}{1 - 2\sqrt{2}} + \frac{22}{(1 - 2\sqrt{2})^2} - 16\frac{\sqrt{2}}{(1 - 2\sqrt{2})^2},$$

rather than the simplified forms

$$q_\sigma = 2\, x - 3, \qquad r_\sigma = 0.$$

The problem here is that some field operations for $\mathbf{Q}(\sqrt{2})$ are not obtained by automatic simplification in these systems. As we shall see in Section 4.2, this is more than just an inconvenience because the greatest common divisor algorithm given in that section gives an incorrect result when it cannot determine that a remainder is zero.

Polynomial Divisors

Definition 4.7. *Let u, v, and q be polynomials in $\mathbf{F}[x]$. A polynomial $v \neq 0$ is a **divisor** of (or **divides**) a polynomial u if there is a polynomial q such that $u = q\, v$. We use the notation $v \mid u$ to indicate that v is a divisor of u and $v \nmid u$ if it does not. The polynomial q is called the **cofactor** of v in u and is represented by $\operatorname{cof}(v, u)$.*

Example 4.8. For the polynomials in $\mathbf{Q}[x]$:

$$u = x^2 + 5x + 6, \qquad v = x + 2, \tag{4.15}$$

then $v \mid u$ with $q = \operatorname{cof}(v, u) = x + 3$. In general, if $v \mid u$, then for $c \neq 0$ in $\mathbf{F}$, $cv \mid u$ as well because

$$u = qv = ((1/c)\, q)\, (cv).$$

Therefore, $2x + 4$ is also a divisor of u in (4.15).   □

Definition 4.9. *Let u, v, and w be polynomials in $\mathbf{F}[x]$.*

1. *A* **common divisor** *(or* **common factor***) of u and v is a polynomial w such that $w \mid u$ and $w \mid v$.*

2. *The polynomials u and v are* **relatively prime** *if they have no common divisor of positive degree.*

Example 4.10. The polynomial $x + 2$ is a common divisor of $u = x^2 - 4$ and $v = x^2 + 5x + 6$. The polynomials $x^2 - 4$ and $x^2 - 1$ are relatively prime.   □

Irreducible Polynomials

An irreducible polynomial is the polynomial analogue of a prime number.

Definition 4.11. *A polynomial u in $\mathbf{F}[x]$ is* **reducible** *if there are polynomials v and w in $\mathbf{F}[x]$ with positive degree such that $u = vw$. A polynomial is* **irreducible** *if it is not reducible.*

Irreducibility depends, of course, on the coefficient field.

Example 4.12. The polynomial $x^2 - 2$ is irreducible when $\mathbf{F} = \mathbf{Q}$ but is reducible as $x^2 - 2 = (x - \sqrt{2})(x + \sqrt{2})$ when $\mathbf{F} = \mathbf{Q}(\sqrt{2})$.   □

Polynomial Expansion

The polynomial expansion of u in terms of v involves the representation of u as a sum whose terms contain non-negative integer powers of v. The following polynomial expansion theorem is based on polynomial division.

Theorem 4.13. *Let u and v be polynomials in $\mathbf{F}[x]$, and suppose that v has positive degree. Then, there are unique polynomials*

$$d_k(x), d_{k-1}(x), \ldots, d_0(x)$$

with $\deg(d_i) < \deg(v)$ *such that*

$$u = d_k v^k + d_{k-1} v^{k-1} + \cdots + d_1 v + d_0. \tag{4.16}$$

The representation for u in Equation (4.16) is called the *polynomial expansion*[3] of u in terms of v. For example, for

$$u = x^5 + 11x^4 + 51x^3 + 124x^2 + 159x + 86, \quad v = x^2 + 4x + 5,$$

then

$$u = (x+3)\,v^2 + (x+2)\,v + (x+1).$$

When $v = x - c$, the polynomial expansion coincides with the Taylor expansion of u about $x = c$ (Exercise 15).

Proof of Theorem 4.13: We show first that the sequence $d_0, d_1, \ldots, d_k$ exists by giving an algorithm to compute it. Using polynomial division, there are polynomials c_0 and d_0 such that

$$u = c_0\, v + d_0, \tag{4.17}$$

where $\deg(d_0) < \deg(v)$. If $c_0 = 0$, there is only one term in the expansion $(k = 0)$ with $u = d_0$. If $c_0 \neq 0$ but $\deg(c_0) < \deg(v)$, then the expansion has two terms $(k = 1)$ with $d_1 = c_0$ and d_0 defined by Equation (4.17). Finally, if $\deg(c_0) \geq \deg(v)$, form the *quotient sequence*

$$
\begin{aligned}
u &= c_0\, v + d_0, \\
c_0 &= c_1\, v + d_1, \\
c_1 &= c_2\, v + d_2, \\
&\;\;\vdots \\
c_{k-2} &= c_{k-1}\, v + d_{k-1}, \\
c_{k-1} &= c_k\, v + d_k, \\
c_k &= 0,
\end{aligned}
\tag{4.18}
$$

where $c_i = \mathrm{quot}(c_{i-1}, v, x)$, $d_i = \mathrm{rem}(c_{i-1}, v, x)$, and $\deg(d_i) < \deg(v)$. Since Theorem 4.1 implies $\deg(c_i) < \deg(c_{i-1})$, the quotient sequence terminates after a finite number of steps with $c_k = 0$. Substituting the expression for c_i into the expression for c_{i-1}, we obtain

---

[3]Don't confuse polynomial expansion with the *Algebraic_expand* operation, which actually destroys a polynomial expansion. The term *expansion* is used here as it is in the Taylor expansion of a function in calculus.

$$
\begin{aligned}
u &= c_0 v + d_0 \\
&= (c_1 v + d_1) v + d_0 = c_1 v^2 + d_1 v + d_0 \\
&= ((c_2 v + d_2) v + d_1) v + d_0 = c_2 v^3 + d_2 v^2 + d_1 v + d_0 \\
&\;\;\vdots \\
&= c_k v^{k+1} + d_k v^k + d_{k-1} v^{k-1} + \cdots + d_1 v + d_0 \\
&= d_k v^k + d_{k-1} v^{k-1} + \cdots + d_1 v + d_0.
\end{aligned}
\tag{4.19}
$$

Now that we know the coefficient sequence d_i exists, we use a degree argument to show that it is the only sequence that satisfies the properties mentioned in the hypothesis of the theorem. Suppose that u had another polynomial expansion:

$$
u = e_{k'} v^{k'} + e_{k'-1} v^{k'-1} + \cdots + e_1 v + e_0,
\tag{4.20}
$$

where the coefficients $e_i(x)$ satisfy $\deg(e_i) < \deg(v)$. We must show that the expansions in Equations (4.16) and (4.20) have the same number of terms ($k = k'$), and that the coefficients of the various powers of v are equal. First, let's show that $e_0 = d_0$. Subtracting Equation (4.20) from Equation (4.16), we obtain

$$
0 = d_k v^k + d_{k-1} v^{k-1} + \cdots + d_1 v + d_0 - (e_{k'} v^{k'} + e_{k'-1} v^{k'-1} + \cdots + e_1 v + e_0),
$$

and moving the terms e_0 and d_0 to the other side of the equation, we have

$$
d_k v^k + d_{k-1} v^{k-1} + \cdots + d_1 v - (e_{k'} v^{k'} + e_{k'-1} v^{k'-1} + \cdots + e_1 v) = e_0 - d_0.
$$

Notice that v is a factor of each term of the left side of the equality. Therefore, if this side of the equality is not zero, it has degree $\geq \deg(v)$. On the other hand, the right side has degree $< \deg(v)$. Since, both of these degree conditions cannot be true simultaneously, both sides of the equation must be zero. Therefore, $e_0 = d_0$ and

$$
d_k v^k + d_{k-1} v^{k-1} + \cdots + d_1 v - (e_{k'} v^{k'} + e_{k'-1} v^{k'-1} + \cdots + e_1 v) = 0.
$$

Dividing both sides of this equation by v and moving the terms containing e_1 and d_1 to the other side, we obtain

$$
d_k v^{k-1} + d_{k-1} v^{k-1} + \cdots + d_2 v - (e_{k'} v^{k'-1} + e_{k'-1} v^{k'-2} + \cdots + e_2 v) = e_1 - d_1.
$$

Using the same degree argument as above, we conclude that $e_1 = d_1$. Continuing in this fashion, we conclude that $e_i = d_i$ for $2 \leq i \leq \min(\{k, k'\})$. If both expansions have the same number of terms ($k = k'$), the proof is complete. If the number of terms are different (say $k > k'$), then after showing $d_{k'} = e_{k'}$, we have

$$d_k v^{(k-k'-1)} + d_{k-1} v^{(k-k'-2)} + \cdots d_{k'+1} = 0.$$

Using the same degree argument as above, we can conclude that all coefficients d_i in this expression are zero, and therefore $k = k'$. $\square$

Example 4.14. Consider the polynomials in $\mathbf{Q}[x]$:

$$u = x^5 + 11x^4 + 51x^3 + 124x^2 + 159x + 86, \quad v = x^2 + 4x + 5.$$

We obtain

$$
\begin{aligned}
c_0 &= x^3 + 7x^2 + 18x + 17, \quad d_0 = x + 1, \\
c_1 &= x + 3, \quad d_1 = x + 2, \\
c_2 &= 0, \quad d_2 = x + 3.
\end{aligned}
$$

The polynomial expansion for u in terms of v is

$$
\begin{aligned}
u &= d_2 v^2 + d_1 v + d_0 \\
&= (x+3)(x^2 + 4x + 5)^2 + (x+2)(x^2 + 4x + 5) + (x+1). \quad \square
\end{aligned}
$$

Definition 4.15. *Let u and v be polynomials in $\mathbf{F}[x]$ with $\deg(v) > 0$, and let t be a symbol. Define the polynomial **expansion polynomial***

$$P_u(t) = d_k t^k + d_{k-1} t^{k-1} + \cdots + d_1 t + d_0, \quad (4.21)$$

where the coefficients $d_i(x)$ are defined by the quotient sequence (4.18).

The polynomial P_u is simply the expansion (4.16) with the polynomial v replaced by the symbol t.

Theorem 4.16. *Suppose that u, v, u_1, u_2, and w polynomials are $\mathbf{F}[x]$ with $\deg(v) > 0$. (All polynomial expansions are in terms of the polynomial v.) The expansion polynomial satisfies the following properties:*

1. $P_0 = 0$.

2. If $\deg(u) < \deg(v)$, then $P_u = u$.

3. $P_{u_1+u_2} = P_{u_1} + P_{u_2}$.

4. $P_{vw} = t P_w$.

Procedure *Polynomial_expansion*(u, v, x, t);
Input
    u : a GPE in x;
    v : a GPE in x with $\deg(v, x) > 0$;
    x, t : symbols;
Output
    The polynomial P_u;
Local Variables
    d, q, r;
Begin
1   **if** $u = 0$ **then**
2   *Return*(0)
3   **else**
4       $d :=$ *Polynomial_division*(u, v, x);
5       $q :=$ *Operand*$(d, 1)$;
6       $r :=$ *Operand*$(d, 2)$;
7   *Return*(*Algebraic_expand*($t *$ *Polynomial_expansion*$(q, v, x, t) + r$))
End

Figure 4.2. An MPL procedure for polynomial expansion. (Implementation: Maple (txt), Mathematica (txt), MuPAD (txt).)

The verification of these properties is straightforward and is left to the reader (Exercise 18).

The next theorem describes a recurrence relation that gives a simple algorithm for computing P_u.

Theorem 4.17. *Let u and v be polynomials in $\mathbf{F}[x]$ with $\deg(v, x) > 0$, and suppose that from polynomial division, $u = q\,v + r$. Then, the expansion polynomial satisfies the recurrence relation*

$$P_u = t\,P_q + r.$$

The proof follows directly from the previous theorem (Exercise 19).

A recursive procedure to compute P_u is given in Figure 4.2. The procedure uses the condition in Theorem 4.16(1) for termination (lines 1–2). The *Algebraic_expand* operator is applied to the recurrence relation (line 7) so that the expression is returned as a polynomial in t rather than as a composite expression.

Notice that the procedure accepts u and v that are GPEs in x, although in some cases more simplification is required to obtain the expansion in a usable form (Exercise 14(b)). Since $\deg(v) > 0$, it follows that $\deg(q) < \deg(u)$, which implies that each successive recursive call involves a polynomial u of smaller degree. Therefore, for some recursive call $\deg(u) < \deg(v)$, which implies $q = 0$ at line 5, and the next call terminates the recursion.

To get the representation for u in Equation (4.16), evaluate the expression

$$\textit{Substitute}(\textit{Polynomial\_expansion}(u, v, x, t), \quad t = v). \tag{4.22}$$

Example 4.18. For the polynomials in Example 4.14

$$u = x^5 + 11\,x^4 + 51\,x^3 + 124\,x^2 + 159\,x + 86, \quad v = x^2 + 4\,x + 5,$$

and so

$$\textit{Polynomial\_expansion}(u, v, x, t) \to x\,t^2 + 3\,t^2 + x\,t + 2\,t + x + 1.$$

Applying the operation (4.22), we have the expansion

$$(x + 3)\,(x^2 + 4\,x + 5)^2 + (x + 2)\,(x^2 + 4\,x + 5) + x + 1. \qquad \square$$

Expansion-Based Substitutions. Using polynomial expansion, we can generalize the notion of substitution for expressions in $\mathbf{F}[x]$ to allow substitutions for "hidden" polynomials. For example, let

$$u = (x + 1)^3 + 2\,(x + 1) + 4,$$

which has the algebraically expanded form

$$u = x^3 + 3x^2 + 5x + 7. \tag{4.23}$$

Now suppose that we would like to substitute t for $x + 1$ in the polynomial (4.23). In this form, $x + 1$ is not a complete sub-expression of u, and so the substitution is not possible with the structural substitution. However, the substitution is readily obtained with the operation

$$\textit{Polynomial\_expansion}(u, x + 1, x, t) \to t^3 + 2\,t + 4.$$

In this sense, polynomial expansion is a generalization of substitution.

Expansion-Based Polynomial Operators. By using polynomial expansion, we can extend the polynomial structural operations (*Degree_gpe*, *Coefficient_gpe*, etc.) to polynomials u that are not (comfortably) polynomials in v in the sense of the definitions in Section 1.4. For example, although

$$u = 3\,x^4 + 5\,x^2 + 7$$

can be viewed as a polynomial in $v = x^2$, the basic structural operators give

$$Degree\_gpe(u, x^2) \rightarrow 1, \quad Coefficient\_gpe(u, x^2, 2) \rightarrow 0.$$

The problem arises because these operators are structure-based, and x^2 is not a complete sub-expression of x^4. The problem can be overcome, however, using polynomial expansion. For example,

$$Degree\_gpe(Polynomial\_expansion(u, x^2, t), t) \rightarrow 2,$$

$$Coefficient\_gpe(Polynomial\_expansion(u, x^2, t), t, 2) \rightarrow 3.$$

In Exercise 20, we describe procedures that perform the basic polynomial structural operations in this setting.

Polynomial Decomposition.   Let $u = 2x^4 + 4x^3 + 9x^2 + 7x + 9$ and $v = x^2 + x + 1$ be polynomials in $\mathbf{Q}[x]$. The polynomial expansion of u in terms of v is given by $u = 2v^2 + 3v + 4$ where, in this case, all the coefficients r_i are in $\mathbf{Q}$. When this occurs, $u(x)$ can be represented as a composition of polynomials $u = f(v(x))$ where $f(x) = 2x^2 + 3x + 4$ is also in $\mathbf{Q}[x]$. In other words, $u(x)$ has a particularly simple structure as a composition of two lower degree polynomials.

Now let's reverse our point of view. Suppose that we are given a polynomial $u(x)$. When is it possible to find polynomials $f(x)$ and $v(x)$ (with $\deg(v) < \deg(u)$ and $\deg(f) < \deg(u)$) so that $u(x) = f(v(x))$? This problem is called the *polynomial decomposition problem* and is considerably more involved than polynomial expansion. We examine the decomposition problem in Chapter 5.

Exercises

1. Let $\mathbf{F}$ be a field. Explain why $\mathbf{F}[x]$ is not a field.

2. Suppose that $u \neq 0$ is in $\mathbf{F}[x]$ and k is a positive integer. Show that

$$\deg(u^k) = k \deg(u).$$

3. Prove Theorem 4.2.

4.   (a) Let u and v be in $\mathbf{F}[x]$. Show that if $u \mid v$ and $v \mid u$, then $u = cv$ for some c in $\mathbf{F}$.

   (b) Suppose that u, v, w, and p are polynomials in $\mathbf{F}[x]$ and $u = v + w$. Show that if p divides any two of the polynomials u, v, and w, then it also divides the third polynomial.

5. Let $u = x^2 + 1$ and $v = x + 2$ be polynomials in $\mathbf{Z}_5[x]$. Find the quotient and remainder of u divided by v.

6. Let u, v, and $w \neq 0$ be polynomials in $\mathbf{F}[x]$, and let c be in $\mathbf{F}$. Show that

$$\text{quot}(u+v, w) = \text{quot}(u, w) + \text{quot}(v, w), \qquad \text{quot}(c\,u, w) = c\,\text{quot}(u, w),$$

$$\text{rem}(u+v, w) = \text{rem}(u, w) + \text{rem}(v, w), \qquad \text{rem}(c\,u, w) = c\,\text{rem}(u, w).$$

7. Let $\mathbf{F}[x] = (\mathbf{Q}(\sqrt{2}))[x]$ and consider the polynomials $u = x$ and $v = x + \sqrt{2}\,x$. Suppose that line 10 in Figure 4.1 is replaced by

$$r := Algebraic\_expand\left(r - v * s * x^{m-n}\right).$$

(See the recurrence relation for r_i in Equation (4.2).) Explain why the Maple, Mathematica, and MuPAD implementations of $Polynomial\_division$ do not terminate using this recurrence relation.

8. Give a recursive procedure for polynomial division. Is it better to use iteration or recursion to implement polynomial division?

9. The polynomial division procedure in Figure 4.1 does not work for polynomials in $\mathbf{Z}_p[x]$ (p a prime) because the operations in lines 8 and 10 are evaluated by automatic simplification using rational number arithmetic. Give a procedure $Poly\_div\_p(u, v, x, p)$ that performs polynomial division for polynomials in $\mathbf{Z}_p[x]$. The procedure returns a list with the quotient and remainder. The procedure $Division\_p$ described in Exercise 11, page 59 is useful in this problem. This exercise is used in the $Poly\_gcd\_p$ procedure described in Exercise 4, page 142.

10. Find all irreducible polynomials in $\mathbf{Z}_3[x]$ with degree $= 2$.

11. Let h be a polynomial in $\mathbf{F}[x]$, and let c and d be distinct members of $\mathbf{F}$. Show that $h - c$ and $h - d$ are relatively prime.

12. Let u be a polynomial in $\mathbf{Q}[x]$. Give a procedure $Polynomial\_divisors(u, x)$ that returns the set of monic divisors of u with positive degree. If u is not a polynomial in x, return the global symbol **Undefined**. Use the factor operator in a CAS to find the factors of u. Each divisor should be returned in expanded form. This procedure is used by the procedures $Poly\_decomp\_2$ in Figure 5.1 and $Complete\_poly\_decomp$ in Figure 5.2.

13. Prove Theorem 4.6.

14. For each of the following, find the polynomial expansion of u in terms of v.

    (a) $u = 2\,x^5 + 13\,x^4 + 41\,x^3 + 68\,x^2 + 63\,x + 11, \quad v = x^2 + 3\,x + 4.$

    (b) In $\mathbf{Q}(\sqrt{2})[x]$, the polynomials

    $$\begin{aligned} u &= (20 + 14\sqrt{2})\,x^3 + (18 + 12\sqrt{2})\,x^2 + (6 + 3\sqrt{2})\,x + 1, \\ v &= (2 + \sqrt{2})\,x + 1. \end{aligned}$$

    In this case, u has a particularly simple expansion in terms of v.

15. Suppose that $v = x - c$. Explain why the polynomial expansion for a polynomial u in terms of v is the same as the Taylor expansion of u about $x = c$.

16. Find the polynomial expansion of $u = x^n$ in terms of $v = x - 1$ for $n = 1, 2, 3, 4, 5$. What pattern do you observe with the coefficients d_i? Explain why the coefficient pattern appears in these expansions.

17. Suppose that u is expressed as a polynomial expansion in terms of v with $\deg(v) > 0$. Show that $\deg(u) = \deg(d_k) + \deg(v) \, k$.

18. Prove Theorem 4.16.

19. Prove Theorem 4.17.

20. Let u and v be polynomials in $\mathbf{Q}[x]$ with $\deg(v, x) \geq 1$. Let's define u to be a polynomial in v if the coefficients d_i in the polynomial expansion of u in terms of v are all rational numbers. For example, $3\,x^4 + 4\,x^2 + 5$ is a polynomial in terms of x^2, while $x^3 + x^2$ is not. Give procedures for the following operators.

    (a) *Polynomial_exp*(u, v, x) which returns **true** if u is a polynomial in terms of v as defined above and **false** otherwise.

    (b) *Degree_exp*(u, v, x) which returns the degree of u in terms of v when u is a polynomial in v or the symbol **Undefined** when u is not a polynomial in v.

    (c) *Coefficient_exp*(u, v, x, i) which returns the coefficient d_i when u is a polynomial in v or the symbol **Undefined** when u is not a polynomial in v.

4.2 Greatest Common Divisors in F[x]

The concept of the *greatest common divisor* of two polynomials u and v is a fundamental one in computer algebra. Loosely speaking, a greatest common divisor of u and v is a divisor of u and v that contains all the factors that u and v have in common. The concept is defined formally in the following definition.

Definition 4.19. *Let u and v be polynomials in $\mathbf{F}[x]$. The **greatest common divisor (gcd)** of u and v (at least one of which is non-zero) is a polynomial d that satisfies the following properties.*

1. *d is a common divisor of u and v.*

2. *If e is any common divisor of u and v, then $e \mid d$.*

3. *d is a monic polynomial.*

The operator $\gcd(u, v, x)$ denotes the greatest common divisor. When there is no chance of confusion, we omit the variable x and use the simpler notation $\gcd(u, v)$.

When both $u = 0$ and $v = 0$, the above definition does not apply. In this case, by definition, $\gcd(0, 0, x) = 0$.

In other words, according to property (2), the greatest common divisor of u and v is "greatest" in the sense that any common divisor of u and v must also divide $\gcd(u, v)$.

Example 4.20. Let

$$
\begin{aligned}
u &= 2x^3 + 12x^2 + 22x + 12 = 2(x+1)(x+2)(x+3), \\
v &= 2x^3 + 18x^2 + 52x + 48 = 2(x+2)(x+3)(x+4).
\end{aligned}
$$

From the factorizations of the two polynomials, it appears that the greatest common divisor is $d = x^2 + 5x + 6$. Let's verify this by showing that d satisfies the three properties in Definition 4.19. First, polynomial division shows that (1) is satisfied. To show (2), we again use polynomial division of u by v to obtain

$$
q_\sigma = \text{quot}(u, v, x) = 1, \quad r_\sigma = \text{rem}(u, v, x) = -6x^2 - 30x - 36,
$$

which implies that

$$
u = q_\sigma v + r_\sigma = (1)v + (-6)d.
$$

Therefore, if $e \mid u$ and $e \mid v$, we have $e \mid d$, and (2) is satisfied. Finally, since d is monic, (3) is satisfied. ☐

The next theorem gives three important properties of the greatest common divisor.

Theorem 4.21. Let u and v be polynomials in $\mathbf{F}[x]$.

1. $\gcd(u, v)$ exists.

2. $\gcd(u, v)$ is unique.

3. If $u \neq 0$, then $\gcd(0, u) = u/\text{lc}(u)$.

Proof: We show (1) by giving an algorithm to compute $\gcd(u, v)$ later in this section. Parts (2) and (3) follow directly from Definition 4.19 (Exercise 2). ☐

Euclid's GCD Algorithm in F[x]

The gcd of two polynomials is obtained with a polynomial version of Euclid's algorithm described in Section 2.1 The algorithm is based on the following theorem.

Theorem 4.22. *Let u and $v \neq 0$ be polynomials in $\mathbf{F}[x]$, and let*

$$r_\sigma = \text{rem}(u, v, x).$$

Then,

$$\gcd(u, v) = \gcd(v, r_\sigma).$$

Proof: The proof of this theorem is the same as the proof for the integer case (see Theorem 2.9, page 21). □

We can find the $\gcd(u, v)$ by repeatedly applying Theorem 4.22. When $v \neq 0$, define a sequence of polynomials $R_{-1}, R_0, R_1, \ldots$ with the following scheme:

$$
\begin{aligned}
R_{-1} &= u, \\
R_0 &= v, \\
R_1 &= \text{rem}(R_{-1}, R_0, x), \\
&\vdots \\
R_{i+1} &= \text{rem}(R_{i-1}, R_i, x), \\
&\vdots
\end{aligned}
\tag{4.24}
$$

The sequence (4.24) is called a *polynomial remainder sequence*. Since

$$\deg(R_{i+1}) < \deg(R_i),$$

some member of the sequence is 0. Let R_ρ be the first such remainder. By Theorem 4.22 and Theorem 4.21(3),

$$
\begin{aligned}
\gcd(u, v) &= \gcd(R_{-1}, R_0) \\
&= \gcd(R_0, R_1) \\
&\vdots \\
&= \gcd(R_{\rho-1}, R_{\rho-2}) \\
&= \gcd(R_{\rho-1}, R_\rho) \\
&= \gcd(R_{\rho-1}, 0) \\
&= R_{\rho-1}/\text{lc}(R_{\rho-1}),
\end{aligned}
\tag{4.25}
$$

where we have used the condition $R_\rho = 0$ to terminate the iteration. This discussion is summarized in the following theorem.

Theorem 4.23. *Let u and v be polynomials in $\mathbf{F}[x]$, at least one of which is not zero. Then, $\gcd(u, v) = R_{\rho-1}/\mathrm{lc}(R_{\rho-1})$.*

Proof: Although the relation was derived by assuming that $v \neq 0$, it is true when $u \neq 0$ and $v = 0$ as well. In this case, $\rho = 0$ and $\gcd(u, 0) = u/\mathrm{lc}(u) = R_{-1}/\mathrm{lc}(R_{-1})$. □

Example 4.24. Let $u = x^7 - 4x^5 - x^2 + 4$ and $v = x^5 - 4x^3 - x^2 + 4$. Then,

$$R_1 = x^4 - 5x^2 + 4, \qquad R_2 = x^3 - x^2 - 4x + 4, \qquad R_3 = 0.$$

Therefore, $\gcd(u, v) = x^3 - x^2 - 4x + 4$. □

A procedure that obtains $\gcd(u, v)$ using the polynomial remainder sequence (4.24) is given in Figure 4.3. Lines 1 and 2 are included so that a division by zero is not encountered at line 6 for the special case when both $u = 0$ and $v = 0$. The *Algebraic_expand* operator at line 9 is included to distribute $1/Leading\_Coefficient\_gpe(U, x)$ when U is a sum.

Notice that in *Polynomial_gcd*, we require that all field operations in $\mathbf{F}$ be obtained with automatic simplification. If this is not so, the procedure may return an incorrect result. The next example illustrates this point.

Example 4.25. Consider the polynomials in $\mathbf{Q}\left(\sqrt{2}\right)[x]$:

$$u = x^2 + \left(-1 - \sqrt{2}\right)x, \qquad v = x^2 + \left(-2 - 2\sqrt{2}\right)x + 3 + 2\sqrt{2}.$$

In this case, $\gcd(u, v) = x + (-1 - \sqrt{2})$. However, if automatic simplification does not perform all operations in $\mathbf{Q}(\sqrt{2})[x]$, we obtain the remainder sequence[4]

$$
\begin{aligned}
R_1 &= x + \sqrt{2}\,x - 3 - 2\sqrt{2}, \\
R_2 &= 3 + 2\sqrt{2} - \frac{14}{\sqrt{2}+1} - \frac{10\sqrt{2}}{\sqrt{2}+1} + \frac{17}{\left(\sqrt{2}+1\right)^2} + \frac{12\sqrt{2}}{\left(\sqrt{2}+1\right)^2}, \\
R_3 &= 0.
\end{aligned}
$$

Since R_2 is free of x, *Polynomial_gcd* returns $\gcd(u, v) = 1$, which is incorrect. In fact, R_2 simplifies to 0, and so the iteration should have terminated with $R_2 = 0$, giving $\gcd(u, v) = R_1/\mathrm{lc}(R_1) = x + (-1 - \sqrt{2})$. □

---

[4]Similar results are obtained with Maple, Mathematica, and MuPAD implementations of *Polynomial_division* and *Polynomial_gcd*.

Procedure *Polynomial_gcd(u, v, x)*;
Input
    u, v : polynomials in $\mathbf{F}[x]$ where all field operations
        in $\mathbf{F}$ are obtained with automatic simplification;
    x : a symbol;
Output
    gcd(u, v, x);
Local Variables
    U, V, R;
Begin
1   **if** $u = 0$ **and** $v = 0$ **then**
2   *Return*(0)
3   **else**
4        $U := u$; $V := v$;
5   **while** $V \neq 0$ **do**
6           $R := Remainder(U, V, x)$;
7           $U := V$;
8           $V := R$;
9   *Return*(*Algebraic_expand*($1/Leading\_Coefficient\_gpe(U, x) * U$))
End

Figure 4.3. An MPL procedure for Euclid's algorithm for polynomials. (Implementation: Maple (txt), Mathematica (txt), MuPAD (txt).)

Algorithms for division and gcd computation for polynomials with algebraic number coefficients are given in Section 4.3.

Computational Difficulties with Euclid's Algorithm

Although Euclid's algorithm is simple and works well for low degree polynomials, it has a property that makes it unsuitable as a general purpose algorithm in a CAS.

Example 4.26. Consider the two polynomials in $\mathbf{Q}[x]$:

$$u = x^8 + 5x^7 + 7x^6 - 3x^5 + 4x^4 + 17x^3 - 2x^2 - 6x + 3,$$

$$v = x^8 + 6x^7 + 3x^6 + x^5 + 10x^4 + 8x^3 + 2x^2 + 9x + 8.$$

The remainder sequence is given by:

$$
\begin{aligned}
R_1 &= -x^7 + 4x^6 - 4x^5 - 6x^4 + 9x^3 - 4x^2 - 15x - 5, \\
R_2 &= 39x^6 - 45x^5 - 41x^4 + 94x^3 - 53x^2 - 146x - 42,
\end{aligned}
$$

$$R_3 = -\frac{896}{507}x^5 - \frac{101}{169}x^4 + \frac{132}{169}x^3 - \frac{655}{169}x^2 - \frac{2749}{507}x - \frac{327}{169},$$

$$R_4 = -\frac{3280121}{802816}x^4 - \frac{3461627}{200704}x^3 - \frac{36161099}{802816}x^2 - \frac{8166587}{802816}x + \frac{17428125}{802816},$$

$$R_5 = -\frac{554215997440}{63663868489}x^3 - \frac{4783122333696}{63663868489}x^2 - \frac{2030741536768}{63663868489}x + \frac{2198164799488}{63663868489},$$

$$R_6 = -\frac{10141289265339652563}{54656781667532800}x^2 - \frac{103051366981906031}{1115444523827200}x + \frac{1272943070806564261}{13664195416883200},$$

$$R_7 = -\frac{12650166638556324421 3043200}{1242653261505575739275 19717}x - \frac{12650166638556324421 3043200}{1242653261505575739275 19717},$$

$$R_8 = 0.$$

Dividing R_7 by its leading coefficient, we have $\gcd(u, v) = x + 1$.   □

Notice that there is a dramatic increase in the number of digits required to represent the coefficients of the remainders, even though the coefficients in the gcd require few digits. This phenomenon, which is known as *coefficient explosion*, occurs frequently with computer algebra algorithms. The culprit here is the recurrence relation for the remainder in polynomial division (see Equation (4.8)). With each application of this formula, a new polynomial is obtained using arithmetic with rational numbers. Unfortunately, the result of an arithmetic operation with rational numbers usually requires more digits than either of the original operands (e.g., $2/3 + 2/5 = 16/15$). For this reason, as the iteration proceeds, there is a tendency for the coefficients to require more and more digits. In fact, for polynomials of degree n with coefficients that are randomly chosen single digit integers (for which it is nearly certain that the polynomials are relatively prime), $R_{\rho-1}$ has coefficients with approximately n^2 digits. On the other hand, if $\deg(\gcd(u, v))$ is close to $\min(\{\deg(u), \deg(v)\})$, few iterations are required, and the coefficient explosion is not nearly as significant.

Coefficient explosion significantly increases both the time and memory requirements of the algorithm. It can be reduced, however, with more sophisticated algorithms that compute the gcd. We return to this problem again in Section 6.3.

The Extended Euclidean Algorithm in F[x]

An extended Euclidean algorithm for polynomials in $\mathbf{F}[x]$ is similar to the algorithm for the integers described in Section 2.1.

Theorem 4.27. *For u and v polynomials in $\mathbf{F}[x]$, there exist polynomials A and B in $\mathbf{F}[x]$ such that*

$$A u + B v = \gcd(u, v). \tag{4.26}$$

The polynomials A and B are given in Equation (4.31).

Proof: Since the development for polynomials is nearly identical to what was done for the integers (page 23), we only summarize the approach. First, for the special case $u = 0$ and $v = 0$, we use $A = 0$ and $B = 0$. When either u or v is not zero, there are polynomials A_i and B_i such that

$$R_i = A_i\, u + B_i\, v. \tag{4.27}$$

These polynomials are given by the recurrence relations:

$$A_{-1} = 1, \quad A_0 = 0, \tag{4.28}$$

$$B_{-1} = 0, \quad B_0 = 1, \tag{4.29}$$

$$A_i = A_{i-2} - Q_i\, A_{i-1}, \quad B_i = B_{i-2} - Q_i\, B_{i-1}, \quad i \geq 1, \tag{4.30}$$

where $Q_i = \mathrm{quot}(R_{i-2}, R_{i-1}, x)$. Now, since $\gcd(u, v) = R_{\rho-1}/\mathrm{lc}(R_{\rho-1})$, we have

$$A = A_{\rho-1}/\mathrm{lc}(R_{\rho-1}), \quad B = B_{\rho-1}/\mathrm{lc}(R_{\rho-1}). \tag{4.31}$$

$$\square$$

Example 4.28. Let

$$u = x^7 - 4\,x^5 - x^2 + 4, \quad v = x^5 - 4\,x^3 - x^2 + 4.$$

The greatest common divisor of these polynomials is given in Example 4.24 as $\gcd(u, v) = R_2 = x^3 - x^2 - 4\,x + 4$. In addition,

$$\begin{aligned}
A_1 &= 1, & B_1 &= -x^2, \\
A_2 &= -x, & B_2 &= x^3 + 1. \\
A_3 &= x^2 + x + 1, & B_3 &= -x^4 - x^3 - x^2 - x - 1.
\end{aligned}$$

Therefore, by Equation (4.31),

$$A = A_2/\mathrm{lc}(R_2) = -x,$$

$$B = B_2/\mathrm{lc}(R_2) = x^3 + 1. \qquad \square$$

A procedure that obtains A and B using the recurrence relations (4.30) with initial conditions (4.28) and (4.29) is given in Figure 4.4. The variables Ap and Bp represent the previous values A_{i-1} and B_{i-1}, and App and Bpp represent A_{i-2} and B_{i-2}.

Theorem 4.27 has many applications in computer algebra including the statements in the next theorem which are used frequently in the remainder of the book.

Theorem 4.29. *Suppose that u, v, and w are in $\mathbf{F}[x]$.*

1. *If $w \mid uv$ and w and u are relatively prime, then $w \mid v$.*

2. *If $w \mid uv$ and w is irreducible, then $w \mid u$ or $w \mid v$.*

3. *If $u \mid w$, $v \mid w$, and $\gcd(u, v) = 1$, then $uv \mid w$.*

Proof: The proofs are similar to the ones for Theorem 2.16 on page 27.$\square$

Using Theorem 4.27, we obtain the following description of relatively prime polynomials.

Theorem 4.30. *Let u and v be polynomials in $\mathbf{F}[x]$ at least one of which is not 0. Then u and v are relatively prime if and only if there are polynomials A and B in $\mathbf{F}[x]$ such that*

$$Au + Bv = 1. \qquad (4.32)$$

Proof: First, if u and v are relatively prime, then Theorem 4.27 implies Equation (4.32).

On the other hand, if Equation (4.32) is true, then since $\gcd(u, v) \mid u$ and $\gcd(u, v) \mid v$, we have $\gcd(u, v) \mid 1$. Therefore, $\deg(\gcd(u, v)) = 0$ which implies that u and v are relatively prime. $\qquad \square$

Properties of A_i and B_i

In the next few theorems, we describe some properties of the polynomial sequences A_i and B_i.

Theorem 4.31. *Let u and v be non-zero polynomials in $\mathbf{F}[x]$. Then, for $i = 1, 2, \ldots, \rho$,*

$$u = (-1)^i (B_i R_{i-1} - B_{i-1} R_i), \qquad (4.33)$$
$$v = (-1)^{i-1} (A_i R_{i-1} - A_{i-1} R_i). \qquad (4.34)$$

Procedure *Extended_Euclidean_algorithm*(u, v, x);
Input
    u, v : polynomials in $\mathbf{F}[x]$ where all field operations
           in $\mathbf{F}$ are obtained with automatic simplification;
    x : a symbol;
Output
       The list $[\gcd(u, v), A, B]$;
Local Variables
    $U, V, Ap, App, Bp, Bpp, q, r, A, B, c$;
Begin

1   **if** $u = 0$ and $v = 0$ **then**
2   *Return*$([0, 0, 0])$
3   **else**
4        $U := u;\ \ V := v;\ \ App := 1;\ \ Ap := 0;\ \ Bpp := 0;\ \ Bp := 1$;
5   **while** $V \neq 0$ **do**
6           $q := Quotient(U, V, x)$;
7           $r := Remainder(U, V, x)$;
8           $A := App\ -\ q * Ap;\ \ \ B := Bpp\ -\ q * Bp$;
9           $App := Ap;\ \ Ap := A;\ \ Bpp := Bp;\ \ Bp := B$;
10         $U := V;\ \ V := r$;
11        $c := Leading\_Coefficient\_gpe(U, x)$;
12        $App := Algebraic\_expand(App/c)$;
13        $Bpp := Algebraic\_expand(Bpp/c)$;
14        $U := Algebraic\_expand(U/c)$;
15   *Return*$([U,\ App,\ Bpp])$
     End

Figure 4.4. An MPL procedure for the extended Euclidean algorithm which returns the list $[\gcd(u, v),\ A,\ B]$. (Implementation: Maple (txt), Mathematica (txt), MuPAD (txt).)

Proof: We verify Equation (4.33) with mathematical induction. For $i = 1$, we obtain using Equations (4.29) and (4.30), $B_1 = -Q_1$. Therefore,

$$
\begin{aligned}
u &= Q_1 v + R_1 = Q_1 R_0 + R_1 \\
&= -B_1 R_0 + B_0 R_1 = (-1)(B_1 R_0 - B_0 R_1)
\end{aligned}
$$

which is Equation (4.33) for $i = 1$. Suppose next that $i \leq \rho$, and assume the induction hypothesis that Equation (4.33) holds for $i - 1$. By definition of the remainder sequence,

$$R_{i-2} = Q_i R_{i-1} + R_i,$$

and from Equation (4.30)

$$B_{i-2} = B_i + Q_i B_{i-1}.$$

Using these relationships and the induction hypothesis, we have

$$
\begin{aligned}
u &= (-1)^{i-1} (B_{i-1} R_{i-2} - B_{i-2} R_{i-1}) \\
&= (-1)^{i-1} (B_{i-1} (Q_i R_{i-1} + R_i) - (B_i + Q_i B_{i-1}) R_{i-1}) \\
&= (-1)^i (B_i R_{i-1} - B_{i-1} R_i)
\end{aligned}
$$

which verifies the relationship for i. $\qquad\square$

One way to find the cofactors $\text{cof}(\gcd(u,v), u)$ and $\text{cof}(\gcd(u,v), v)$ is to divide u and v by $\gcd(u,v)$. It is also possible to obtain them using the relationships in the next theorem.

Theorem 4.32. *Let u and v be non-zero polynomials in $\mathbf{F}[x]$. Then*

$$
\begin{aligned}
\text{cof}(\gcd(u,v), u) &= (-1)^\rho B_\rho \, \text{lc}(R_{\rho-1}), \\
\text{cof}(\gcd(u,v), v) &= (-1)^{\rho-1} A_\rho \, \text{lc}(R_{\rho-1}).
\end{aligned}
$$

Proof: Since $R_\rho = 0$ and $\gcd(u,v) = R_{\rho-1}/\text{lc}(R_{\rho-1})$, these relationships follow directly from Equations (4.33) and (4.34). $\qquad\square$

Example 4.33. Let $u = x^7 - 4x^5 - x^2 + 4$ and $v = x^5 - 4x^3 - x^2 + 4$. The greatest common divisor of these polynomials is given in Example 4.24 as $\gcd(u,v) = R_2 = x^3 - x^2 - 4x + 4$ with $\rho = 3$. The sequences A_i and B_i are given in Example 4.28 where

$$A_3 = x^2 + x + 1, \qquad B_3 = -x^4 - x^3 - x^2 - x - 1.$$

Therefore,

$$
\begin{aligned}
\text{cof}(\gcd(u,v), u) &= (-1)^3 B_3 \, \text{lc}(R_2) = x^4 + x^3 + x^2 + x + 1, \\
\text{cof}(\gcd(u,v), v) &= (-1)^2 A_3 \, \text{lc}(R_2) = x^2 + x + 1.
\end{aligned}
$$
$\qquad\square$

In Example 4.28, the degrees of the sequences A_i and B_i increase with i. In the next theorem we substantiate this observation.

Theorem 4.34. *Let u and v be non-zero polynomials in $\mathbf{F}[x]$.*

1. If $\rho \geq 2$, then $\deg(A_1) < \deg(A_2) < \cdots < \deg(A_\rho)$.

2. *If $\rho \geq 2$ and $Q_1 \neq 0$, then $\deg(B_1) < \deg(B_2) < \cdots < \deg(B_\rho)$.*

3. *If $\rho \geq 3$ and $Q_1 = 0$, then $\deg(B_2) < \deg(B_3) < \cdots < \deg(B_\rho)$.*

Proof: We prove the theorem for the sequence A_i using mathematical induction. First, according to Exercise 3, $\deg(Q_i) > 0$ for $i = 2, 3, \ldots, \rho$. Now, since $A_1 = 1$ and $A_2 = -Q_2$, we have $\deg(A_1) < \deg(A_2)$, and so the degree relation holds for the base case $i = 2$. Let's suppose that $3 \leq i \leq \rho$, and assume the induction hypothesis that $\deg(A_{i-2}) < \deg(A_{i-1})$. This inequality together with Theorem 4.1(2) implies that

$$\deg(A_{i-2}) < \deg(Q_i) + \deg(A_{i-1}) = \deg(Q_i\, A_{i-1}),$$

which in turn implies that

$$\deg(A_{i-2} - Q_i\, A_{i-1}) = \max(\{\deg(A_{i-2}), \deg(Q_i\, A_{i-1})\}) = \deg(Q_i\, A_{i-1}).$$

Finally, this relation together with the recurrence relation (4.30) gives

$$
\begin{aligned}
\deg(A_{i-1}) \quad &< \quad \deg(Q_i) + \deg(A_{i-1}) \\
&= \quad \deg(Q_i\, A_{i-1}) \\
&= \quad \deg(A_{i-2} - Q_i\, A_{i-1}) \\
&= \quad \deg(A_i)
\end{aligned}
$$

which is the degree relation for i.

The proof of the degree relationships for the sequence B_i is left to the reader (Exercise 7). In this case the initial step for the induction depends on whether or not $Q_1 = 0$. $\square$

Using the previous theorem we have the following important result.

Theorem 4.35. *Let u and v be polynomials in $\mathbf{F}[x]$ with positive degree. Then, $\deg(A) < \deg(v)$ and $\deg(B) < \deg(u)$.*

Proof: First assume that $\rho = 1$. In this case $v \mid u$, and so $\gcd(u, v) = v/\mathrm{lc}(v)$. Therefore, $A = 0$ and $B = 1/\mathrm{lc}(v)$ which implies that $\deg(A) < \deg(v)$ and $\deg(B) < \deg(u)$.

Next suppose that $\rho \geq 2$. Since $A = A_{\rho-1}/\mathrm{lc}(R_{\rho-1})$, we have by Theorem 4.32 and Theorem 4.34:

$$
\begin{aligned}
\deg(A) \quad &= \quad \deg(A_{\rho-1}) \\
&< \quad \deg(A_\rho) \\
&\leq \quad \deg((-1)^{\rho-1} A_\rho\, \mathrm{lc}(R_{\rho-1})\gcd(u, v)) \\
&= \quad \deg(v).
\end{aligned}
$$

A similar argument shows that $\deg(B) < \deg(v)$ when $\rho \geq 2$, although the case with $\rho = 2$ and $Q_1 = 0$ is handled separately.   □

Uniqueness of A and B

The polynomials $A = A_{\rho-1}/\mathrm{lc}(R_{\rho-1})$ and $B = B_{\rho-1}/\mathrm{lc}(R_{\rho-1})$ are not the only ones that satisfy Equation (4.26). Indeed, if w is any polynomial, define

$$A' = A + w\,v, \qquad B' = B - w\,u.$$

Then,

$$A'\,u + B'\,v = (A + w\,v)\,u + (B - w\,u)\,v = A\,u + B\,v = \gcd(u,v).$$

However, when u and v are relatively prime, we have the following uniqueness result.

Theorem 4.36. *Let u and v be polynomials in $\mathbf{F}[x]$ with positive degree, and suppose that $\gcd(u,v) = 1$. Then the polynomials A and B obtained with the extended Euclidean algorithm are the only polynomials with $\deg(A) < \deg(v)$ and $\deg(B) < \deg(u)$ such that $A\,u + B\,v = 1$.*

Proof: The existence of A and B is guaranteed by Theorem 4.35. To show the uniqueness property, suppose that there are two pairs of polynomials A, B and A', B' that satisfy the degree properties. Then, $A\,u + B\,v = 1 = A'\,u + B'\,v$, and

$$(A - A')\,u = (B' - B)\,v.$$

Since u divides the left side of this equation, it must also divide the right side. Therefore, since $\gcd(u,v) = 1$, Theorem 4.29(1) implies that $u \mid B' - B$. However, since

$$\deg(B' - B) < \deg(u),$$

we have $B' - B = 0$. In a similar way $A' - A = 0$.   □

Example 4.37. The preceding theorem is not true when u and v have a common factor with positive degree. For example, for $u = x^2 - 1$ and $v = x^2 + 2\,x + 1$, then

$$\gcd(u,v) = R_1/\mathrm{lc}(R_1) = x + 1, \qquad A = -1/2, \qquad B = 1/2.$$

However, the polynomials $A' = x/2$ and $B' = -x/2 + 1$ also satisfy the degree requirements, and $A'\,u + B'\,v = x + 1$. In fact, there are infinitely many pairs of polynomials A' and B' that satisfy these conditions (Exercise 13). This example is not just a rare occurrence but is indicative of the underlying theory. We show in Chapter 7 that whenever u and v are not relatively prime, this non-uniqueness property appears. $\qquad\square$

Factorization of Polynomials

An important application of Theorem 4.27 is the following factorization theorem for polynomials.[5]

Theorem 4.38. *Let u be a polynomial in $\mathbf{F}[x]$ with $\deg(u) > 0$. Then, u can be factored as*

$$u = c\, p_1 p_2 \cdots p_r, \qquad (4.35)$$

where c is in $\mathbf{F}$ and each p_i is a monic, irreducible polynomial in $\mathbf{F}[x]$ with $\deg(p_i) > 0$. The factorization is unique up to the order of the polynomials. The factorization (4.35) is called the **irreducible factorization** *of u in $\mathbf{F}[x]$.*

In other words, the uniqueness property states that two factorizations with the factors p_i listed in different orders are considered the same factorization.

Example 4.39. In $\mathbf{Q}[x]$ we have the unique factorization

$$2\,x^2 + 10\,x + 12 = 2\,(x+2)\,(x+3).$$

An irreducible factorization depends on the field of coefficients. For example, $u = x^2 - 2$ is irreducible in $\mathbf{Q}[x]$ but can be factored as $x^2 - 2 = (x + \sqrt{2})\,(x - \sqrt{2})$ in $\mathbf{Q}\left(\sqrt{2}\right)[x]$. Another example is $x^2 + 1$ which is irreducible in $\mathbf{Q}[x]$ and $\mathbf{Z}_3[x]$, but can be factored as

$$
\begin{aligned}
x^2 + 1 &= (x + i)(x - i), & \text{in } \mathbf{Q}(i)[x], \\
x^2 + 1 &= (x + 1)^2, & \text{in } \mathbf{Z}_2[x], \\
x^2 + 1 &= (x + 2)(x + 3), & \text{in } \mathbf{Z}_5[x]. \qquad\square
\end{aligned}
$$

Proof of Theorem 4.38: First, if u is irreducible, then its irreducible factorization is given by

$$u = \mathrm{lc}(u)\, p,$$

where p is the monic polynomial with coefficients obtained from the coefficients of u divided by $\mathrm{lc}(u)$.

---

[5]Theorem 4.27 is applied through Theorem 4.29(2) which follows from Theorem 4.27.

Next, if u is reducible, it can be factored $u = v\,w$ where $0 < \deg(v) < \deg(u)$ and $0 < \deg(w) < \deg(u)$. If v and w are irreducible, the process terminates. If not, the process continues by factoring v and w. Theorem 4.1(2) implies that the process eventually terminates with the factorization

$$u = q_1\, q_2 \cdots q_r, \qquad (4.36)$$

where each q_i is irreducible and $0 < \deg(q_i) < \deg(u)$. In addition,

$$u = q_1\, q_2 \cdots q_r = c\, p_1\, p_2 \cdots p_r$$

where

$$c = \mathrm{lc}(q_1)\, \mathrm{lc}(q_2) \cdots \mathrm{lc}(q_r),$$

and p_i is an irreducible monic polynomial with coefficients obtained from the coefficients of q_i divided by $\mathrm{lc}(q_i)$.

To show the uniqueness of the factorization, suppose that u has two irreducible factorizations

$$u = c\, p_1\, p_2 \cdots p_r = c'\, p_1'\, p_2' \cdots p_s'. \qquad (4.37)$$

Since $c = \mathrm{lc}(u) = c'$, we divide both factorizations by the leading coefficient to obtain

$$p_1\, p_2 \cdots p_r = p_1'\, p_2' \cdots p_{r'}'. \qquad (4.38)$$

Now, since the irreducible polynomial p_1 divides the both sides of Equation (4.38), it divides some factor p_j' on the right (Theorem 4.29(2)). In fact, since both polynomials are monic and irreducible, $p_i = p_j'$. Dividing both sides of Equation (4.38) by this polynomial, we obtain

$$p_2\, p_3 \cdots p_r = p_1'\, p_2' \cdots p_{j-1}'\, p_{j+1}' \cdots p_{r'}'.$$

Repeating this argument for each successive polynomial p_i, $i = 2, \ldots, r$, it follows that $r = r'$ and each polynomial p_i corresponds to a unique polynomial on the right side of Equation (4.37). Therefore, the two factorizations are the same except for a possible rearrangement of the factors.   □

By combining like factors in the previous theorem we obtain the following theorem.

Theorem 4.40. *Let u be a polynomial in $\mathbf{F}[x]$ with $\deg(u) > 0$. Then, u can be factored uniquely as*

$$u = c\, p_1^{n_1}\, p_2^{n_2} \cdots p_s^{n_s},$$

where the polynomials p_i are monic, irreducible, and relatively prime.

Theorems 4.38 and 4.40 are "non-constructive" mathematical statements because they assert the existence of a unique factorization but do not give an algorithm to compute it. In Chapter 9 we describe algorithms that obtain the irreducible factorization of polynomials in $\mathbf{Q}[x]$.

Chinese Remainder Problem in F[x]

The next theorem, which is also based on the extended Euclidean algorithm, is the polynomial version of the Chinese Remainder Theorem.

Theorem 4.41. *Let $u_1, u_2, \ldots, u_r$ be polynomials in $\mathbf{F}[x]$ with positive degree that are pairwise relatively prime, and let $a_1, a_2, \ldots, a_r$ be polynomials in $\mathbf{F}[x]$ with $\deg(a_i) < \deg(u_i)$. Then, there is exactly one h in $\mathbf{F}[x]$ with*

$$\deg(h) < \deg(u_1) + \cdots + \deg(u_r) \tag{4.39}$$

that satisfies the remainder equations

$$\mathrm{rem}(h, u_i) = a_i, \quad i = 1, 2, \ldots, r. \tag{4.40}$$

Proof: We prove the theorem with mathematical induction. First, for the base case $r = 1$, the division property for polynomials implies that $h = a_1$. Let's assume now that there is a polynomial z that satisfies the remainder equations

$$\mathrm{rem}(z, u_i) = a_i, \quad i = 1, \ldots, r-1, \tag{4.41}$$

with $\deg(z) < \deg(u_1) + \cdots + \deg(u_{r-1})$ and show how to extend the process one step further to find a polynomial h that satisfies the conditions in the theorem. Observe that the condition (4.41) implies that

$$z = q_i u_i + a_i, \quad i = 1, \ldots, r-1, \tag{4.42}$$

where $q_i = \mathrm{quot}(z, u_i)$. In addition, for $v = u_1 \cdots u_{r-1}$, the pairwise relatively prime hypothesis in the theorem implies that $\gcd(v, u_r) = 1$. Now, using the extended Euclidean algorithm, we obtain polynomials A and B such that

$$A v + B u_r = 1. \tag{4.43}$$

Let

$$\begin{aligned} g &= A v a_r + B u_r z, \tag{4.44} \\ u &= v u_r = u_1 \cdots u_r. \end{aligned}$$

Although it can be shown that g satisfies the remainder equations, it does not satisfy the degree condition (4.39). To obtain a solution with the proper degree, divide g by u to obtain

$$g = q\,u + h, \tag{4.45}$$

where the remainder h satisfies (4.39).

We show now that h satisfies all the remainder equations. First, for $1 \leq i \leq r - 1$, we use Equation (4.43) to eliminate $B\,u_r$ from h:

$$
\begin{aligned}
h &= g - q\,u \\
&= A\,v\,a_r + B\,u_r\,z - q\,u \\
&= A\,v\,a_r + (1 - A\,v)\,z - q\,u \\
&= A\,v\,a_r - A\,v\,z - q\,u + z. \tag{4.46}
\end{aligned}
$$

Using Equation (4.42) to eliminate the z on the far right, we obtain for $1 \leq i \leq r - 1$,

$$h = (A\,v\,a_r - A\,v\,z - q\,u + q_i\,u_i) + a_i.$$

Observe that u_i divides each of the terms in parentheses, and therefore, the uniqueness property for polynomial division implies that $\operatorname{rem}(h, u_i) = a_i$ (for $1 \leq i \leq r - 1$).

For $i = r$, we use Equation (4.43) to eliminate $A\,v$ from h:

$$
\begin{aligned}
h &= g - q\,u \\
&= A\,v\,a_r + B\,u_r\,z - q\,u \\
&= (B\,u_r\,z - B\,u_r\,a_r - q\,u) + a_r.
\end{aligned}
$$

Since u_r divides each term in parentheses, the uniqueness property for polynomial division implies that $\operatorname{rem}(h, u_r) = a_r$, and therefore, h satisfies all the remainder equations.

To show the uniqueness of h, suppose that h and h' both satisfy the conditions in the theorem. Then, for $1 \leq i \leq r$, we represent h and h' as

$$h = f_i\,u_i + a_i, \qquad h' = g_i\,u_i + a_i$$

which implies that $h - h' = (f_i - g_i)\,u_i$. Therefore, $u_i \mid h - h'$, and since $u_1, \ldots, u_r$ are pairwise relatively prime, Theorem 4.29(3) implies that

$$u \mid h - h'. \tag{4.47}$$

However, since $\deg(h - h') < \deg(u)$, we have $h - h' = 0$ which proves the uniqueness property. $\qquad\square$

Example 4.42. Let

$$u_1 = x^2 + 3x + 2, \quad u_2 = x + 3,$$

$$a_1 = 4x + 5, \quad a_2 = 2,$$

be polynomials in $\mathbf{Q}[x]$. Then, using the notation in the proof of the
theorem, when $r = 2$, we have $z = a_1$ and $v = u_1$. Applying the extended
Euclidean algorithm to v and u_2, we obtain $A = 1/2$, $B = (-1/2)x$.
Therefore Equation (4.44) gives $g = -2x^3 - 15/2x^2 - 9/2x + 2$, which
satisfies the remainder equations but not the degree condition. However,
$h = \mathrm{rem}(g,\, u_1 u_2) = 9/2x^2 + 35/2x + 14$ satisfies the remainder equations
and the degree condition.   □

A procedure that finds the polynomial h in the Chinese remainder theorem is similar to the one for the integer version of the theorem described
in Section 2.1. We leave the details to the reader (Exercise 21).

Exercises

1. Use Euclid's algorithm to find the gcd of each of the pairs of polynomials.

   (a) $u = x^2 + 6x + 9, \quad v = x^2 + 5x + 6$.

   (b) $u = x^3 + 9x^2 + 26x + 24, \quad v = x^3 - 2x^2 - 11x + 12$.

   (c) $u = x^3 + 2x^2 + 10x + 8, \quad v = x^3 - 1$.

   (d) $u = x^3 + 5x^2 + \sqrt{2}x^2 + 6x + 5\sqrt{2}x + 6\sqrt{2}, \quad v = x^3 + \sqrt{2}x^2 - 4x - 4\sqrt{2}$.

   (e) $u = 2x^2 + 3x + 1, \quad v = 2x^2 + 4x + 2$, in $\mathbf{Z}_5[x]$.

2. Use Definition 4.19 to give proofs for properties (2) and (3) in Theorem 4.21.

3. Let u and $v \neq 0$ be polynomials in $\mathbf{F}[x]$, and consider the remainder sequence (4.24). Let $Q_i = \mathrm{quot}(R_{i-2}, R_{i-1}, x)$ and suppose that $\rho \geq 2$.
   Explain why $\deg(Q_i) > 0$ for $i = 2, 3, \ldots, \rho$. Give an example where
   $\deg(Q_1) = 0$.

4. Let p be a prime number, and let u and v be polynomials in $\mathbf{Z}_p[x]$.

   (a) Give a procedure
   $$Poly\_gcd\_p(u, v, x, p)$$
   that computes the gcd of u and v in $\mathbf{Z}_p[x]$.

   (b) Give a procedure

   $$Ext\_Euclidean\_alg\_p(u, v, x, p)$$

   that returns the list $[A, B, \gcd(u, v)]$ generated by the extended Euclidean algorithm in $\mathbf{Z}_p[x]$.

The procedure *Poly_div_p* described in Exercise 9 on page 125 and the procedure *Multiplicative_inverse_p* described in Exercise 11 on page 59 are useful in this exercise.

5. Compute A and B obtained with the extended Euclidean algorithm for each pair of polynomials in Exercise 1.

6. In this exercise we describe an approach for deriving the integral reduction formula

$$\int \frac{dx}{(x^2+c)^n} = \frac{1}{2c(n-1)} \left(\frac{x}{(x^2+c)^{n-1}} + (2n-3) \int \frac{dx}{(x^2+c)^{n-1}} \right)$$
(4.48)

   where the constant $c \neq 0$ and $n \geq 1$ is an integer. The derivation uses Theorem 4.27.

   (a) Let $s = x^2 + c$ and so $s' = 2x$. Using the extended Euclidean algorithm, find polynomials A and B such that $As + Bs' = 1$.

   (b) Show that

   $$\int \frac{dx}{(x^2+c)^n} = \int \frac{dx}{s^n} = \int \frac{A\,dx}{s^{n-1}} + \int \frac{B\,s'}{s^n}\,dx.$$
   (4.49)

   (c) Using integration by parts, show that the second integral in the sum (4.49) is given by

   $$\int \frac{B\,s'}{s^n}\,dx = \frac{-B}{(n-1)s^{n-1}} + \int \frac{B'}{(n-1)s^{n-1}}\,dx.$$
   (4.50)

   (d) Combine Equations (4.49) and (4.50) to obtain Equation (4.48).

7. (a) Prove Theorem 4.34(2),(3).

   (b) Give an example where $Q_1 = 0$ and $\deg(B_1) = \deg(B_2)$.

8. Let u and v be in $\mathbf{F}[x]$.

   (a) Show that if $\gcd(u, v) \neq 1$, then $\deg(A_\rho) < \deg(v)$ and $\deg(B_\rho) < \deg(u)$.

   (b) Give an example that shows that the conclusion in (a) is not true when u and v are relatively prime.

9. Let u, v, and w be in $\mathbf{F}[x]$ with $\gcd(u, v) = 1$ and $\deg(w) > 0$. Suppose that w is irreducible and $w^n \mid uv$. Show that either $w^n \mid u$ or $w^n \mid v$ but w^n does not divide both u and v. Give an example that shows that this statement is false if $\gcd(u, v) \neq 1$.

10. Let u, v, and w be in $\mathbf{F}[x]$ with $\gcd(u, v) = 1$, w a monic polynomial, and $w \mid uv$. Show that

    $$w = \gcd(w, u)\,\gcd(w, v).$$

    Give an example that shows that this statement is false if $\gcd(u, v) \neq 1$.

11. Let u, v, and $w \neq 0$ be in $\mathbf{F}[x]$. Show that

$$\gcd(w\,u, w\,v) = w/\mathrm{lc}(w)\,\gcd(u, v).$$

12. Let u v, and w be in $\mathbf{F}[x]$ with $\gcd(u, v) = 1$. Show that

$$\gcd(w, u\,v) = \gcd(w, u)\,\gcd(w, v).$$

Give an example that shows that this statement is false if $\gcd(u, v) \neq 1$.

13. Suppose that $u = x^2 - 1$ and $v = x^2 + 2\,x + 1$ where $\gcd(u, v) = x + 1$. Show that there are infinitely many pairs of polynomials A and B where $A\,u + B\,v = x + 1$ and $\deg(A) < \deg(v)$, $\deg(B) < \deg(u)$.

14. Let u and v be relatively prime and have positive degree, and suppose that S and T are two polynomials such that $S\,u + T\,v = 1$. (S and T are not necessarily the ones generated by the extended Euclidean algorithm.) Let $A = \mathrm{rem}(S, v)$ and $B = \mathrm{rem}(T, u)$. Show that A and B are the polynomials obtained with the extended Euclidean algorithm.

15. Let u and v be polynomials in $\mathbf{F}[x]$ with positive degree. Show that $\gcd(u, v) \neq 1$ if and only if there are polynomials C and D with $\deg(C) < \deg(v)$ and $\deg(D) < \deg(u)$ such that $C\,u = D\,v$.

16. In this problem, p and u are polynomials in $\mathbf{Q}[x]$ where p is irreducible and u' is the derivative of u.

(a) Suppose that $u = p^n\,w$. Show that $p^{n-1} \mid \gcd(u, u')$.

(b) Now show the converse of Part (a). Suppose that $p^{n-1} \mid \gcd(u, u')$. Show that $p^n \mid u$.

17. Let u, v and w be in $\mathbf{F}[x]$ where u and v have positive degree, $\gcd(u, v) = 1$, and $\deg(w) < \deg(u) + \deg(v)$. Show there are unique polynomials C and D in $\mathbf{F}[x]$ with $\deg(C) < \deg(v)$ and $\deg(D) < \deg(u)$ such that $C\,u + D\,v = w$.

18. In this exercise we describe a simple version of a rational simplification operator. The operator is applied to expressions in the class RS-IN (where IN refers to "input") that satisfy one of the following rules.

RS-IN-1. u is an integer.

RS-IN-2. u is a fraction.

RS-IN-3. u is the symbol x.

RS-IN-4. u is a sum with operands that are in RS-IN.

RS-IN-5. u is a product with operands that are in RS-IN.

RS-IN-6. $u = v^n$, where v is in RS-IN, and n is an integer.

The goal is to transform u to a rational expression v in $\mathbf{Q}(x)$ in a *standard form* that satisfies one of the following rules.

RS-1. v is a polynomial in $\mathbf{Q}[x]$.

RS-2. For $n = Numerator(v)$ and $d = Denominator(v)$, n and d are monic polynomials in $\mathbf{Q}[x]$ which satisfy the following conditions.

(a) d has positive degree.

(b) n and d are relatively prime as polynomials in $\mathbf{Q}[x]$.

RS-3. $v = cw$ where c is a rational number and w satisfies RS-2.

(a) Give a procedure for $Rational\_simplify\_sv(u, x)$ (where the suffix "sv" stands for "single variable") that either returns the standard form of u or the global symbol **Undefined** if a division by zero is encountered. For example,

$$Rational\_simplify\_sv\left(\frac{x^2 - 1}{x + 1}, x\right) \to x - 1,$$

$$Rational\_simplify\_sv\left(\frac{x + 1}{x^2 - 1 - (x + 1) * (x - 1)}, x\right) \to$$
$$\textbf{Undefined},$$

$$Rational\_simplify\_sv\left(\frac{1}{1 + \frac{1}{x+1}} + \frac{2}{x + 2}, x\right) \to \frac{x + 3}{x + 2},$$

$$Rational\_simplify\_sv\left(\frac{2x + 4}{3x + 9}, x\right) \to (2/3)\frac{x + 2}{x + 3}.$$

(b) For a CAS, the standard form for a rational expression may be different from the one given above. Experiment with a CAS to determine the standard form for rationally simplified expressions in $\mathbf{Q}(x)$. (Use **normal** in Maple and MuPAD and **Together** in Mathematica.) On page 257 we describe another standard form for rational expressions.

19. Find the irreducible factorization of each of the following polynomials in $\mathbf{Z}_5[x]$.

(a) $x^2 + 2x + 2$.

(b) $x^3 + 1$.

20. Consider the polynomial $(-1 + 5i)x^2 + (9 + 8i)x + (3 - 6i)$ in $(\mathbf{Q}(i))[x]$. Factor the polynomial in the form (4.35).

21. Let $A = [a_1, \ldots, a_r]$ and $U = [u_1, \ldots, u_r]$ be lists of polynomials in $\mathbf{Q}[x]$ that satisfy the conditions in the Chinese Remainder Theorem (Theorem 4.41). Give a procedure $Poly\_Chinese\_rem(A, U, x)$ that obtains the polynomial h described in the theorem.

4.3   Computations in Elementary Algebraic Number Fields

Although the polynomial procedures described in Sections 4.1 and 4.2 terminate for expressions that are GPEs in x, they may return inappropriate or incorrect results when automatic simplification does not handle all of the field operations. In this section we describe procedures for field operations, polynomial division, and polynomial gcd calculations when the coefficient field is a simple algebraic number field. This section is both an application and an extension of the previous sections. It is an application in the sense that polynomial division and the extended Euclidean algorithm are used to perform arithmetic with algebraic numbers, and an extension in the sense that the polynomial procedures in Sections 4.1 and 4.2 are extended to this more general setting.

Simplification of Algebraic Number Polynomial Expressions

Consider the polynomial

$$u = 2 + 3\,\imath + 4\,\imath^2 + 5\,\imath^3 + 6\,\imath^4$$

where the symbol i satisfies

$$\imath^2 = -1. \tag{4.51}$$

Although u is a polynomial in the symbol i, it differs from a polynomial in a variable x because it can be simplified using the side relation (4.51). One way to simplify u is to apply the structural substitutions

$$\imath^2 = -1, \qquad \imath^3 = -i, \qquad \imath^4 = 1$$

which simplifies the expression to $4 - 2\,\imath$. Since structural substitution replaces only complete sub-expressions, all three substitutions are required.

A better way to obtain the simplification involves polynomial division. Dividing u by the polynomial $p = \imath^2 + 1$ we obtain

$$u = \quad \mathrm{quot}(u,\,p,\,\imath) \cdot p + \mathrm{rem}(\imath,\,p,\,\imath) = (6\,\imath^2 + 5\,\imath - 2)\,(\imath^2 + 1) + (4 - 2\,\imath).$$

Since $\imath^2 + 1$ simplifies to 0, this expression simplifies to the remainder $4 - 2\,\imath$. Simplification by polynomial division is used for the remainder of this section.

Example 4.43. Consider the expression

$$u = 3\,(2 + \sqrt{2})^3 + (2 + \sqrt{2})^2 - 4,$$

and let's view u as a polynomial in the algebraic number $x = 2 + \sqrt{2}$, which satisfies the side relation $f(x) = x^2 - 4x + 2 = 0$. The simplified form of u is given by

$$u = \operatorname{rem}(u(\alpha), f(\alpha), \alpha) = -30 + 46\,\alpha \qquad (4.52)$$
$$= -30 + 46\,(2 + \sqrt{2}). \qquad (4.53)$$

In fact, the general relation in Equation (4.52) holds for any root of $f(x) = 0$. For example, for the root $\alpha = 2 - \sqrt{2}$, we also have the simplification

$$3\,(2 - \sqrt{2})^3 + (2 - \sqrt{2})^2 - 4 = -30 + 46\,(2 - \sqrt{2}). \qquad \square$$

The Minimal Polynomial of an Algebraic Number

Let α be an algebraic number that is a root of the polynomial equation $f(x) = 0$, where f is in $\mathbf{Q}[x]$ and $\deg(f) \geq 2$. In order to simplify a polynomial $u(\alpha)$, we must divide u by the simplest polynomial that has α as a root. This polynomial is described in the following theorem.

Theorem 4.44. *Let α be an algebraic number. Then, there is a unique, monic, irreducible polynomial $p(x) \neq 0$ in $\mathbf{Q}[x]$ such that $p(\alpha) = 0$. The polynomial p is called the* **minimal polynomial** *of α.*

Proof: Suppose that α is a root of the polynomial equation $f(x) = 0$. Then f can be factored as $f(x) = c \cdot p_1(x) \cdot p_2(x) \cdots p_n(x)$, where each p_i is a monic and irreducible polynomial in $\mathbf{Q}[x]$. Since $f(\alpha) = 0$, α is the root of one of these polynomials, say $p_{i'}$. Let $p = p_{i'}$.

To show that p is the only polynomial with these properties, we first show that p is a divisor of any (non-zero) polynomial $g(x)$ in $\mathbf{Q}[x]$ that has α as a root. Since p is irreducible, the two polynomials p and g are either relatively prime or $p \mid g$. We show that the first of these two options is impossible.

If p and g were relatively prime, then, by the extended Euclidean algorithm, there are polynomials $A(x)$ and $B(x)$ in $\mathbf{Q}[x]$ such that

$$A(x)\,p(x) + B(x)\,g(x) = \gcd(g, p) = 1. \qquad (4.54)$$

Substituting $x = \alpha$ into this expression, we obtain

$$A(\alpha)\,p(\alpha) + B(\alpha)\,g(\alpha) = A(\alpha) \cdot 0 + B(\alpha) \cdot 0 = 0. \qquad (4.55)$$

Since Equations (4.54) and (4.55) cannot both be true ($0 \neq 1$), p and g cannot be relatively prime, and so $p \mid g$. Therefore, if g is a polynomial which satisfies the three properties in the theorem (monic, irreducible, $g(\alpha) = 0$), we must have $p = g$.   □

Example 4.45. In order to find the simplest form of $u(\alpha)$ using polynomial division, the side relation associated with α must be the minimal polynomial. For example, let $\alpha = i$ and

$$u = 2 + 3i + 4i^2 + 5i^3 + 6i^4.$$

The minimal polynomial of i is $p(x) = x^2 + 1$. Of course i is also a root of any multiple of $p(x)$ such as

$$f(x) = (x^2 + 1)(x - 1) = x^3 - x^2 + x - 1.$$

However,

$$\operatorname{rem}(u, f(i), i) \to 9i^2 - 2i + 13,$$

rather than the simplified form

$$\operatorname{rem}(u, i^2 + 1, i) = 4 - 2i.$$ □

Arithmetic in Algebraic Number Fields

In Section 2.3 we described a number of fields that are constructed by adjoining an algebraic number to the rational numbers $\mathbf{Q}$ (e.g., $\mathbf{Q}(i)$). In the next definition and theorem we describe a systematic approach for arithmetic in this setting.

Definition 4.46. *Let α be an algebraic number with minimal polynomial p such that $\deg(p, x) = n \geq 2$, and let $\mathbf{Q}(\alpha)$ be the set of polynomials u in $\mathbf{Q}[\alpha]$ such that $\deg(u, \alpha) < n$. For polynomials $u(\alpha)$ and $v(\alpha)$ in $\mathbf{Q}(\alpha)$, define an addition operation as ordinary polynomial addition*

$$u + v,$$

and a multiplication operation as

$$\operatorname{rem}(u \cdot v, p(\alpha), \alpha).$$

Notice that with these operations both the sum and product of polynomials in $\mathbf{Q}(\alpha)$ are in simplified form with respect to the side relation $p(\alpha) = 0$.

Example 4.47. Let $\alpha = 2^{1/3}$. Since the minimal polynomial of α is $p(x) = x^3 - 2$, the set $\mathbf{Q}(\alpha)$ consists of polynomials of the form $a + b\alpha + c\alpha^2$, where a, b, and c are rational numbers. Let

$$u = 1 + \alpha + 2\alpha^2, \quad v = 2 - \alpha + 3\alpha^2.$$

Using the operations described in Definition 4.46, the sum of u and v is

$$u + v = 3 + 5\alpha^2,$$

and the product is

$$
\begin{aligned}
\mathrm{rem}(u \cdot v, p(\alpha), \alpha) &= \mathrm{rem}(2 + \alpha + 6\alpha^2 + \alpha^3 + 6\alpha^4, \alpha^3 - 2, \alpha) \\
&= 4 + 13\alpha + 6\alpha^2.
\end{aligned}
$$
$\quad\square$

Theorem 4.48. *The set $\mathbf{Q}(\alpha)$ with the addition and multiplication operations described in Definition 4.46 is a field.*

Proof: The addition and multiplication operations are defined so that the closure axioms (F-1 and F-2) are automatically satisfied, and using the properties of polynomials, the field axioms F-3 through F-11 are also easy to verify.

To show that $\mathbf{Q}(\alpha)$ is a field, we need only show that each field element $u(\alpha) \neq 0$ has a multiplicative inverse (Axiom F-12). Since the minimal polynomial p is irreducible with $\deg(u) < \deg(p)$, and since $u(\alpha) \neq 0$, we have $\gcd(u(\alpha), p(\alpha)) = 1$. By the extended Euclidean algorithm, there are polynomials $A(\alpha)$ and $B(\alpha)$ such that

$$A(\alpha)\, u(\alpha) + B(\alpha)\, p(\alpha) = 1 \tag{4.56}$$

where, by Theorem 4.36, $\deg(A) < \deg(p)$. To show that $A(\alpha)$ is the multiplicative inverse of $u(\alpha)$, we compute the product of $A(\alpha)$ and $u(\alpha)$ in $\mathbf{Q}(\alpha)$. Using Equation (4.56), we have

$$\mathrm{rem}(A(\alpha)\, u(\alpha), p(\alpha), \alpha) = \mathrm{rem}(1 - B(\alpha)\, p(\alpha), p(\alpha), \alpha) = 1.$$

Therefore, A is the multiplicative inverse of u, and $\mathbf{Q}(\alpha)$ is a field. $\quad\square$

Example 4.49. Consider the field $\mathbf{Q}(2^{1/3})$, where the minimal polynomial of $\alpha = 2^{1/3}$ is $p(x) = x^3 - 2$. Let $u = 2 + 3 \cdot 2^{1/3}$. Applying the extended Euclidean algorithm we obtain

$$u^{-1} = A = 9/62\,\alpha^2 - 3/31\,\alpha + 2/31,$$

and therefore,

$$\left(2 + 3 \cdot 2^{1/3}\right)^{-1} = 9/62 \cdot 2^{2/3} - 3/31 \cdot 2^{1/3} + 2/31.$$

In fact, A is the inverse for all solutions α of $x^3 - 2 = 0$. For example, another solution is

$$\alpha = 2^{1/3}\left(-\frac{1}{2} + \frac{3^{1/2}\,\imath}{2}\right),$$

and so

$$\left(2^{1/3}\left(-\frac{1}{2} + \frac{3^{1/2}\,\imath}{2}\right)\right)^{-1}$$
$$= \quad 9/62 \cdot \left(2^{1/3}\left(-\frac{1}{2} + \frac{3^{1/2}\,\imath}{2}\right)\right)^2 - 3/31 \cdot \left(2^{1/3}\left(-\frac{1}{2} + \frac{3^{1/2}\,\imath}{2}\right)\right)$$
$$+\,2/31. \qquad\qquad\qquad\qquad\qquad\qquad\qquad\qquad\qquad\qquad\qquad \Box$$

Procedures for the Field Operations in $\mathbf{Q}(\alpha)$.   Addition of u and v in $\mathbf{Q}(\alpha)$ is obtained with automatic simplification, and the additive inverse of u is obtained with[6]

$$Algebraic\_expand(-1 \cdot u).$$

Multiplication in $\mathbf{Q}(\alpha)$ is given by

$$Remainder(Algebraic\_expand(u \cdot v),\, p,\, \alpha),$$

where the minimal polynomial is expressed in terms of the symbol α instead of the symbol x.

Procedures for the multiplicative inverse and division operations are given in Figure 4.5. The multiplicative inverse is obtained in the procedure *Alg_mult_inverse* with the extended Euclidean algorithm (line 1).

A general simplifier for expressions constructed from integers, fractions, the algebraic number α, sums, products, and powers with integer exponents is described in Exercise 4.

Procedures for Polynomial Operations in $\mathbf{Q}(\alpha)[x]$.   Let's consider now the extension of the polynomial operations described in Sections 4.1 and 4.2 for polynomials in $\mathbf{Q}(\alpha)[x]$.

---

[6]In a computer algebra system where -1 is distributed over a sum by automatic simplification, the *Algebraic_expand* operator can be omitted.

Procedure   *Alg_mult_inverse*(v, p, α);
Input
    α : a symbol that represents an algebraic number;
    p : a monic, irreducible polynomial in $\mathbf{Q}[\alpha]$ with $\deg(p, \alpha) \geq 2$;
    v : a non-zero polynomial in $\mathbf{Q}(\alpha)$ with $\deg(v) < \deg(p)$;
Output
    The multiplicative inverse of v in $\mathbf{Q}(\alpha)$;
Local Variables
    w;
Begin
1    $w :=$ *Extended_Euclidean_algorithm*(v, p, α);
2   *Return*$($*Operand*$(w, 2))$
End

Procedure   *Alg_divide*(u, v, p, α);
Input
    α : a symbol that represents an algebraic number;
    p : a monic, irreducible polynomial in $\mathbf{Q}[\alpha]$ with $\deg(p, \alpha) \geq 2$;
    u, v : polynomials in $\mathbf{Q}(\alpha)$ with degree $< \deg(p)$ and $v \neq 0$;
Output
    The quotient of u divided by v in $\mathbf{Q}(\alpha)$;
Local Variables
    w;
Begin
1    $w :=$ *Alg_mult_inverse*(v, p, α);
2   *Return*$($*Remainder*$($*Algebraic_expand*$(u * w), p, \alpha))$
End

Figure 4.5. MPL procedures for the division and multiplicative inverse operations in $\mathbf{Q}(\alpha)$. In both procedures, the minimal polynomial is expressed in terms of the symbol α rather than the symbol x. (Implementation: Maple (txt), Mathematica (txt), MuPAD (txt).)

The polynomial division procedure[7] shown in Figure 4.6 is a modification of the one given in Figure 4.1 on page 116.

Observe that at line 8, we obtain the quotient lcr/lcv with the *Alg_divide* operator. At line 10, the product of the algebraic number s with the coefficients of v may give coefficients that are unsimplified with respect to $p(\alpha) = 0$ (see Example 4.50). For this reason, the operator *Alg_coeff_simp* is included at line 11 to simplify each of the (algebraic number) coefficients of the polynomial (Exercise 2).

---

[7]See the comment concerning the MuPAD implementation in footnote 2 on page 115.

Procedure $Alg\_polynomial\_division(u, v, x, p, \alpha)$;
Input
 u, v : polynomials in $\mathbf{Q}(\alpha)[x]$ with $v \neq 0$;
 x : a symbol;
 α : a symbol that represents an algebraic number;
 p : a monic, irreducible polynomial in $\mathbf{Q}[\alpha]$ with degree ≥ 2;
Output
 A list $[q, r]$ with the quotient and remainder;
Local Variables
 q, r, m, n, lcv, lcr, s;
Begin
1 $q := 0$;
2 $r := u$;
3 $m := Degree\_gpe(r, x)$;
4 $n := Degree\_gpe(v, x)$.
5 $lcv := Leading\_Coefficient\_gpe(v, x)$;
6 **while** $m \geq n$ **do**
7 $lcr := Leading\_Coefficient\_gpe(r, x)$;
8 $s := Alg\_divide(lcr, lcv, p, \alpha)$;
9 $q := q + s * x^{m-n}$;
10 $r := Algebraic\_expand\left((r - lcr * x^m) - (v - lcv\ x^n) * s * x^{m-n}\right)$;
11 $r := Alg\_coeff\_simp(r, x, p, a)$;
12 $m := Degree\_gpe(r, x)$;
13 $Return([q, r])$
End

$$Alg\_quotient(u, v, x, p, \alpha) \overset{\text{function}}{:=}$$
$$Operand(Alg\_polynomial\_division(u, v, x, p, \alpha), 1);$$
$$Alg\_remainder(u, v, x, p, \alpha) \overset{\text{function}}{:=}$$
$$Operand(Alg\_polynomial\_division(u, v, x, p, \alpha), 2);$$

Figure 4.6. An MPL procedure for polynomial division in $\mathbf{Q}(\alpha)[x]$ and functions that extract the quotient and remainder. (Implementation: Maple (txt), Mathematica (txt), MuPAD (txt).)

Example 4.50. Let α be a root of $p(x) = x^2 - 2 = 0$, and consider the polynomials in $\mathbf{Q}(\alpha)[x]$:

$$u = 2\,x^2 + \alpha\,x, \qquad v = \alpha\,x + \alpha.$$

Let's consider the division of u by v using the *Alg_polynomial_division* procedure. The quotient and remainder are initialized at lines 1–2 with

$$q = 0, \qquad r = 2\,x^2 + \alpha\,x.$$

On the first pass through the loop, we have

$$s = \frac{2}{\alpha} = \alpha, \qquad q = \alpha\,x,$$

and

$$r = x\,\alpha - \alpha^2\,x = (\alpha - 2)\,x,$$

where the middle expression is the unsimplified form from line 10 and the expression on the right is the simplified form from line 11. On the second pass, we have

$$s = \frac{\alpha - 2}{\alpha} = 1 - \alpha, \qquad q = \alpha\,x + 1 - \alpha,$$

and

$$r = -\alpha + \alpha^2 = -\alpha + 2,$$

where again the middle expression is the unsimplified form and the expression on the right is the simplified form. At this point, the process terminates with quotient $q = \alpha\,x + 1 - \alpha$ and remainder $-\alpha + 2$. $\square$

A procedure that obtains the greatest common divisor[8] in $\mathbf{Q}(\alpha)[x]$ based on Euclid's algorithm is given in Figure 4.7. The operator *Alg_monic* in line 6 assures that the gcd is a monic polynomial (Exercise 3).

Example 4.51. Let α be a root of $p(x) = x^2 - 2 = 0$ and consider the polynomials in $\mathbf{Q}(\alpha)[x]$:

$$u = x^2 + (-1 - \alpha)\,x, \qquad v = x^2 + (-2 - 2\,\alpha)\,x + 3 + 2\,\alpha.$$

The remainder sequence produced by *Alg_polynomial_gcd* at line 3 is

$$\begin{aligned} R &= x + \alpha\,x - 3 - 2\,\alpha, & (4.57) \\ R &= 0. \end{aligned}$$

[8] Another approach is suggested in Exercise 10, page 347.

Procedure $Alg\_polynomial\_gcd(u, v, x, p, \alpha)$;
Input
    u, v : polynomials in $\mathbf{Q}(\alpha)[x]$;
    x : a symbol;
    α : a symbol that represents an algebraic number;
    p : a monic, irreducible polynomial in $\mathbf{Q}[\alpha]$ with degree ≥ 2;
Output
    $\gcd(u, v)$;
Local Variables
    U, V, R;
Begin
1    $U := u$; $V := v$;
2   **while** $V \neq 0$ **do**
3        $R := Alg\_remainder(U, V, x, p, \alpha)$;
4        $U := V$;
5        $V := R$;
6    $Return(Alg\_monic(U, x, p, \alpha))$
End

Figure 4.7. An MPL procedure for the Euclidean algorithm that computes $\gcd(u, v)$ in $\mathbf{Q}(\alpha)[x]$. (Implementation: Maple (txt), Mathematica (txt), MuPAD (txt).)

Therefore, using the remainder (4.57), at line 6 we obtain

$$\gcd(u, v) = R/\mathrm{lc}(R) = x - 1 - \alpha. \qquad \square$$

In a similar way, we can modify the procedures for polynomial expansion and the extended Euclidean algorithm to handle polynomials in $\mathbf{Q}(\alpha)[x]$ (Exercises 5 and 6).

Multiple Algebraic Extensions

We consider now the arithmetic and polynomial operations when a finite set $\{\alpha_1, \alpha_2, \ldots, \alpha_s\}$ of algebraic numbers is adjoined to $\mathbf{Q}$. We begin with some examples.

Example 4.52. Consider the simplification problem for a polynomial expression in the two positive algebraic numbers $\alpha_1 = \sqrt{2}$ and $\alpha_2 = \sqrt{3}$, which satisfy the two side relations

$$p_1(x_1) = x_1^2 - 2 = 0, \quad p_2(x_2) = x_2^2 - 3 = 0.$$

To simplify

$$u = 2 + 3\alpha_1 + 4\alpha_1^2 + \alpha_2 - \alpha_1\alpha_2 + \alpha_1^2\alpha_2 + 3\alpha_2^2 - \alpha_1\alpha_2^2 - 2\alpha_1^2\alpha_2^2,$$

we view u as a polynomial in $\mathbf{Q}(\alpha_1)[\alpha_2]$:

$$u = (2 + 3\alpha_1 + 4\alpha_1^2) + (1 - \alpha_1 + \alpha_1^2)\alpha_2 + (3 - \alpha_1 - 2\alpha_1^2)\alpha_2^2.$$

Simplifying the coefficients in parentheses by replacing each of them with the remainder of division by $p_1(\alpha_1)$, we obtain

$$u = (10 + 3\alpha_1) + (3 - \alpha_1)\alpha_2 + (-1 - \alpha_1)\alpha_2^2.$$

The final simplification is the remainder obtained by dividing this polynomial (considered in $\mathbf{Q}(\alpha_1)[\alpha_2]$) by $p_2(\alpha_2)$:

$$u = \operatorname{rem}(u, p_2(\alpha_2), \alpha_2) \quad \rightarrow \quad 7 + (3 - \alpha_1)\alpha_2$$
$$= \quad 7 + \left(3 - \sqrt{2}\right)\sqrt{3}. \qquad \square$$

Example 4.53. Consider the simplification problem for a polynomial expression in the two positive algebraic numbers $\alpha_1 = \sqrt{2}$ and $\alpha_2 = \sqrt{8}$, which satisfy the two side relations

$$p_1(x_1) = x_1^2 - 2 = 0, \quad p_2(x_2) = x_2^2 - 8 = 0.$$

Let

$$u = \alpha_1\alpha_2 - 4,$$

and consider the problem of simplifying u. If we proceed as in the last example by dividing u by $p_2(\alpha_2)$, we obtain the remainder $u = \alpha_1\alpha_2 - 4$ rather than the simplified form $u = \sqrt{2}\sqrt{8} - 4 = 0$. The difficulty arises because there is a dependence relation between $\sqrt{2}$ and $\sqrt{8}$:

$$2\alpha_1 - \alpha_2 = 2\sqrt{2} - \sqrt{8} = 0. \qquad \square$$

The last example shows that when there is an algebraic relationship between algebraic numbers, simplification with polynomial division by the minimal polynomials may not determine that an expression simplifies to zero. In the next definition we describe the form of these dependence relations.

Definition 4.54. *Let $S = \{\alpha_1, \alpha_2, \ldots, \alpha_s\}$ be algebraic numbers, and let $p_i(x_i)$ be the minimal polynomial of α_i. The set S is **algebraically dependent** if there exists a non-zero polynomial $h(x_1, x_2, \ldots, x_s)$ with rational number coefficients such that*

$$\deg(h, x_i) < \deg(p_i, x_i), \text{ for } i = 1, 2, \ldots, s \qquad (4.58)$$

and

$$h(\alpha_1, \alpha_2, \ldots, \alpha_s) = 0. \tag{4.59}$$

If the set S is not dependent, it is called **algebraically independent**.

In other words, a set S of algebraic numbers is algebraically dependent if there is a polynomial relation between the numbers. The degree condition (4.58) is essential because without it the definition is meaningless. Indeed,

$$h(x_1, \ldots, x_s) = p_1(x_1) \cdots p(x_s)$$

satisfies Equation (4.59) but not the inequality (4.58).

Example 4.55. Consider the positive algebraic numbers $\alpha_1 = \sqrt{2}$ and $\alpha_2 = \sqrt{8}$ with minimal polynomials $p_1(x_1) = x_1^2 - 2$ and $p_2(x_2) = x_2^2 - 8$. These numbers are algebraically dependent with

$$h(x_1, x_2) = 2\,x_1 - x_2.$$

Notice that the dependence relation depends on which roots of the minimal polynomials are considered. For example, for $\alpha_1 = -\sqrt{2}$ and $\alpha_2 = \sqrt{8}$, the dependence relation is

$$h(x_1, x_2) = 2\,x_1 + x_2.$$

The positive algebraic numbers $\alpha_1 = \sqrt{2}$, $\alpha_2 = \sqrt{3}$, and $\alpha_3 = \sqrt{6}$ are algebraically dependent with

$$h(x_1, x_2, x_3) = x_1\,x_2 - x_3.$$

The positive algebraic numbers

$$\alpha_1 = \sqrt{2}, \quad \alpha_2 = \sqrt{3}, \quad \alpha_3 = \sqrt{30 + 12\sqrt{6}}$$

are algebraically dependent with

$$h(x_1, x_2, x_3) = 3\,x_1 + 2\,x_2 - x_3.$$

The dependence relationship $h(\alpha_1, \alpha_2, \alpha_3) = 0$ can be verified by squaring both sides of the equation

$$3\sqrt{2} + 2\sqrt{3} = \sqrt{30 + 12\sqrt{6}}. \qquad \square$$

Example 4.56. The positive algebraic numbers $\alpha_1 = \sqrt{2}$ and $\alpha_2 = \sqrt{3}$ are algebraically independent. To show this we must show that it is impossible to find a non-zero polynomial

$$h(x_1, x_2) = a + b\,x_1 + c\,x_2 + d\,x_1\,x_2, \qquad (4.60)$$

with a, b, c, and d rational numbers such that

$$h(\alpha_1, \alpha_2) = a + b\,\sqrt{2} + c\,\sqrt{3} + d\,\sqrt{2}\,\sqrt{3} = 0. \qquad (4.61)$$

First, if such a polynomial did exist, then both c and d cannot be zero because, if this were so, Equation (4.61) becomes

$$a + b\sqrt{2} = 0. \qquad (4.62)$$

In this expression, $b \neq 0$ because, if this were not so, then $a = 0$ as well, and $h(x_1, x_2)$ is the zero polynomial. On the other hand, if $b \neq 0$, Equation (4.62) implies that $\sqrt{2}$ is a rational number which, of course, is not true.

Therefore, we assume that either c or d is not zero which implies, using the above reasoning, that $c + d\,\sqrt{2} \neq 0$. When this is so, Equation (4.61) implies

$$\sqrt{3} = \frac{-a - b\sqrt{2}}{c + d\sqrt{2}},$$

and by rationalizing the denominator, we obtain

$$\sqrt{3} = e + f\sqrt{2},$$

where e and f are rational numbers. Squaring both sides of this expression, we obtain

$$3 = e^2 + 2\,e\,f\,\sqrt{2} + 2\,f^2. \qquad (4.63)$$

If both $e \neq 0$ and $f \neq 0$, then again we would conclude that $\sqrt{2}$ is a rational number, and so either e or f is zero.

Suppose that $e = 0$ and $f = f_1/f_2$ where f_1 and f_2 are integers with no common factor. Then from Equation (4.63), we have

$$3 f_2^2 = 2 f_1^2. \qquad (4.64)$$

Since 3 divides the left side of the equality, Theorem 2.16(2) implies that $3 \mid f_1^2$, and then Theorem 2.16(1) implies that $3 \mid f_1$. Therefore, $3^2 \mid f_1^2$, and by Equation (4.64), $3 \mid f_2^2$. Again, Theorem 2.16(1) implies that $3 \mid f_2$, and so we conclude that f_1 and f_2 have a common factor, which is a contradiction. Therefore, $e \neq 0$. In a similar way we can conclude that $f \neq 0$.

Finally, since both e and f cannot be non-zero, the polynomial (4.60) does not exist, and so $\sqrt{2}$ and $\sqrt{3}$ are independent algebraic numbers. $\square$

The last example is a special case of the next theorem which gives a large set of independent algebraic numbers.[9]

Theorem 4.57. *Let $n_1, n_2, \ldots, n_s$ be distinct prime numbers, and let m_1, $m_2, \ldots, m_s$ be integers that are ≥ 2. Then $\{n_1^{1/m_1}, n_2^{1/m_2}, \ldots, n_s^{1/m_s}\}$ is an set of independent algebraic numbers.*

Arithmetic with Two Independent Algebraic Numbers

Definition 4.58. *Let α_1 and α_2 be independent algebraic numbers with minimal polynomials $p_1(x_1)$ and $p_2(x_2)$ with rational number coefficients and degree ≥ 2. Let*

$$\mathbf{Q}(\alpha_1, \alpha_2)$$

be the set of polynomials u in $\mathbf{Q}(\alpha_1)[\alpha_2]$ with $\deg(u, \alpha_2) < \deg(p_2(x_2), x_2)$. For u and v in $\mathbf{Q}(\alpha_1, \alpha_2)$, the addition operation is defined as ordinary polynomial addition, and the multiplication operation is

$$\mathrm{rem}\big(u \cdot v, \, p_2(\alpha_2), \, \alpha_2\big),$$

where the remainder operation is performed in $\mathbf{Q}(\alpha_1)[\alpha_2]$.

For $u \neq 0$, the multiplicative inverse is obtained with the extended Euclidean algorithm in $\mathbf{Q}(\alpha_1)[x_2]$ using an approach similar to what was done in the proof of Theorem 4.48. (See Example 4.60 and Exercise 6.) This approach requires that the minimal polynomial $p_2(x_2)$ be irreducible in $\mathbf{Q}(\alpha_1)[x_2]$. Although $p_2(x_2)$ is irreducible as a polynomial in $\mathbf{Q}[x_2]$, it is only irreducible in the larger setting $\mathbf{Q}(\alpha_1)[x_2]$ when α_1 and α_2 are independent algebraic numbers. This is the content of the next theorem.

Theorem 4.59. *Let α_1 and α_2 be algebraic numbers with minimal polynomials $p_1(x_1)$ and $p_2(x_2)$. The polynomial $p_2(x_2)$ is irreducible in $\mathbf{Q}(\alpha_1)[x_2]$ if and only if α_1 and α_2 are algebraically independent.*

Proof: First assume that α_1 and α_2 are algebraically independent. Let's suppose that p_2 were reducible as

$$p_2 = f\,g$$

---

[9]Theorem 4.57 is proved in Besicovitch [9].

where f and g are monic polynomials in $\mathbf{Q}(\alpha_1)[x_2]$ and

$$0 < \deg(f, x_2) < \deg(p_2, x_2), \quad 0 < \deg(g, x_2) < \deg(p_2, x_2). \quad (4.65)$$

Since α_2 is a root of p_2, it is a root of one of the factors (say f), and since p_2 is the (unique) minimal polynomial of α_2, at least one coefficient of f must depend on α_1. (Otherwise f would be the minimal polynomial for α_2 over $\mathbf{Q}$.) Define a multivariate polynomial with

$$h(x_1, x_2) = Substitute(f, \alpha_1 = x_1)$$

where

$$h(\alpha_1, \alpha_2) = f(\alpha_2) = 0. \quad (4.66)$$

In addition, since f is in $\mathbf{Q}(\alpha_1)[x_2]$, we have

$$\deg(h, x_1) < \deg(p_1, x_1), \quad (4.67)$$

and (4.65) implies

$$\deg(h, x_2) = \deg(f, x_2) < \deg(p_2, x_2). \quad (4.68)$$

Observe that the conditions (4.66), (4.67), and (4.68) imply that h is a dependence relation for α_1 and α_2. However, since α_1 and α_2 are independent, a dependence relation does not exist, and so p_2 irreducible in $\mathbf{Q}(\alpha_1)[x_2]$.

Conversely, let's assume next that $p_2(x_2)$ is irreducible in $\mathbf{Q}(\alpha_1)[x_2]$. If there were a dependence relation $h(x_1, x_2)$ that satisfies the conditions of Definition 4.54, then, since $\deg(h, x_2) < \deg(p_2, x_2)$, we have

$$\gcd(h(\alpha_1, x_2), p_2(x_2), x_2) = 1$$

in $\mathbf{Q}(\alpha_1)[x_2]$. By the extended Euclidean algorithm, there are polynomials A and B in $\mathbf{Q}(\alpha_1)[x_2]$ such that

$$A(x_2)\, h(\alpha_1, x_2) + B(x_2)\, p_2(x_2) = 1.$$

However, when $x_2 = \alpha_2$, the left side of this equations is zero. Therefore, $h(x_1, x_2)$ cannot exist, and so α_1 and α_2 are algebraically independent. $\square$

Example 4.60. Consider the field $\mathbf{Q}(\alpha_1, \alpha_2)$ where

$$p_1 = x_1^2 - 2 = 0, \quad p_2 = x_2^3 - 3 = 0.$$

By Theorem 4.57, α_1 and α_2 are algebraically independent. Let

$$u = (6\,\alpha_1 + 5)\,\alpha_2^2 + (4\,\alpha_1 + 3)\,\alpha_2 + 2\,\alpha_1 + 1.$$

To find u^{-1}, we view u as a polynomial in $\mathbf{Q}(\alpha_1)[\alpha_2]$ and apply the extended Euclidean algorithm (Exercise 6).[10] Since u and $p_2(\alpha_2)$ are relatively prime, the extended Euclidean algorithm obtains polynomials A and B in $\mathbf{Q}(\alpha_1)[\alpha_2]$, such that

$$A\,u + B\,p_2 = 1.$$

In addition, the product of A and u in $\mathbf{Q}(\alpha_1, \alpha_2)$ is

$$\operatorname{rem}(A \cdot u,\, p_2,\, \alpha_2) = \operatorname{rem}((1 - B\,p_2) \cdot u,\, p_2,\, \alpha_2) = 1,$$

and so

$$
\begin{aligned}
u^{-1} &= A \\
&= \left(\frac{227}{182354}\alpha_1 - \frac{109}{182354}\right)\alpha_2^2 + \left(\frac{17289}{364708}\alpha_1 - \frac{7565}{182354}\right)\alpha_2 \\
&\quad - \frac{12985}{364708}\alpha_1 + \frac{3065}{91177}.
\end{aligned}
$$

$\square$

The General Approach

Definition 4.61. *Let* $\{\alpha_1, \alpha_2, \ldots, \alpha_s\}$ *be a set of independent algebraic numbers with minimal polynomials* $p_1(x_1), p_2(x_2), \ldots, p_s(x_s)$ *with rational number coefficients and degree* ≥ 2. *Let*

$$\mathbf{Q}(\alpha_1, \alpha_2, \ldots, \alpha_s)$$

be the set of all the polynomials u *in* $\mathbf{Q}(\alpha_1, \alpha_2, \ldots, \alpha_{s-1})[\alpha_s]$ *such that*

$$\deg(u,\, \alpha_s) < \deg(p_s(x_s),\, x_s).$$

For u *and* v *in* $\mathbf{Q}(\alpha_1, \alpha_2, \ldots, \alpha_s)$, *the addition operation is defined as ordinary polynomial addition, and the multiplication operation is*

$$\operatorname{rem}(u \cdot v,\, p_s(\alpha_s),\, \alpha_s)$$

where the remainder operation is performed in $\mathbf{Q}(\alpha_1, \alpha_2, \ldots, \alpha_{s-1})[\alpha_s]$.

As above, the independence of the algebraic numbers implies that $p_s(x_s)$ is irreducible in $\mathbf{Q}(\alpha_1, \alpha_2, \ldots, \alpha_{s-1})[x_s]$ (Exercise 10). In addition, we obtain the multiplicative inverse using the extended Euclidean algorithm as was done in Example 4.60 above.

This previous discussion is summarized in the following theorem.

Theorem 4.62. *Suppose that* $\{\alpha_1, \alpha_2, \ldots, \alpha_s\}$ *is a set of independent algebraic numbers. Then* $\mathbf{Q}(\alpha_1, \alpha_2, \ldots, \alpha_s)$ *is a field.*

---

[10] Another(Exercise 6). approach is suggested in Exercise 11, page 347.

Dependent Algebraic Numbers. Our description of the field $\mathbf{Q}(\alpha_1, \alpha_2, \ldots, \alpha_s)$ requires that the algebraic numbers are independent. When there are dependence relations between the algebraic numbers, we can still obtain a field as long as the simplifications are defined using division by both the minimal polynomials and the dependence relations.

Example 4.63. Consider the dependent algebraic numbers $\alpha_1 = \sqrt{2}$ and $\alpha_2 = \sqrt{8}$ with minimal polynomials $p_1(x_1) = x_1^2 - 2$, $\quad p_2(x_2) = x_2^2 - 8$, and dependence relation

$$h(x_1, x_2) = 2\,x_1 - x_2 = 0, \tag{4.69}$$

and consider the expression

$$u = \alpha_1^2 + 2\,a_2^2 + a_1^2\,a_2^2 - 8\,a_1\,a_2 - 2$$

which simplifies to 0. Simplification of u with division by the minimal polynomials $p_1(\alpha_1)$ and $p_2(\alpha_2)$ obtains

$$u = -\alpha_1\,\alpha_2 + 32$$

rather than the simplified form $u = 0$. However, simplification with division by one of the minimal polynomials (say $p_1(\alpha_1)$) and $h(\alpha_1, \alpha_2)$ obtains the simplified form $u = 0$. $\qquad\square$

The simplification problem with dependent algebraic numbers is best handled using the techniques in Chapter 8. We consider the above simplification again in Example 4.63 on page 161.

Procedures. When two independent algebraic numbers α_1 and α_2 are involved, the field computations in $\mathbf{Q}(\alpha_1, \alpha_2)$ and polynomial computations in $\mathbf{Q}(\alpha_1, \alpha_2)[x]$ are based on the procedures when only one algebraic number is involved. We leave these procedures to the reader (Exercise 11).

Procedures for arithmetic and polynomial calculations with more than two independent algebraic numbers are based on the recursive structure of the field described in the discussion preceding Theorem 4.62. These procedures are also left to the reader (Exercise 12).

Manipulations with Explicit Algebraic Numbers. In most applications, the explicit algebraic numbers that appear in an expression are radicals of integers for which the minimal polynomials are easy to obtain. For more complicated algebraic numbers, however, the minimal polynomials have high degree and coefficients with many digits. For example, the algebraic number

$$\sqrt{2+\sqrt{3}+\cfrac{1}{\sqrt{5+\sqrt{7}+\left(11+\sqrt{13}\right)^{-1}}}}$$

has the minimal polynomial[11]

$$
\begin{aligned}
p(x) \;=\; & x^{32} - 32\,x^{30} + \frac{66085344}{145609}\,x^{28} \\
& - \frac{545732992}{145609}\,x^{26} + \frac{421691755033884}{21201980881}\,x^{24} \\
& - \frac{1494907538191008}{21201980881}\,x^{22} + \frac{103051255680436160}{614857445549}\,x^{20} \\
& - \frac{159162134521396736}{614857445549}\,x^{18} + \frac{4196565298931285686}{17830865920921}\,x^{16} \\
& - \frac{1505578046340828000}{17830865920921}\,x^{14} - \frac{746238387428625184}{17830865920921}\,x^{12} \\
& + \frac{809449751146726016}{17830865920921}\,x^{10} - \frac{86721370491939780}{17830865920921}\,x^{8} \\
& - \frac{97493394843118048}{17830865920921}\,x^{6} + \frac{14479220688074368}{17830865920921}\,x^{4} \\
& + \frac{4517575807284480}{17830865920921}\,x^{2} + \frac{229036045032825}{17830865920921}.
\end{aligned}
$$

In these situations, the algorithms described in this section involve considerable computation, and the expressions for multiplicative inverses are so involved, they are not useful in practice.

The Denesting Problem. Another simplification problem for explicit algebraic numbers is known as the *denesting problem*. The goal here is to determine if an algebraic number that is defined using nested radicals can also be expressed using fewer nested radicals. For example, each of the following nested radicals on the left can be completely denested:

$$
\begin{aligned}
\sqrt{5+2\sqrt{2}\sqrt{3}} &= \sqrt{2}+\sqrt{3}, \\
\left(2^{1/3}-1\right)^{1/3} &= (1/9)^{1/3} - (2/9)^{1/3} + (4/9)^{1/3}, \\
\sqrt{5^{1/3}-4^{1/3}} &= 1/3\left(2^{1/3}+20^{1/3}-25^{1/3}\right).
\end{aligned}
$$

---

[11]In Section 7.2 we describe a procedure that obtains a multiple of the minimal polynomial of an explicit algebraic number although, in most cases, it involves considerable computation. This minimal polynomial was obtained with this procedure.

On the other hand, some nested radicals such as $\sqrt{1 + 2\sqrt{2}\sqrt{3}}$ cannot be denested.

The denesting problem is a very involved problem which has only been partially solved.[12]

Exercises

1. Find the multiplicative inverse of each of the following using the extended Euclidean algorithm.

   (a) $4 + 3\sqrt{2}$.

   (b) $1 - 5^{1/3} + 2 \cdot 5^{2/3}$.

   (c) $1 + \sqrt{2} + \sqrt{3}$.

2. Let α be an algebraic number with minimal polynomial p (in terms of α), and let u be a polynomial in x that has coefficients that are polynomials in $\mathbf{Q}[\alpha]$. Give a procedure $Alg\_coeff\_simp(u, x, p, \alpha)$ that simplifies each coefficient of u using the side relation $p(\alpha) = 0$. Return an expression in collected form with respect to x. For example,

$$Alg\_coeff\_simp\left(\alpha^3 x + \alpha^2 + 3,\ x,\ p,\ x\right) \rightarrow 2\alpha x + 5,$$
$$Alg\_coeff\_simp\left((\alpha^2 + \alpha)\ x - 2x - \alpha x,\ x,\ p,\ x\right) \rightarrow 0.$$

3. Let α be an algebraic number with minimal polynomial p (in terms of α), and let u be a polynomial in $\mathbf{Q}(\alpha)[x]$. Give a procedure

$$Alg\_monic(u, x, p, \alpha)$$

that transforms a polynomial u to monic form.

4. Let α be an algebraic number with minimal polynomial p (in terms of α). Consider the class $\mathbf{G}$ of expressions defined by the following rules.

   G-1. u is an integer.

   G-2. u is a fraction.

   G-3. u is the symbol α.

   G-4. u is sum with operands that are in $\mathbf{G}$.

   G-5. u is a product with operands that are in $\mathbf{G}$.

   G-6. u is a power with a base that is in $\mathbf{G}$ and an exponent that is an integer.

   Give a procedure

$$Alg\_gen\_simp(u, p, \alpha)$$

---

[12]See Landau [60] and Jeffrey and Rich [49].

that transforms u to either an expression in $\mathbf{Q}(\alpha)$ or the global symbol **Undefined** if a division by zero is encountered. For example, if $p = \alpha^5 - \alpha + 1$ and

$$u = 2 + \cfrac{\alpha}{3 + \alpha + \cfrac{4}{2\alpha + \alpha^5}},$$

then

$$Alg\_gen\_simp(u, p, \alpha) \quad \rightarrow \quad -\frac{1457}{27821}\alpha^4 + \frac{2191}{27821}\alpha^3 - \frac{5357}{27821}\alpha^2$$
$$+\frac{13555}{27821}\alpha + \frac{50559}{27821},$$

$$Alg\_gen\_simp(1/p, p, \alpha) \quad \rightarrow \quad \textbf{Undefined}.$$

5. Let α be an algebraic number with minimal polynomial p (in terms of α), and let u and v be polynomials in $\mathbf{Q}(\alpha)[x]$ with $\deg(v, x) \geq 1$. Give a procedure

$$Alg\_poly\_expansion(u, v, x, t, p, \alpha)$$

for polynomial expansion in $\mathbf{Q}(\alpha)[x]$.

6. Let α be an algebraic number with minimal polynomial p (in terms of α), and let u and v be polynomials in $\mathbf{Q}(\alpha)[x]$. Give a procedure

$$Alg\_extend(u, v, x, p, \alpha)$$

for the extended Euclidean algorithm in $\mathbf{Q}(\alpha)[x]$.

7. This exercise describes another example of a finite field. Let p be a prime number, and let $f(x)$ be an irreducible polynomial with positive degree in $\mathbf{Z}_p[x]$. The field $\mathbf{Z}_p(\alpha)$ contains all polynomials u in $\mathbf{Z}_p[\alpha]$ which satisfy $\deg(u) < \deg(f)$, where the addition and multiplication operations are defined as in Definition 4.46.

   (a) Let $f(x) = x^2 + x + 1$ in $\mathbf{Z}_2[x]$. List the polynomials in $\mathbf{Z}_2(\alpha)$, and give the addition and multiplication tables.

   (b) Suppose that $\deg(f) = n$. Show that $\mathbf{Z}_p(\alpha)$ contains p^n polynomials. This field is called the *Galois field* of order p^n.

8. (a) Let α be an algebraic number with minimal polynomial p (in terms of α). This exercise is similar to Exercise 18 on page 144. Give a rational simplification procedure $Alg\_rational\_simplify\_sv(u, x, p, \alpha)$ where the coefficient field is now $\mathbf{Q}(\alpha)$ instead of $\mathbf{Q}$.

   (b) Let $\alpha = \sqrt{2}$, and consider

$$u = \frac{x^2 + (6 + 8\alpha)x + 38 + 22\alpha}{x^2 + (8 + 10\alpha)x + 54 + 32\alpha}.$$

      i. Find the simplified form of this expression using the procedure described in Part (a).

ii. Try to simplify u using the following simple approach. Since both the numerator and denominator of u are quadratic in x, we can find factorizations for them using the quadratic formula for the roots. What problems do you encounter when trying to simplify the expression with this approach?

9. Determine whether the following sets of algebraic numbers are algebraically dependent or independent. If the numbers are dependent, find the polynomial h which gives the dependence relation between the numbers.

(a) $S = \{\sqrt{2}, i\}$.

(b) $S = \left\{\sqrt{2}, \sqrt{3}, \sqrt{5 + 2\sqrt{2}\sqrt{3}}\right\}$.

(c) $S = \{\sqrt{2}, \sqrt{3}, \sqrt{5}\}$.

(d) $S = \{\sqrt{6}, \sqrt{10}, \sqrt{15}\}$.

(e) $S = \{\sqrt{3} + \sqrt{2}, \sqrt{3} - \sqrt{2}\}$.

(f) $S = \left\{2^{1/3}, -1/2 + \sqrt{3}/2\,i\right\}$.

(g) $S = \left\{2^{1/3}, 2^{1/3}(-1/2 + \sqrt{3}/2\,i)\right\}$.

10. Let $\alpha_1, \alpha_2, \ldots, \alpha_s$ be a set of independent algebraic numbers, and let p_i be the minimal polynomial of α_i. Show that p_s is irreducible as a polynomial in $\mathbf{Q}(\alpha_1, \alpha_2, \ldots, \alpha_{s-1})[x_s]$.

11. Consider two independent algebraic numbers α_1 and α_2 with minimal polynomials p_1 and p_2.

(a) Let a and $b \neq 0$ be in $\mathbf{Q}(\alpha_1, \alpha_2)$. Give procedures for the following operations.

i. $Alg\_mult\_inverse\_2(a, p_1, \alpha_1, p_2, \alpha_2)$ for the multiplicative inverse of a.

ii. $Alg\_divide\_2(a, b, p_1, \alpha_1, p_2, \alpha_2)$ for the division of a by b.

(b) Let u and v be in $\mathbf{Q}(\alpha_1, \alpha_2)[x]$. Give procedures for the following operations.

i. For $v \neq 0$, $Alg\_polynomial\_division\_2(u, v, x, p_1, \alpha_1, p_2, \alpha_2)$ for polynomial division.

ii. $Alg\_polynomial\_gcd\_2(u, v, x, p_1, \alpha_1, p_2, \alpha_2)$ for the polynomial greatest common divisor.

iii. $Alg\_extend\_2(u, v, x, p_1, \alpha_1, p_2, \alpha_2)$ for the extended Euclidean algorithm.

12. Let $\{\alpha_1, \alpha_2, \ldots, \alpha_s\}$ be a set of independent algebraic numbers with minimal polynomials $p_1, p_2, \ldots, p_s$. Define a list of two element lists

$$L = [[\alpha_1, p_1], [\alpha_2, p_2], \ldots, [\alpha_s, p_s]].$$

(a) Let a and $b \neq 0$ be in $\mathbf{Q}(\alpha_1, \alpha_2, \ldots, \alpha_s)$. Give procedures for the following operations.

    i. *Multiple_alg_mult_inverse*(b, L) for the multiplicative inverse of a.

    ii. *Multiple_alg_divide*(a, b, L) for division of a by b.

(b) Let u and v be in $\mathbf{Q}(\alpha_1, \alpha_2, \ldots, \alpha_s)[x]$. Give procedures for the following operations.

    i. For $v \neq 0$, *Multiple_alg_polynomial_division*(u, v, x, L) for polynomial division.

    ii. *Multiple_alg_gcd*(u, v, x, L) for the polynomial greatest common divisor.

    iii. *Multiple_alg_extend*(u, v, x, L) for the extended Euclidean algorithm.

4.4 Partial Fraction Expansion in F(x)

Partial fraction expansion provides a way to decompose an involved rational expression into a sum of simpler rational expressions. The technique is important for indefinite integration, the solution of differential equations, and many areas of engineering. In this section we describe an approach that is based on the extended Euclidean algorithm and polynomial expansion.

Let $\mathbf{F}$ be a field. In this section, a rational expression u/v is the ratio of polynomials u and $v \neq 0$, each of which is in $\mathbf{F}[x]$. The notation $\mathbf{F}(x)$ represents the set of all rational expressions.

Definition 4.64. *A rational expression $w = u/v$ in $\mathbf{F}(x)$ is* **proper** *if*

$$\deg(u, x) < \deg(v, x).$$

If w is not proper, it is called **improper**.

Notice that, based on the definitions of the *Numerator* and *Denominator* operators,[13] this definition includes the expression 0 because

$$Numerator(0) \rightarrow 0, \qquad Denominator(0) \rightarrow 1,$$

and

$$\deg(0) = -\infty < \deg(1) = 0.$$

---

[13]Formal definitions of the MPL *Numerator* and *Denominator* operators are given on the CD that accompanies this book. The corresponding operators in Maple, Mathematica, and MuPAD perform in this way (see page 14).

Theorem 4.65. *A rational expression* $w = u/v$ *in* $\mathbf{F}(x)$ *can be represented uniquely as*

$$\frac{u}{v} = q + \frac{r}{v}$$

where q *and* r *are in* $\mathbf{F}[x]$ *and* r/v *is proper.*

Proof: The theorem follows from Theorem 4.1 (page 116) on polynomial division (Exercise 1). $\qquad\square$

The next theorem contains the fundamental relation for partial fraction expansion.

Theorem 4.66. *Let* u, $v_1 \neq 0$, *and* $v_2 \neq 0$ *be polynomials in* $\mathbf{F}[x]$ *such that* $\gcd(v_1, v_2) = 1$, *and suppose that*

$$\frac{u}{v_1 v_2}$$

is a proper rational expression. Then, there are unique polynomials u_1 *and* u_2 *in* $\mathbf{F}[x]$ *such that*

$$\frac{u}{v_1 v_2} = \frac{u_1}{v_1} + \frac{u_2}{v_2}$$

where u_1/v_1 *and* u_2/v_2 *are proper rational expressions.*

Proof: The proof, which is constructive, is based on the extended Euclidean algorithm. Since $\gcd(v_1, v_2) = 1$, using the extended Euclidean algorithm, there are polynomials $A(x)$ and $B(x)$ such that $A v_1 + B v_2 = 1$. Therefore,

$$\frac{u}{v_1 v_2} = \frac{(A v_1 + B v_2) \cdot u}{v_1 v_2} = \frac{A u}{v_2} + \frac{B u}{v_1}. \tag{4.70}$$

The proof is not complete, however, because the two rational expressions in the last sum may not be proper (see Example 4.67). By performing the divisions in the sum on the right, we obtain

$$B u = q_1 v_1 + u_1, \quad A u = q_2 v_2 + u_2,$$

with $\deg(u_1) < \deg(v_1)$ and $\deg(u_2) < \deg(v_2)$, and so Equation (4.70) becomes

$$\frac{u}{v_1 v_2} = q_2 + \frac{u_2}{v_2} + q_1 + \frac{u_1}{v_1} \tag{4.71}$$

$$= q_1 + q_2 + \frac{v_1 u_2 + v_2 u_1}{v_1 v_2}. \tag{4.72}$$

Since both of the rational expressions on the right in Equation (4.71) are proper, their sum in Equation (4.72) is also proper (Exercise 2). By Theorem 4.65, the proper rational expression $u/(v_1 v_2)$ can only be represented in one way: as the sum of a polynomial and a proper rational expression. Since the polynomial part of this expression is 0, $q_1 + q_2$ in Equation (4.72) is also zero. Therefore,

$$\frac{u}{v_1 v_2} = \frac{u_1}{v_1} + \frac{u_2}{v_2}.$$

To show the uniqueness property, suppose that there were two representations

$$\frac{u}{v_1 v_2} = \frac{u_1}{v_1} + \frac{u_2}{v_2} = \frac{w_1}{v_1} + \frac{w_2}{v_2} \tag{4.73}$$

$$= \frac{u_1 v_2 + u_2 v_1}{v_1 v_2} = \frac{w_1 v_2 + w_2 v_1}{v_1 v_2} \tag{4.74}$$

where all the rational expressions in Equation (4.73) are proper. Since both expressions in Equation (4.74) have the same denominator, their numerators are equal, and so

$$(u_1 - w_1)\, v_2 = (w_2 - u_2)\, v_1.$$

Therefore, since $\gcd(v_1, v_2) = 1$, Theorem 4.29(1) (page 133) implies that

$$v_2 \mid (w_2 - u_2). \tag{4.75}$$

However, since all rational expressions in Equation (4.73) are proper,

$$\deg(w_2 - u_2) < \deg(v_2), \tag{4.76}$$

and therefore, the conditions (4.75) and (4.76) imply that $w_2 - u_2 = 0$. In a similar way, $u_1 - w_1 = 0$, and so the representation is unique. □

Example 4.67. Let

$$u = x^2 + 2, \qquad v_1 = x + 2, \qquad v_2 = x^2 - 1.$$

Therefore,

$$\frac{u}{v_1 v_2} = \frac{x^2 + 2}{x^3 + 2\,x^2 - x - 2}$$

is a proper rational expression with $\gcd(v_1, v_2) = 1$. By the extended Euclidean algorithm, we obtain polynomials A and B such that

$$A\, v_1 + B\, v_2 = (-1/3\, x + 2/3)\, v_1 + (1/3)\, v_2 = 1.$$

Therefore,

$$
\begin{aligned}
\frac{u}{v_1\,v_2} &= \frac{A\,u}{v_2} + \frac{B\,u}{v_1} \\
&= \frac{(-1/3)\,x^3 + (2/3)\,x^2 - (2/3)\,x + 4/3}{v_2} \\
&\quad + \frac{(1/3)\,x^2 + 2/3}{v_1}
\end{aligned}
\tag{4.77}
$$

where the two rational expressions in Equation (4.77) are improper. Performing the divisions in Equation (4.77), we obtain

$$
\begin{aligned}
\frac{u}{v_1\,v_2} &= (-1/3)\,x + 2/3 + \frac{(-x+2)}{v_2} + (1/3)\,x - 2/3 + \frac{2}{v_1} \\
&= \frac{2}{v_1} + \frac{(-x+2)}{v_2}.
\end{aligned}
\qquad\qquad \square
$$

A procedure that performs the manipulations in the proof of Theorem 4.66 is shown in Figure 4.8.

The next theorem describes the partial fraction expansion u/v in terms of the irreducible factorization in $\mathbf{F}[x]$ of the denominator v.

Theorem 4.68. *Let u/v be a proper rational expression, and let*

$$
v = c \cdot p_1^{n_1} \cdot p_2^{n_2} \cdots p_s^{n_s}
$$

be the irreducible factorization of v in $\mathbf{F}[x]$. Then, there are unique polynomials $u_1, u_2, \ldots, u_s$ in $\mathbf{F}[x]$, with

$$
\deg(u_i) < n_i \cdot \deg(p_i),
$$

such that

$$
\frac{u}{v} = (1/c) \sum_{i=1}^{s} \frac{u_i}{p_i^{n_i}}.
\tag{4.78}
$$

Proof: Since the polynomials $p_1, \ldots, p_s$ are relatively prime, the two polynomials

$$
p_1^{n_1} \quad \text{and} \quad p_2^{n_2} \cdot p_3^{n_3} \cdots p_s^{n_s}
\tag{4.79}
$$

are relatively prime as well. Therefore, by Theorem 4.66, there are polynomials u_1 and w such that

$$
\frac{u}{v} = \frac{1}{c}\left(\frac{u_1}{p_1^{n_1}} + \frac{w}{p_2^{n_2} p_3^{n_3} \cdots p_s^{n_s}} \right)
\tag{4.80}
$$

Procedure *Partial_fraction_1* $(u, v1, v2, x)$;
Input
    $u, v1, v2$: polynomials in $\mathbf{F}[x]$ where all field operations
            in $\mathbf{F}$ are obtained with automatic simplification,
            $\gcd(v1, v2) = 1$, and $u/(v1\, v2)$ is proper;
    x : a symbol;
Output
    the list $[u1, u2]$;
Local Variables
    $s, A, B, u1, u2$;
Begin

```
1     s := Extended_Euclidean_algorithm(v1, v2, x);
2     A := Operand(s, 2);
3     B := Operand(s, 3);
4     u1 := Remainder(Algebraic_expand(B * u), v1, x);
5     u2 := Remainder(Algebraic_expand(A * u), v2, x);
6     Return([u1, u2])
```

End

Figure 4.8. An MPL procedure for the two term partial fraction expansion described in Theorem 4.66. (Implementation: Maple (txt), Mathematica (txt), MuPAD (txt).)

where both rational expressions on the right are proper. The expansion is completed by reapplying this process (recursively) to

$$\frac{w}{p_2^{n_2} p_3^{n_3} \cdots p_s^{n_s}}.$$

The proof of the uniqueness is left to the reader (Exercise 5).   □

A recursive procedure that returns the expansion in Theorem 4.68 is given in Figure 4.9. The statement at line 1 is the termination condition for the recursion. When the condition is **true**, v has only one factor and so u/v is returned. The statements at lines 4–5 separate the first factor of v from the remaining factors as is done in (4.79). In lines 6–7, a check is done to see if this factor is not part of c in Equation (4.78). If f is not free of x, then lines 9–11 obtain Equation (4.80) using procedure *Partial_fraction_1*. Notice that both f and r are expanded because both may be in a factored form. At line 12, the first term is added to the partial fraction expansion of the remaining terms of the sum which is determined recursively.

The most general form of the partial fraction expansion is described in the next theorem.

Procedure *Partial_fraction_2(u, v, x)*
Input
    u : a polynomial in $\mathbf{F}[x]$ where all field operations
       in $\mathbf{F}$ are obtained with automatic simplification;
    v : the irreducible factorization of a polynomial in $\mathbf{F}[x]$ with
       positive degree and u/v proper;
    x : a symbol;
Output
    the partial fraction expansion described in Theorem 4.68;
Local Variables
    $f, r, s, u1, w$;
Begin

```
1      if  Kind(v) ≠ " * " then
2          Return(u/v)
3      else
4          f := Operand(v, 1);
5          r := v/f;
6          if  Free_of(f, x) then
7              Return((1/f) * Partial_fraction_2(u, r, x))
8          else
9              s := Partial_fraction_1(u, Algebraic_expand(f),
                                          Algebraic_expand(r), x);
10             u1 := Operand(s, 1);
11             w := Operand(s, 2);
12             Return(u1/f + Partial_fraction_2(w, r, x))
    End
```

Figure 4.9. An MPL procedure for the partial fraction expansion described in Theorem 4.68. (Implementation: Maple (txt), Mathematica (txt), MuPAD (txt).)

Theorem 4.69. *Let $w = u/v$ be a rational expression in $\mathbf{F}(x)$, and let*

$$v = c \cdot p_1^{n_1} \cdot p_2^{n_2} \cdots p_s^{n_s}$$

be the irreducible factorization of v in $\mathbf{F}[x]$. Then, there are unique polynomials q and

$$g_{i,j}, \qquad i = 1, \ldots, s, \qquad j = 1, \ldots, n_i,$$

with $\deg(g_{i,j}) < \deg(p_i)$, such that

$$w = \frac{u}{v} = q + (1/c) \sum_{i=1}^{s} \sum_{j=1}^{n_i} \frac{g_{i,j}}{p_i^j}. \tag{4.81}$$

Proof: By Theorems 4.65 and 4.68, a rational expression can be represented as

$$\frac{u}{v} = q + \frac{r}{v} = q + (1/c) \sum_{i=1}^{s} \frac{u_i}{p_i^{n_i}} \tag{4.82}$$

where $u_i/p_i^{n_i}$ is a proper rational expression. By Theorem 4.13 on polynomial expansion, each polynomial u_i, can be represented by the polynomial expansion

$$u_i = \sum_{j=0}^{n_i-1} g_{i,j} p_i^j, \tag{4.83}$$

with $\deg(g_{i,j}) < \deg(p_i)$. (Notice that since $u_i/p_i^{n_i}$ is proper, the upper limit of the sum is $n_i - 1$.) Therefore,

$$\frac{u_i}{p_i^{n_i}} = \sum_{j=1}^{n_i} \frac{g_{i,j}}{p_i^j},$$

and substituting this expression into Equation (4.82) we obtain Equation (4.81).

The proof of the uniqueness of the representation is left to the reader (Exercise 6). $\square$

The procedure that determines the representation in Theorem 4.69 is left to the reader (Exercise 8).

Example 4.70. Consider the rational expression

$$f = \frac{x^3 + 7x^2 + 26x - 105}{5x^3 - 25x^2 - 40x + 240}$$

in $\mathbf{Q}(x)$. Since f is not proper, dividing the denominator into the numerator we obtain

$$f = 1/5 + \frac{12x^2 + 34x - 153}{5(x+3)(x-4)^2}$$

where we have represented the denominator of the rational part in factored form. By Theorem 4.68, we can represent this expression as

$$f = 1/5 + (1/5)\left(\frac{(-3)}{x+3} + \frac{15x - 35}{(x-4)^2}\right). \tag{4.84}$$

We modify the last term in the sum by finding the polynomial expansion of $15x - 35$ in terms of $x - 4$:

$$15x - 35 = 15(x - 4) + 25. \tag{4.85}$$

Substituting Equation (4.85) into Equation (4.84), we obtain the complete expansion

$$f = 1/5 + (1/5)\left(\frac{(-3)}{x+3} + \frac{15}{x-4} + \frac{25}{(x-4)^2}\right).$$ □

Exercises

1. Prove Theorem 4.65.

2. Show that the sum of two proper rational expressions is a proper rational expression.

3. Show that the derivative of a proper rational expression is a proper rational expression.

4. Find a partial fraction expansion (in $\mathbf{Q}(x)$) for each of the following using the algorithm described in this section.

    (a) $\dfrac{1}{x^3 + 2x^2 + x}$.

    (b) $\dfrac{x^3 + 2x^2 - x - 3}{x^4 - 4x^2 + 4}$.

    (c) $\dfrac{x^3}{x^2 + 3x + 2}$.

5. Prove that the representation in Theorem 4.68 is unique.

6. Prove that the representation in Theorem 4.69 is unique.

7. Consider the proper rational expression u/v, and suppose that u and v have a common factor of positive degree. How is this reflected in the partial fraction expansion? Justify your answer.

8. In this exercise we describe procedures that obtain the partial fraction expansion of a rational expression f in $\mathbf{Q}(x)$.

    (a) Let u/v be a proper rational expression where u is in expanded form and v is the irreducible factorization of a polynomial in $\mathbf{Q}(x)$. Give a procedure $Partial\_fraction\_3(u, v, x)$ that finds the partial fraction expansion described in Theorem 4.69. (The quotient q is zero for this case.)

    (b) Let f be a rational expression in $\mathbf{Q}(x)$ (which may be proper or improper) where the numerator and the denominator of f are in algebraically-expanded form. Give a procedure $Partial\_fraction(f, x)$ that obtains the full partial fraction expansion described in Theorem 4.69. Use the factor operator in a CAS to obtain the factorization of the denominator of f.

9. Find the partial fraction expansion in $\mathbf{Q}\left(\sqrt{2}\right)(x)$ for

$$\frac{x^4 + \left(2\sqrt{2}+1\right)x^2 + \left(5+9\sqrt{2}\right)x + 9\sqrt{2} - 12}{x^3 + 3\,x^2 - 2\,x - 6}.$$

10. In this exercise we describe procedures that can evaluate the anti-derivative of the partial fraction expansion in Equation (4.81) for rational expressions in $\mathbf{Q}(x)$. The approach can obtain the anti-derivative when either $g_{i,j}/p_i^j$ in Equation (4.81) satisfies

$$0 < \deg(p_i, x) \le 2,$$

or $g_{i,j}/p_i^j$ can be transformed using a substitution based on polynomial expansion to a similar expression that satisfies this degree condition. The algorithm proceeds in a number of steps depending on the forms of $g_{i,j}$ and p_i.

(a) Suppose that $g_{i,j}$ is a rational number and p_i is linear in x. This case is handled by

$$\int \frac{g_{i,j}}{(a\,x+b)^j}\,dx = \begin{cases} -\dfrac{g_{i,j}}{(j-1)\,a\,(a\,x+b)^{j-1}}, & j > 1, \\[2ex] \dfrac{g_{i,j}\,\ln(a\,x+b)}{a}, & j = 1. \end{cases}$$

(b) Suppose that $g_{i,j} = 1$, and $p_i = a\,x^2 + b\,x + c$ with $a \ne 0$ and $d = b^2 - 4\,a\,c \ne 0$. (The case $b^2 - 4\,a\,c = 0$ does not occur here because this implies $a\,x^2 + b\,x + c = a\,(x+b/(2\,a))^2$, which is handled by the partial fraction expansion using case (a) with $p_i = x + b/(2\,a)$. For this case, the integral is given by the reduction formula

$$\int \frac{1}{(a\,x^2 + b\,x + c)^n}\,dx = \qquad\qquad\qquad (4.86)$$

$$\begin{cases} \begin{cases} 2\dfrac{\arctan\left(\frac{2\,a\,x+b}{\sqrt{4\,a\,c-b^2}}\right)}{\sqrt{4\,a\,c - b^2}}, & \text{if } b^2 - 4\,a\,c < 0, \\[3ex] -2\dfrac{\operatorname{arctanh}\left(\frac{2\,a\,x+b}{\sqrt{b^2-4\,a\,c}}\right)}{\sqrt{b^2 - 4\,a\,c}}, & \text{if } b^2 - 4\,a\,c > 0, \\[3ex] -\dfrac{2}{2\,a\,x+b}, & \text{if } b^2 - 4\,a\,c = 0, \end{cases} & n = 1, \\[8ex] \dfrac{-(b+2\,a\,x)}{(n-1)\,d\,(a\,x^2+b\,x+c)^{n-1}} \\[2ex] \quad + \dfrac{-(4\,n-6)\,a}{(n-1)\,d}\displaystyle\int \frac{1}{(a\,x^2+b\,x+c)^{n-1}}\,dx, & n > 1. \end{cases}$$

(A similar formula is derived in Exercise 6 on page 143.) Give a procedure

$$Rational\_red\_1\,(a, b, c, n, x)$$

which evaluates the integral using Equation (4.86).

(c) Suppose that $g_{i,j} = r\,x + s$ and $p_i = a\,x^2 + b\,x + c$ with $a \neq 0$. Then,

$$\int \frac{r\,x + s}{(a\,x^2 + b\,x + c)^n}\, dx = \qquad (4.87)$$

$$\begin{cases} (r/(2\,a))\,\ln(a\,x^2 + b\,x + c) \\ +(s - r\,b/(2\,a)) \displaystyle\int \frac{dx}{a\,x^2 + b\,x + c}, & n = 1, \\[3em] \dfrac{-r}{2\,(n-1)\,a\,(a\,x^2 + b\,x + c)^{n-1}} \\ + \dfrac{-b\,r + 2\,a\,s}{(2\,a)} \displaystyle\int \frac{1}{(a\,x^2 + b\,x + c)^n}\, dx, & n > 1 \end{cases}$$

where the integrals are evaluated using $Rational\_red\_1$. Give a procedure

$$Rational\_red\_2\,(r, s, a, b, c, n, x)$$

that evaluates the integral using Equation (4.87).

(d) Suppose that $\deg(p_i) > 2$. For this case we only evaluate the antiderivative when $g_{i,j}(x)/p_i^j(x)$ can be transformed using a substitution $v = f(x)$ to the form $1/q(v)^j$ with

$$1 \leq \deg(q(v), v) \leq 2. \qquad (4.88)$$

For example, the expression $2\,x/(x^4 + 1)$ is transformed by the substitution $v = x^2$ to $1/(v^2 + 1)$, and

$$\int \frac{2\,x}{x^4 + 1}\, dx = \int \frac{dv}{v^2 + 1} = \arctan(v) = \arctan(x^2).$$

Notice that the polynomial v satisfies $v' = g_{i,j}$. In general, to find if a substitution $v = f(x)$ is possible, let $f(x) = \int g_{i,j}\, dx$ and

$$q(v) = Polynomial\_expansion(p_i, f(x), x, v).$$

If $q(v)$ is a polynomial in v with rational number coefficients, then the substitution is $v = f(x)$. (If $q(v)$ has coefficients that are not free of x, then the substitution does not exist.) When

$$1 \leq \deg(q(v), v) \leq 2, \qquad (4.89)$$

$\int dv/q(v)^j$ is evaluated by either Part (a) or Equation (4.87), and

$$\int \frac{g_{i,j}}{p_i^j}\, dx = Substitute\left(v = f(x), \int \frac{dv}{q^j}\right).$$

If $\deg(q(v), v) > 2$, the anti-derivative is not evaluated by the techniques described here. Give a procedure

$$Rational\_red\_3(g, p, j, x)$$

that evaluates the anti-derivative of g/p^j by

    i. integrating g,

    ii. obtaining the polynomial expansion with *Polynomial_expansion*,

    iii. checking if the substitution is possible, and

    iv. evaluating the anti-derivative when the degree condition (4.89) is satisfied.

If the anti-derivative cannot be evaluated with this approach, return the global symbol **Fail**.

(e) Give a procedure

$$Rational\_red\_4(g, p, j, x)$$

that evaluates the anti-derivative of g/p^j by evaluating the degrees of g and p and applying one of the cases (a), (b), (c), or (d). If the anti-derivative cannot be evaluated using these techniques, return the global symbol **Fail**.

(f) Let f be an algebraic expression. Give a procedure

$$Rational\_integrate(f, x)$$

that checks if f is a rational expression in $\mathbf{Q}(x)$ and when this is so evaluates $\int f \, dx$ using the approach described in this exercise. If f does not have the proper form or if $\int f \, dx$ cannot be evaluated using these techniques, return the global symbol **Fail**.

11. Give a procedure

$$Numerical\_part\_frac(r)$$

that expresses a rational number $0 \le r < 1$ in terms of rational numbers with prime denominators. For example, $r = 1/6 = 1/2 - 1/3$. To bypass the automatic simplification of rational number expressions, use the function representations described in Figure 2.6 on page 39.

Use the `ifactor` operator in Maple and MuPAD and the `FactorInteger` operator in Mathematica to obtain the prime factorization of the denominator of r. (Implementation: Maple (mws), Mathematica (nb), MuPAD (mnb).)

Further Reading

4.2 Greatest Common Divisors. See Akritas [2], Brown and Traub [16], Davenport et al. [29], Geddes et al. [39], Knuth [55], Yap [105], Mignotte [66], von zur Gathen and Gerhard [96], and Zippel [108] for modern approaches in greatest common divisor calculation.

4.3 Computations in Algebraic Number Fields. See Buchberger et al. [17] and Smedley [91] for discussions of computations in algebraic number fields. Landau [60] describes an algorithm for denesting radicals based on Galois theory.

4.4 Partial Fraction Expansion. von zur Gathen and Gerhard [96] discuss partial fraction expansion.

5

Polynomial Decomposition

A polynomial $u(x)$ is called a *composite* polynomial if there are polynomials $v(w)$ and $w(x)$, both with degree greater than one, such that u is represented by the functional composition $u(x) = v(w(x))$. The composite expression $v(w(x))$ is also called a *decomposition* of u in terms of the components $v(w)$ and $w(x)$, and the process for finding v and w is called *decomposing* a polynomial. For example,

$$u(x) = x^4 - 5\,x^2 + 6 \tag{5.1}$$

is a composite polynomial with decomposition given by

$$v(w) = w^2 - 5\,w + 6, \quad w(x) = x^2.$$

Decomposition plays an important role in the solution of polynomial equations. For example, to solve

$$u(w(x)) = (x^2)^2 - 5\,(x^2) + 6 = 0,$$

first solve the equation for $w(x) = x^2$ to obtain $x^2 = 2, 3$, and then solve these equations to obtain $x = \pm\sqrt{2}, \pm\sqrt{3}$. Most computer algebra systems apply decomposition as part of the solution of polynomial equations.

It is easy to recognize that $u(x)$ in Equation (5.1) is a composite polynomial. With more involved examples, however, a decomposition may not be apparent or even possible. For example,

$$u(x) = x^4 + 4\,x^3 + 3\,x^2 - 2\,x + 3 \tag{5.2}$$

179

can be decomposed as

$$(x^2 + 2x + 1)^2 - 3(x^2 + 2x + 1) + 5, \qquad (5.3)$$

while $u(x) + x$ cannot be decomposed at all.

How can we determine if a polynomial can be decomposed? Surprisingly, there are some simple algorithms that do the job. In Section 5.1 we describe the basic theoretical concepts of polynomial decomposition, and in Section 5.2 we describe an algorithm that either finds a decomposition for $u(x)$ (if one exists) or determines that no such decomposition is possible.

5.1 Theoretical Background

In this chapter all polynomials $u(x)$ are in $\mathbf{Q}[x]$. First of all, it is convenient to simplify our notation. We denote a composite polynomial $u(x) = v(w(x))$ using the operator notation $\circ$ as

$$u = v \circ w,$$

and to emphasize the dependence on a variable x, we represent the composite as

$$u(x) = (v \circ w)(x).$$

For example, the composite polynomial

$$u(x) = (x^2 + 2x + 1)^2 - 3(x^2 + 2x + 1) + 5,$$

is expressed as

$$u(x) = (w^2 - 3w + 5) \circ (x^2 + 2x + 1). \qquad (5.4)$$

It also simplifies matters if we use the same variable name x to describe all polynomial formulas. With this convention, Equation (5.4) becomes

$$u(x) = (x^2 - 3x + 5) \circ (x^2 + 2x + 1).$$

Using the $\circ$ notation we can also represent more complex composite expressions such as

$$f(x) = (u \circ v \circ w)(x) = u(v(w(x))).$$

Actually, the composition symbol $\circ$ is nothing more than a shorthand notation for the *Substitute* operator

$$(v \circ w)(x) = Substitute(v(x),\ x = w(x)).$$

The composition operator satisfies the following properties.

Theorem 5.1. *Let u, v, and w be polynomials in $\mathbf{Q}[x]$.*

1. *$u \circ (v \circ w) = (u \circ v) \circ w$ (associative property).*

2. *$u \circ x = x \circ u = u$.*

3. *For $u \neq 0$ and $v \neq 0$, $\deg((u \circ v)) = \deg(u)\deg(v)$.*

Proof: Statement (1) is a general statement about mathematical functions. Statement (2), which is easy to verify, says that the polynomial $I(x) = x$ acts as an identity element with respect to composition. Statement (3), which is also easy to verify, is left to the reader (Exercise 2). □

Example 5.2. In general, the composition operation is not commutative. If

$$v(x) = 3\,x^2 - x, \qquad w(x) = x^2 + 2\,x,$$

then

$$(v \circ w)(x) = 3\,x^4 + 12\,x^3 + 11\,x^2 - 2\,x,$$

while

$$(w \circ v)(x) = 9\,x^4 - 6\,x^3 + 7\,x^2 - 2\,x.$$

There are, however, some polynomials for which the operation is commutative. (See Equation (5.7) and Exercise 8.) □

Definition 5.3. *A polynomial in $\mathbf{Q}[x]$ is **decomposable** if there is a finite sequence of polynomials $g_1(x), \dots, g_n(x)$, each with degree greater than one, such that*

$$u = g_n \circ g_{n-1} \circ \dots \circ g_1, \quad n > 1. \tag{5.5}$$

*Equation (5.5) is called a **functional decomposition** (or just a **decomposition**) of $u(x)$, and the polynomials $g_i(x)$ are the **components** of the decomposition. If no such decomposition exists, then $u(x)$ is **indecomposable**. If each of the g_i is also indecomposable, then Equation (5.5) is called a **complete decomposition** of $u(x)$.*

Most computer algebra systems have a command that obtains a polynomial decomposition (`compoly` in Maple, `Decompose` in Mathematica, `polylib::decompose` in MuPAD). (Implementation: Maple, Mathematica, MuPAD.)

The definition requires that the decomposition be non-trivial in the sense that all components g_i must have degree greater than one. This means

$$4x^2 + 18x + 22 = (2x+3)^2 + 3(2x+3) + 4$$

is not a decomposition because $2x+3$ has degree one.

Example 5.4. The following polynomial has a decomposition with three components:

$$
\begin{aligned}
u(x) &= x^8 + 4x^7 + 10x^6 + 16x^5 + 29x^4 + 36x^3 + 40x^2 + 24x + 39 \\
&= \left((x^2+x)^2 + 2(x^2+x)\right)^2 + 12\left((x^2+x)^2 + 2(x^2+x)\right) + 39 \\
&= (x^2 + 12x + 39) \circ (x^2 + 2x) \circ (x^2 + x).
\end{aligned}
$$

Since none of the components can be decomposed, the decomposition is complete (Exercise 3).   □

Example 5.5. A complete decomposition of a polynomial is not unique. For example, here are two different decompositions of a polynomial:

$$
\begin{aligned}
u(x) &= x^6 + 2x^4 + x^2 + 1 \\
&= (x^3 + 2x^2 + x + 1) \circ (x^2) \qquad\qquad\qquad\qquad (5.6) \\
&= (x^3 - x^2 + 1) \circ (x^2 + 1).
\end{aligned}
$$
                                                                            □

In fact, any polynomial that can be decomposed has infinitely many decompositions. This fact is based on the observation that the two linear polynomials $f(x) = mx+b$ and $g(x) = x/m - b/m$ are inverses with respect to composition because they satisfy the property

$$(f \circ g)(x) = (g \circ f)(x) = x. \qquad\qquad\qquad (5.7)$$

Therefore, if u has a decomposition $u = v \circ w$, another decomposition is obtained (using Theorem 5.1) in the following way:

$$u = v \circ w = v \circ x \circ w = v \circ (f \circ g) \circ w = (v \circ f) \circ (g \circ w). \qquad (5.8)$$

For Equation (5.6), with $f = x - 1$, $g = x + 1$, this construction takes the form:

$$
\begin{aligned}
u(x) &= x^6 + 2x^4 + x^2 + 1 \\
&= (x^3 + 2x^2 + x + 1) \circ x^2 \\
&= (x^3 + 2x^2 + x + 1) \circ x \circ x^2 \qquad\qquad\qquad (5.9) \\
&= (x^3 + 2x^2 + x + 1) \circ \left((x-1) \circ (x+1)\right) \circ x^2 \\
&= \left((x^3 + 2x^2 + x + 1) \circ (x-1)\right) \circ \left((x+1) \circ x^2\right) \\
&= (x^3 - x^2 + 1) \circ (x^2 + 1).
\end{aligned}
$$

The next theorem plays an important role in our decomposition algorithm.

Theorem 5.6. *Let u be a polynomial that can be decomposed. Then, there exists a decomposition $u = v \circ w$ such that $w(0) = 0$.*

Proof: Suppose that $u = v \circ w$. In case $w(0) \neq 0$, define the new polynomials

$$v_2(x) = v(x + w(0)), \qquad w_2(x) = w(x) - w(0).$$

Then $w_2(0) = 0$, and

$$
\begin{aligned}
(v_2 \circ w_2)(x) &= v_2(w_2(x)) \\
&= v_2(w(x) - w(0)) \\
&= v(w(x) - w(0) + w(0)) \\
&= v(w(x)) \\
&= u(x). \qquad \qquad \square
\end{aligned}
$$

Ritt's Decomposition Theorem

Suppose that we have two distinct decompositions for a polynomial

$$u = v_n \circ v_{n-1} \circ \ldots v_1 = w_m \circ w_{m-1} \circ \ldots w_1. \tag{5.10}$$

What is the relationship between the two decompositions? The theorem that describes this connection is known as Ritt's theorem.

Theorem 5.7. [Ritt's Theorem] *Let $u(x)$ be a polynomial in $\mathbf{Q}[x]$.*

1. *Any two complete decompositions of u contain the same number of components.*

2. *The degrees of the components in one complete decomposition are the same as the those in another complete decomposition, except possibly for the order in which they occur.*

The proof of Ritt's theorem utilizes some advanced mathematics and is well beyond the scope of our discussion.[1]

Ritt's theorem implies that the decomposition problem is not as clean as the polynomial factorization problem. The factorization of a polynomial

---

[1]For a proof of Ritt's theorem, see the references at the end of the chapter.

into irreducible factors is unique up to the order of the factors (Theorem 4.38, page 138). On the other hand, two complete decompositions of a polynomial can look quite different. For example,

$$x^6 + 2\,x^4 + x^2 + 1 \;=\; (x^3 + 2\,x^2 + x + 1) \circ (x^2) \qquad (5.11)$$
$$=\; (x^2 + 1) \circ (x^3 + x).$$

Notice that each of these complete decompositions contains two polynomials, one with degree two and one with degree three, but the degrees appear in opposite order.

We have seen that one way two decompositions of a polynomial can differ is through the application of linear polynomials and their inverses as is done in Equation (5.8). In fact, in some cases, it is possible to go from one decomposition to another by simply inserting linear polynomials and their inverses at appropriate places in the first decomposition (see Equation (5.9)). When this happens, the two decompositions are called *equivalent* decompositions. However, the situation can be more complex. For example, the two decompositions in Equation (5.11) are not equivalent in this sense because it is impossible to change the order of the degrees in a decomposition by simply inserting a linear polynomial and its inverse (Exercise 4).

The relationship between different decompositions of a polynomial can be described in more detail than we have done in this brief discussion. However, it is not necessary to understand these relationships to describe a decomposition algorithm, and because the algorithm is our primary goal, we will not pursue this matter further. The reader who is interested in pursuing these questions should consult the references at the end of the chapter.

Sufficiency of Rational Decompositions

We conclude this section with an important theoretical point. Suppose that u is a polynomial in $\mathbf{Q}[x]$ that is indecomposable if we require that the components are also in $\mathbf{Q}[x]$. Might it be possible to decompose u in terms of polynomials with coefficients in a larger number field? After all, a polynomial's irreducible factorization depends on the coefficient field. Could the same thing happen for decompositions? Fortunately, for polynomial decomposition the answer is no.

Theorem 5.8. *Suppose that u is a polynomial in $\mathbf{Q}[x]$ that has a decomposition $u = v \circ w$. Then, u has an equivalent decomposition in terms of polynomials with rational number coefficients.*

Proof: Suppose that $u = v \circ w$. By Theorem 5.6, there is an equivalent decomposition with $w(0) = 0$, so we assume that w satisfies this property. In a similar way, we can also find an equivalent decomposition so that w is a monic polynomial (Exercise 11(a)); therefore, we assume that w satisfies this property as well. We show that if u has rational coefficients and if w satisfies these two properties, then both w and v must have rational coefficients as well.

Let

$$u = u_m x^m + u_{m-1} x^{m-1} + \cdots + u_1 x + u_0,$$

$$v = v_n x^n + v_{n-1} x^{n-1} + \cdots + v_1 x + v_0, \quad n > 1,$$

$$w = x^p + w_{p-1} x^{p-1} + \cdots + w_1 x, \quad p > 1, \tag{5.12}$$

where the degrees of the polynomials satisfy the relation $m = np$. (Notice that w in Equation (5.12) is monic and $w(0) = 0$.) Then

$$\begin{aligned} u & = v \circ w \\ & = v_n \left(x^p + w_{p-1} x^{p-1} + \cdots + w_1 x \right)^n \\ & \quad + v_{n-1} \left(x^p + w_{p-1} x^{p-1} + \cdots + w_1 x \right)^{n-1} \\ & \quad + \cdots + v_1 \left(x^p + w_{p-1} x^{p-1} + \cdots + w_1 x \right) + v_0. \end{aligned} \tag{5.13}$$

In this representation, v_n is the coefficient of x^{np}, and since $np = m = \deg(u)$, we have

$$v_n = u_m \tag{5.14}$$

which shows that v_n is a rational number. In addition, the largest power of x that is contributed by the second term of the representation (5.13) is $x^{p(n-1)}$, and so all of the powers

$$x^{pn}, x^{pn-1}, \ldots, x^{p(n-1)+1}$$

come from the first term

$$v_n \left(x^p + w_{p-1} x^{p-1} + \cdots + w_1 x \right)^n. \tag{5.15}$$

By examining the form of the coefficients in the expression (5.15), we will show that all the coefficients of w are rational.

For example, let's look at the coefficient of x^{pn-1}. Expanding the expression (5.15) with the binomial theorem, we have

$$\begin{aligned} & v_n \left(x^p + \left(w_{p-1} x^{p-1} + \cdots + w_1 x \right) \right)^n \\ & = v_n \left((x^p)^n + n (x^p)^{n-1} \left(w_{p-1} x^{p-1} + w_{p-2} x^{p-2} + \cdots + w_1 x \right) \right. \\ & \quad \left. + \frac{n(n-1)}{2} (x^p)^{n-2} \left(w_{p-1} x^{p-1} + w_{p-2} x^{p-2} + \cdots + w_1 x \right)^2 + \cdots \right). \end{aligned} \tag{5.16}$$

Observe that the largest power of x that is contributed by the second term in Equation (5.16) is

$$x^{p(n-1)+(p-1)} = x^{pn-1},$$

and the largest power contributed by the third term is

$$x^{p(n-2)+2(p-1)} = x^{pn-2}.$$

From these observations, the coefficient of x^{pn-1} in Equation (5.16) is $v_n\, n\, w_{p-1}$, and since $np - 1 = m - 1$, we have from Equation (5.13)

$$u_{m-1} = v_n\, n\, w_{p-1}.$$

Therefore, since u_{m-1} and v_n are both rational, w_{p-1} is rational as well.

Let's consider next the coefficient of x^{pn-2} in Equation (5.16). This term is obtained with

$$v_n \left(n\, x^{p(n-1)}\, w_{p-2}\, x^{p-2} + \frac{n(n-1)}{2}\, x^{p(n-2)}\, w_{p-1}^2\, x^{2(p-1)} \right)$$

$$= v_n \left(n\, w_{p-2} + \frac{n(n-1)}{2}\, w_{p-1}^2 \right) x^{pn-2}.$$

Therefore, since $np - 2 = m - 2$, we have from Equation (5.13)

$$u_{m-2} = v_n \left(n\, w_{p-2} + \frac{n(n-1)}{2}\, w_{p-1}^2 \right),$$

and since u_{m-2}, v_n, and w_{p-1} are rational, w_{p-2} is also rational. Continuing in this fashion, we show that all the coefficients of w are rational.

We can also show that the coefficients of v are rational by comparing coefficients in Equation (5.13). We have shown above that v_n is rational (see Equation (5.14)), and to show that v_{n-1} is rational, consider the coefficients of $x^{p(n-1)}$. From Equation (5.13), we have

$$u_{p(n-1)} = v_n\, r + v_{n-1},$$

where r is obtained by combining coefficients of $x^{p(n-1)}$ from the term

$$(x^p + w_{p-1}\, x^{p-1} + \cdots + w_1\, x)^n.$$

Since, all the coefficients of w are rational, r is rational, and since $u_{p(n-1)}$ and v_n are rational, v_{n-1} is rational as well. In a similar way, by comparing coefficients of $x^{p(n-2)}$, we can show that v_{n-2} is rational. Continuing in this fashion, we can show that all of the coefficients of v are rational. $\square$

Exercises

1. For the polynomial in Equation (5.2), solve the equation $u(x) = 0$ with the help of polynomial decomposition.

2. Prove Theorem 5.1(3).

3. Explain why a polynomial of degree two or degree three cannot be decomposed.

4. Explain why the order of the degrees in a decomposition of a polynomial $u(x)$ cannot be changed by the insertion of a linear polynomial and its inverse between two components.

5. Is every indecomposable polynomial irreducible? Is every irreducible polynomial indecomposable?

6. Find six distinct decompositions of $u(x) = x^{30}$ such that the powers of the components appear in different orders.

7. Find two different decompositions for the polynomial $u = x^4 + 2x^3 + x^2$. *Hint:* there is a decomposition with one component x^2. Do not use computer algebra software for this problem.

8. The Chebyshev polynomial $T_n(x)$ of degree n is defined (for $-1 \leq x \leq 1$) by the trigonometric relations

$$T_n(x) = \cos(n\,\theta), \qquad x = \cos(\theta).$$

For example, $T_0(x) = \cos(0) = 1$ and $T_1(x) = \cos(\theta) = x$.

   (a) Show that $T_2(x) = 2x^2 - 1$.

   (b) Derive the recurrence relation

$$T_{n+1}(x) = 2x\,T_n(x) - T_{n-1}(x).$$

   Hint: Use the identity $\cos(a + b) + \cos(a - b) = 2\cos(a)\cos(b)$.

   (c) Show that $T_{mn} = T_m \circ T_n$.

   (d) Show that $T_n \circ T_m = T_m \circ T_n$. In other words, the polynomials T_n and T_m commute under composition.

9. Consider the two monic polynomials $u = x^2 + bx + c$ and $v = x^2 + ex + f$. Show that if $u \circ v = v \circ u$, then $u = v$. (This implies that two distinct quadratic polynomials cannot commute with respect to composition.)

10. Consider the following four polynomials:

$$u_1 = 2x^2 + 3x + 4, \qquad v_1 = x^2 - 5,$$

$$u_2 = 8x^2 + 30x + 31, \qquad v_2 = \frac{x^2}{2} - 4.$$

   (a) Show that $u_1 \circ v_1 = u_2 \circ v_2$.

(b) Find polynomials $f = m\,x + b$ and $g = x/m - b/m$ such that $u_1 \circ f = u_2$ and $g \circ v_1 = v_2$.

11. Suppose that u has a complete decomposition

$$u = g_p \circ g_{p-1} \circ \cdots \circ g_1.$$

(a) Show that it is possible to modify the decomposition so that all of the components except g_p are monic.

(b) Show that it is possible to modify the decomposition so that the only component that has a constant term is g_p.

12. In this exercise we describe a collection of polynomials that illustrates the second statement in Ritt's Theorem 5.7 (page 183). Let $g(x)$ be a non-zero polynomial, and consider the polynomials

$$u(x) = x^n, \qquad v(x) = x^r\, g(x^n), \qquad w(x) = x^r\, g(x)^n.$$

(a) Show that $\deg(v, x) = \deg(w, x) = r + \deg(g, x)\, n$.

(b) Show that $u \circ v = w \circ u$.

(c) Show that, if n and $r + \deg(g, x)\, n$ are prime numbers, then u, v, and w are indecomposable.

This implies, if n and $r + \deg(g, x)\, n$ are distinct as well as prime, that $u \circ v$ and $v \circ w$ are two distinct complete decompositions of the same polynomial with degrees that occur in opposite order. For example, for $n = 3$, $r = 2$, and $g(x) = x^3 - 1$, we have $v = x^{11} - x^2$, $w = x^{11} - 3\,x^8 + 3\,x^5 - x^2$, and $u \circ v = w \circ u = x^{33} - 3\,x^{24} + 3\,x^{15} - x^6$.

13. The polynomial

$$u = x^4 - 10\,x^3 + 27\,x^2 - 10\,x + 3$$

can be decomposed as

$$u = \left(2\,x^2 + \left(\frac{2\sqrt{6}}{3} + 2\sqrt{2}\right) x + \frac{10}{3} + \frac{2\sqrt{3}}{3}\right) \circ \left(\frac{\sqrt{2}}{2}\,x^2 - \frac{5\sqrt{2}}{2}\,x - \frac{\sqrt{6}}{6}\right).$$

Find an equivalent decomposition that involves polynomials with rational coefficients.

5.2   A Decomposition Algorithm

In this section we describe an algorithm that finds the complete decomposition of a polynomial $u(x)$ (if it exists) or determines that no such decomposition is possible. We begin, however, by looking at the simpler problem of finding a two-level decomposition $u = v \circ w$. In some cases, it will be possible to decompose v or w.

Let's assume that $u = v \circ w$, and let

$$u(x) = u_m x^m + u_{m-1} x^{m-1} + \cdots + u_1 x + u_0,$$

$$v(x) = v_n x^n + v_{n-1} x^{n-1} + \cdots + v_1 x + v_0.$$

The algorithm is based on the properties of a decomposition that are described in Theorems 5.9 and 5.10.

Theorem 5.9. *Let u be a polynomial in $\mathbf{Q}[x]$ that can be decomposed. Then, there is a decomposition $u = v \circ w$ such that w satisfies the following properties.*

1. $1 < \deg(w, x) < \deg(u, x)$.

2. $\deg(w, x) \mid \deg(u, x)$.

3. $w(0) = 0$.

4. $w(x) \mid (u(x) - u_0)$.

Proof: Property (1), which follows from Theorem 5.1(3), is just a restatement of the property that both v and w must have degree greater than 1. Property (2) also follows from Theorem 5.1(3). Property (3) is a restatement of Theorem 5.6.

To show property (4), we have, using property (3),

$$u_0 = u(0) = v(w(0)) = v(0) = v_0.$$

Therefore,

$$\begin{aligned} u(x) - u_0 &= u(x) - v_0 \\ &= v_n w(x)^n + v_{n-1} w(x)^{n-1} + \cdots + v_1 w(x), \end{aligned}$$

and since $w(x)$ divides each term in the sum, it also divides $u(x) - u_0$. $\square$

The property $w(0) = 0$ is a sufficient condition to guarantee that

$$w(x) \mid (u(x) - u_0)$$

but it is not a necessary condition. There can be other divisors of $u(x) - u_0$ that don't satisfy this condition but lead to a decomposition (see Example 5.11).

Although we don't know which divisors of $u(x) - u_0$ lead to a decomposition, we do know that there is a decomposition $v(x) \circ w(x)$ for which

$w(x)$ is one of these divisors. The next theorem gives a way to compute $v(x)$ and provides a test that determines whether or not some trial divisor $w(x)$ of $u(x) - u_0$ leads to a decomposition.

Theorem 5.10. *Suppose that a polynomial u in $\mathbf{Q}[x]$ can be decomposed as $u = v \circ w$. Then v is given by the polynomial expansion of u in terms of w, and this expansion has coefficients that are free of x.*

Proof: Since $u = v \circ w$,

$$u(x) = v_n\, w(x)^n + v_{n-1}\, w(x)^{n-1} + \cdots + v_1\, w(x) + v_0 \qquad (5.17)$$

where the coefficients v_i are rational numbers. But the representation (5.17) is just the (unique) polynomial expansion u in terms of w (see Theorem 4.13 on page 118). In general, the coefficients in a polynomial expansion can be polynomials in x, but, in this case, the v_i are rational numbers and therefore free of x. $\qquad\qquad\square$

The two previous theorems suggest the following approach to find a decomposition for $u(x)$. First, find the set of distinct polynomial divisors of $u(x) - u_0$. These divisors are found by first finding all irreducible (non-constant) factors of $u(x) - u_0$ and then forming all possible products of these factors. In this context, multiplication of a divisor w by a rational number c is not another distinct divisor, and so, it is sufficient to consider only monic divisors. For example, the set of distinct polynomial divisors of $3\,(x-1)\,(x-2)^2$ is (in factored form)

$$S = \left\{ x-1,\ \ x-2,\ \ (x-2)^2,\ \ (x-1)\,(x-2),\ \ (x-1)\,(x-2)^2 \right\}.$$

Each divisor that satisfies properties (1), (2), and (4) of Theorem 5.9 is a potential candidate for $w(x)$.

Next, for each such candidate $w(x)$ in S, form the polynomial expansion (5.17). If all the coefficients v_i are free of x, then v and w give a decomposition of u. On the other hand, if none of the divisors in S leads to an expansion with this property, then u is indecomposable.

Example 5.11. Consider the polynomial $u(x) = x^6 + 6\,x^4 + 3\,x^3 + 9\,x^2 + 9\,x + 5$. Since $u(x) - 5$ can be factored as

$$u(x) - 5 = x\,(x^2 + 3)\,(x^3 + 3x + 3),$$

the divisor set is

$$\begin{aligned} S \ =\ & \left\{ x,\ x^2 + 3,\ x(x^2 + 3),\ x^3 + 3\,x + 3,\ x\,(x^3 + 3\,x + 3), \right. \\ & \left. (x^2 + 3)\,(x^3 + 3\,x + 3),\ x\,(x^2 + 3)\,(x^3 + 3\,x + 3) \right\}. \qquad (5.18) \end{aligned}$$

However, the only polynomial that satisfies all the properties of Theorem 5.9 is $w(x) = x^3 + 3\,x$. To see if this expression gives a decomposition, we find the polynomial expansion[2] of u in terms of w:

$$v(t) = Polynomial\_expansion(u,\ w,\ x,\ t) \to t^2 + 3\,t + 5.$$

Since all of the coefficients of $v(t)$ are free of x, we have the decomposition

$$u = v(x) \circ w(x) = (x^2 + 3\,x + 5) \circ (x^3 + 3\,x).$$

There are two other divisors in (5.18) that do not satisfy Theorem 5.9 (3) but do satisfy Parts (1), (2), and (4):

$$x^2 + 3, \quad x^3 + 3\,x + 3.$$

For these cases, the polynomial expansions are

$$w(x) = x^3 + 3\,x + 3, \quad v(t) = t^2 - 3\,t + 5, \qquad (5.19)$$
$$w(x) = x^2 + 3, \quad v(t) = t^3 - 3\,t^2 + 3\,x\,t + 5. \qquad (5.20)$$

Since $v(t)$ in (5.19) is free of x, we have another decomposition

$$u = v(x) \circ w(x) = (x^2 - 3\,x + 5) \circ (x^3 + 3\,x + 3)$$

even though $w(0) \neq 0$. On the other hand, since $v(t)$ in (5.20) is not free of x, we do not obtain an expansion with $w(x) = x^2 + 3$.   □

Example 5.12. Consider the polynomial $u(x) = x^4 + x^3 + 3$. Since

$$u(x) - 3 = x^3\,(x + 1),$$

the divisor set is

$$S = \left\{ x,\ x + 1,\ x^2,\ x^2 + x,\ x^3,\ x^3 + x^2,\ x^4 + x^3 \right\}.$$

The only divisors that satisfy all the properties of Theorem 5.9 are $w_1(x) = x^2 + x$ and $w_2(x) = x^2$. The polynomial expansion of u in terms of $w_1(x)$ is

$$3 - x\,t + t^2$$

and the expansion in terms of $w_2(x)$ is

$$3 + x\,t + t^2.$$

---

[2]The *Polynomial_expansion* operator is given in Figure 4.2 on page 122.

Procedure *Poly_decomp_2*(*u*, *x*);
Input
    u : a polynomial in **Q**[*x*];
    x : a symbol;
Output
    Either a decomposition [*v*, *w*] or *u* if no decomposition exists;
Local Variables *U*, *S*, *i*, *w*, *v*;
Global *t*;
Begin
1   *U* := *u* − *Coefficient_gpe*(*u*, *x*, 0);
2   *S* := *Polynomial_divisors*(*U*, *x*);
3   *i* := 1;
4   **while** *i* ≤ *Number_of_operands*(*S*) **do**
5   *w* := *Operand*(*S*, *i*);
6   **if** *Degree_gpe*(*w*, *x*) ≠ *Degree_gpe*(*u*, *x*) **and** *Degree_gpe*(*w*, *x*) > 1
            and *Irem*(*Degree_gpe*(*u*, *x*), *Degree_gpe*(*w*, *x*)) = 0 **then**
7   *v* := *Polynomial_expansion*(*u*, *w*, *x*, *t*);
8   **if** *Free_of*(*v*, *x*) **then**
9   *Return*[(*Substitute*(*v*, *t* = *x*), *w*]);
10   *i* := *i* + 1;
11   *Return*(*u*)
End

Figure 5.1. An MPL procedure for a two-level polynomial decomposition. (Implementation: Maple (txt), Mathematica (txt), MuPAD (txt).)

In both cases, the expansion has a coefficient that depends on *x*, and so, *u*(*x*) cannot be decomposed in terms of these polynomials. Therefore, *u*(*x*) cannot be decomposed. □

A procedure for the decomposition process is shown in Figure 5.1. If *u* is decomposable, the procedure returns a list [*v*, *w*] that contains the polynomials that make up a two-level decomposition, and if *u* is not decomposable, then the procedure returns *u*. The global symbol *t* is a mathematical symbol that is used in line 7 as an auxiliary variable in the operator *Polynomial_expansion*. We have used a global symbol to avoid using a local variable that would not be assigned before it was used (see page 3).
At line 2, the operator *Polynomial_divisors* is used to obtain the divisors[3] of *u* − u_0. At line 7, the operator *Polynomial_expansion* returns a polynomial in terms of the fourth input parameter *t* with coefficients that

---

[3]The *Polynomial_divisors* procedure is described in Exercise 12 on page 125. The factorization of *u* − u_0 is obtained by the factor operator of a CAS in this procedure.

may depend on x. We check for this possibility by applying the *Free_of* operator to the entire polynomial (line 8). If the **while** loop (lines 4–10) does not find a decomposition, the polynomial u cannot be decomposed. In this case, the algorithm simply returns u (line 11).

Notice that the algorithm does not check the condition $w(0) = 0$ and may return a decomposition without this property.

The Complete Decomposition Algorithm

Next, we describe an algorithm that finds a complete decomposition for u. The algorithm either finds a sequence of indecomposable polynomials $g_1, g_2, \ldots, g_p$ (with $p > 1$) such that

$$u = g_p \circ g_{p-1} \circ \cdots \circ g_1$$

or determines that u is indecomposable. The general idea is to build up the decomposition, one component at a time, taking care not to miss any intermediate components. We show below that it is possible to arrange things so that each of the partial compositions

$$g_i \circ g_{i-1} \circ \cdots \circ g_1, \quad 1 \le i < p,$$

is a divisor of $u - u_0$ where the components g_i are obtained with polynomial expansions (Theorems 5.13 and 5.14). We insure that each g_i is indecomposable by applying a minimal degree condition to the divisors of $u - u_0$ (Theorem 5.15).

Let's see how to find the component g_i (for $i < p$) assuming that the components $g_1, g_2, \ldots, g_{i-1}$ have already been found. (The last component g_p is computed in a different way.) To simplify the notation, let

$$
\begin{aligned}
u &= g_p \circ g_{p-1} \circ \cdots \circ g_{i+1} \circ g_i \circ g_{i-1} \circ \cdots \circ g_1 \\
&= R \circ g_i \circ C,
\end{aligned}
$$

where

$$C = g_{i-1} \circ \cdots \circ g_1 \tag{5.21}$$

is the decomposition so far and

$$R = g_p \circ g_{p-1} \circ \cdots \circ g_{i+1} \tag{5.22}$$

is the decomposition that remains to be found after g_i has been found. (For $i = 1$ we initialize C to x.) Since $i < p$, we can assume that $\deg(R, x) > 0$.

Theorem 5.13. *Let u be a polynomial in $\mathbf{Q}[x]$ that can be decomposed as*

$$u = R \circ g_i \circ C.$$

Then, for $i < p$, it is possible to choose g_i with the following properties:

1. $\deg(C, x) < \deg(g_i \circ C, x) < \deg(u, x)$.

2. $\deg(g_i \circ C, x) \mid \deg(u, x)$.

3. $(g_i \circ C)(0) = 0$.

4. $(g_i \circ C)(x) \mid (u(x) - u_0)$.

Proof: The proof of this theorem is similar to the proof for Theorem 5.9 (Exercise 5). □

The theorem implies that we can find $g_i \circ C$ by examining the divisors of $u(x) - u_0$. Notice in this case that we do not get g_i directly as some divisor of $u - u_0$, but instead obtain the composite expression $g_i \circ C$. This is consistent with what was done in the two-level algorithm because when computing g_1, C is initialized to x and we get g_1 directly.

The next theorem gives a way to compute g_i from $g_i \circ C$ and provides a way to tell if some trial divisor of $u - u_0$ contributes to the decomposition.

Theorem 5.14. *Suppose that u in $\mathbf{Q}[x]$ can be decomposed as $u = R \circ g_i \circ C$.*

1. *The polynomial R is found by computing the polynomial expansion of u in terms of $g_i \circ C$, and this expansion has coefficients that are free of x.*

2. *The polynomial g_i is found by computing the polynomial expansion of $g_i \circ C$ in terms of C, and this expansion has coefficients that are free of x.*

Proof: To verify (1), let

$$R = r_s\, x^s + r_{s-1}\, x^{s-1} + \cdots + r_0.$$

Observe that

$$u = R \circ g_i \circ C = r_s\, (g_i(C))^s + r_{s-1}\, (g_i(C))^{s-1} + \cdots + r_0,$$

which is just the unique polynomial expansion of u in terms of $g_i(C)$. Since the coefficients $r_s, r_{s-1}, \ldots, r_0$ are rational numbers, they are free of x. Although statement (1) is similar to Theorem 5.10 in the two-level case, we shall see that R only becomes part of the decomposition when finding the final component g_p. For the other components g_i, $i < p$, we need only check that the coefficients of R are free of x.

To verify (2), let

$$g_i(x) = a_p\, x^p + a_{p-1}\, x^{p-1} + \cdots + a_1\, x + a_0.$$

Observe that

$$(g_i \circ C)(x) = a_p\, (C(x))^p + a_{p-1}\, (C(x))^{p-1} + \cdots + a_1\, C(x) + a_0, \quad (5.23)$$

which is just the unique polynomial expansion of $g_i \circ C$ in terms of C. In order for g_i to be a component, its coefficients must be rational numbers and hence free of x. □

The next theorem provides a way to choose $g_i \circ C$ so that g_i is indecomposable.

Theorem 5.15. *Let $u = R \circ g_i \circ C$, and suppose that $g_i \circ C$ is the divisor of $u(x) - u_0$ of minimal degree that satisfies the conditions in Theorem 5.13 and Theorem 5.14. Then, g_i is indecomposable.*

Proof: If g_i were decomposable, then we can show that $g_i \circ C$ does not satisfy the minimal degree condition. To see why, suppose that g_i is decomposable as

$$g_i = v \circ w$$

with $\deg(v, x) > 1$ and $\deg(w, x) > 1$. (This implies that $\deg(w, x) < \deg(g_i, x)$.) Then

$$\begin{aligned} u &= R \circ g_i \circ C = R \circ v \circ w \circ C \\ &= (R \circ v) \circ (w \circ C). \end{aligned}$$

By an argument similar to the one in the proof of Theorem 5.9, we can assume that $(w \circ C)(0) = 0$, and therefore $w \circ C$ is a divisor of $u - u_0$. Since

$$\begin{aligned} \deg(w \circ C,\ x) &= \deg(w, x)\deg(C, x) \\ &< \deg(g_i, x)\deg(C, x) \\ &= \deg(g_i \circ C,\ x), \end{aligned}$$

$g_i \circ C$ does not satisfy the minimal degree condition. Therefore g_i must be indecomposable. □

The above theorems suggest the following approach for finding a complete decomposition of $u(x)$. First, find all the divisors of $u - u_0$. To find g_1, initialize C to x, and find a divisor w of $u - u_0$ of smallest degree that

satisfies the conditions in Theorems 5.13 and 5.14. Since $C = x$, we immediately have $g_1 = w$ as was done in the two-level case.[4] If no such divisor
w exists, u is indecomposable.

To compute g_i $(1 < i < p)$, let $C = g_{i-1} \circ \cdots \circ g_1$ be the decomposition
so far, and let w be a divisor of $u - u_0$ of minimal degree that satisfies the
conditions in Theorems 5.13 and 5.14. In this case, we find g_i using the
polynomial expansion of w in terms of C as is done in Equation (5.23).
If no such divisor w exists, then C includes the first $p - 1$ components in
the decomposition and we need only compute the last component g_p. This
component can be found by computing the polynomial expansion of u in
terms of C. It is not difficult to show that g_p computed in this way is
indecomposable (Exercise 6).

Example 5.16. Consider the polynomial

$$u := x^8 + 4\,x^7 + 6\,x^6 + 4x^5 + 3x^4 + 4x^3 + 2\,x^2 + 1.$$

The polynomial $u - 1$ can be factored as

$$u - 1 = x^2 \,(x^4 + 2\,x^3 + x^2 + 2)\,(x + 1)^2.$$

Although there are 17 divisors of $u - 1$, the degree requirements of Theorem 5.13 eliminate all but five of them:

$$x^2, \quad x^2 + x, \quad x^2 + 2x + 1, \quad x^4 + 2x^3 + x^2, \quad x^4 + 2x^3 + x^2 + 2. \quad (5.24)$$

To compute g_1, let $C = x$, and first consider the three divisors of degree
two in (5.24). There are three possibilities for $g_1 \circ C$:

$$g_1 \circ C \;=\; x^2, \quad R = t^4 + (4\,x + 6)\,t^3 + (4\,x + 3)\,t^2$$
$$+\,2t + 1, \qquad\qquad\qquad (5.25)$$
$$g_1 \circ C \;=\; x^2 + x, \quad R = t^4 + 2\,t^2 + 1, \qquad\qquad (5.26)$$
$$g_1 \circ C \;=\; x^2 + 2\,x + 1, \quad R = t^4 + (2 - 4\,x)\,t^3 + (-1 - 4\,x)\,t^2$$
$$+\,(-2 - 4\,x)\,t + 1. \qquad (5.27)$$

Again, we have expressed the polynomial expansions for R in terms of
a variable t to clearly distinguish the coefficients that may depend on x.
(When $C = x$, we have $g_1 \circ C = g_1$, and so we have not computed the
polynomial expansion of $g_1 \circ C$ in terms of C.) Since R in (5.26) is the only
one with coefficients that are free of x, we choose

$$g_1 \circ C = x^2 + x,$$

---

[4]For the sake of consistency, the algorithm finds g_1 using polynomial expansion even
though it can be found directly.

and since $C = x$,

$$g_1 = x^2 + x.$$

Now to choose g_2, let $C = g_1$ and observe there are two divisors with degree four in (5.24), each of which is a candidate for $g_2 \circ C$. Computing the polynomial expansions in Theorem 5.14 for each of them, we obtain

$$g_2 \circ C = x^4 + 2x^3 + x^2, \quad g_2 = t^2, \tag{5.28}$$
$$R = t^2 + 2t + 1, \tag{5.29}$$

or

$$g_2 \circ C = x^4 + 2x^3 + x^2 + 2, \quad g_2 = t^2 + 2, \tag{5.30}$$
$$R = t^2 - 2t + 1. \tag{5.31}$$

In both cases, both g_2 and R are free of x and so either divisor will contribute to the decomposition. Using Equation (5.28) (with x substituted for t), we have $g_2 = x^2$. Since there are no more divisors to consider, we obtain g_3 by substituting x for t in R in Equation (5.29). Therefore, a complete decomposition is

$$u = (x^2 + 2x + 1) \circ (x^2) \circ (x^2 + x). \qquad \square$$

A procedure that obtains the complete decomposition of a polynomial in $\mathbf{Q}[x]$ is shown in Figure 5.2. The procedure returns either a list $[g_p, g_{p-1}, \ldots, g_1]$ that represents the decomposition or u if the polynomial is indecomposable. The **while** loop (lines 5–14) examines the divisors of $U = u - u_0$ and tries to choose an appropriate one to build up the decomposition. At line 6, the *Find_min_deg* operator (Exercise 4) chooses a trial divisor w of minimal degree from the set S, and the statement at line 8 checks the degree requirements mentioned in Theorem 5.13. If w passes these tests, the polynomial expansions referred to in Theorem 5.14 are computed (lines 9–10), and the conditions mentioned in this theorem are checked in line 11. (Recall that the operator *Polynomial_expansion* returns a polynomial in terms of the fourth input parameter t with coefficients that may depend on x.) If both expansions are free of x, we have found the next component which is adjoined to the list *decomposition* (line 12). At line 13, C is assigned the value w in preparation for searching for the next component.

At line 14, the polynomial expansion R is assigned to *final_component*. At this point, R has been used only to check the validity of w for computing the next component (line 11). However, when the next to last component (g_{p-1}) is computed, the current value of R is the polynomial expansion

Procedure $Complete\_poly\_decomp(u, x)$;
Input
    u : a polynomial in $\mathbf{Q}[x]$;
    x : a symbol;
Output
    A decomposition $[g_p, g_{p-1}, \ldots, g_1]$ or u if no decomposition exists;
Local Variables
    $U, S, decomposition, C, w, g, R, final\_component$;
Global t;
Begin

```
1      U := u − Coefficient_gpe(u, x, 0);
2      S := Polynomial_divisors(U, x);
3      decomposition := [ ];
4      C := x;
5      while  S ≠ ∅ do
6          w := Find_min_deg(S, x);
7          S := S − {w};
8          if  Degree_gpe(C, x) < Degree_gpe(w, x)
               and Degree_gpe(w, x) < Degree_gpe(u, x)
               and Irem(Degree_gpe(u, x), Degree_gpe(w, x)) = 0 then
9              g := Polynomial_expansion(w, C, x, t);
10             R := Polynomial_expansion(u, w, x, t);
11             if  Free_of(g, x) and Free_of(R, x) then
12                 decomposition := Adjoin(Substitute(g, t = x),  decomposition);
13                 C := w;
14                 final_component := R;
15     if  decomposition = [ ] then
16         Return(u)
17     else
18         Return(Adjoin(Substitute(final_component, t = x),  decomposition))
   End
```

Figure 5.2. An MPL procedure for complete polynomial decomposition. (Implementation: Maple (txt), Mathematica (txt), MuPAD (txt).)

of u in terms of $g_{p-1} \circ \cdots \circ g_1$ which is the last component g_p in the decomposition.

If we go through the entire loop without finding any components, the polynomial is indecomposable and we return the input polynomial (line 16). Otherwise, we add the *final_component* to the *decomposition* list (line 18).

Notice that nowhere in the algorithm do we use the condition $w(0) = 0$ (property (3) in Theorem 5.13). In fact, no harm is done by choosing a divisor that does not satisfy this condition as long as all the other conditions in Theorems 5.13 and 5.14 are satisfied. The condition $w(0) = 0$ in Theorem 5.13 simply guarantees that we can limit our search to the divisors of $u - u_0$. If a divisor in S without this property is used, we simply get another decomposition.

Exercises

1. Find a decomposition for the polynomial $u(x) = x^4 + 6x^3 + 14x^2 + 15x + 9$.

2. Find a decomposition for the polynomial $u(x) = x^8 + 8x^7 + 48x^6 + 176x^5 + 503x^4 + 988x^3 + 1456x^2 + 1320x + 749$.

3. Show that $u(x) = x^4 + x + 1$ cannot be decomposed.

4. Let S be a set of polynomials in $\mathbf{Q}[x]$. Give a procedure $Find\_min\_deg(S, x)$ that returns a polynomial of smallest degree in S. If $S = \emptyset$, return the global symbol **Undefined**.

5. Prove Theorem 5.13.

6. Show that final component g_p selected by the algorithm in Figure 5.2 is indecomposable.

7. Let $u(x) = (x^2 + 1) \circ (x^2 + 1) \circ (x^2 + 1)$. What decomposition is produced for this polynomial by the algorithm in Figure 5.2? Explain why the above decomposition cannot be found by the algorithm.

8. Another approach that finds the complete decomposition of a polynomial $u(x)$ is to modify the two-level procedure (Figure 5.1) so that it attempts to further decompose v and w by recursively calling itself with each of these polynomials as input. What are the advantages and disadvantages of this approach compared with the approach described in the text?

Further Reading

5.1 Theoretical Background. Barbeau [4] discusses many interesting questions related to composition of polynomials. Theorem 5.7 was first proved by J. F. Ritt in [85]. This article also describes in detail the relationship between two different decompositions of a polynomial. Other proofs of Theorem 5.7 are given in Dorey and Whaples [33] and Engstrom [34].

5.2 A Decomposition Algorithm. The algorithms given in this section are similar to those given in Barton and Zippel [5]. Another algorithm that does not require polynomial factorization is given in Kozen and Landau [59].

6

Multivariate Polynomials

Most symbolic computation involves mathematical expressions that contain more than one symbol or generalized variable. In order to manipulate these expressions, we must extend the concepts and algorithms described in Chapter 4 to polynomials with several variables. This generalization is the subject of this chapter. It includes a description of coefficient domains and a recursive view of multivariate polynomials (Section 6.1), versions of polynomial division and expansion (Section 6.2), greatest common divisor algorithms (Section 6.3), and the extended Euclidean algorithm (Section 6.3).

6.1 Multivariate Polynomials and Integral Domains

A *multivariate polynomial* u in the set of distinct symbols $\{x_1, x_2, \ldots, x_p\}$ is a finite sum with (one or more) monomial terms of the form

$$c\, x_1^{n_1}\, x_2^{n_2} \cdots x_p^{n_p}$$

where the coefficient c is in a coefficient domain $\mathbf{K}$ and the exponents n_j are non-negative integers. (The axioms for $\mathbf{K}$ (which may not be a field) are given in Definition 6.2 below.) The notation $\mathbf{K}[x_1, x_2, \ldots, x_p]$ represents the set of polynomials in the symbols $x_1, x_2, \ldots, x_p$ with coefficients in $\mathbf{K}$. For example, $\mathbf{Z}[x, y]$ represents all polynomials in x and y with coefficients that are integers.

A particularly important instance when $\mathbf{K}$ is not a field has to do with the recursive representation of multivariate polynomials. For example, let

$\mathbf{Q}[x, y]$ be the polynomials in x and y with rational number coefficients. By collecting coefficients of powers of x, a polynomial $u(x, y)$ is represented as

$$u(x, y) = u_m(y)x^m + u_{m-1}(y)x^{m-1} + \cdots + u_0(y),$$

where the coefficients $u_i(y)$ are in $\mathbf{Q}[y]$. In this sense, $\mathbf{Q}[x, y]$ is equivalent to $\mathbf{K}[x]$ where the coefficient domain $\mathbf{K} = \mathbf{Q}[y]$. In general, we can view $\mathbf{K}[x_1, x_2, \ldots, x_p]$ recursively as

$$\mathbf{K}[x_1, x_2, \ldots, x_p] = \mathbf{K}_1[x_1], \tag{6.1}$$

where

$$\mathbf{K}_i = \mathbf{K}_{i+1}[x_{i+1}], \quad 1 \leq i \leq p-1, \quad \text{and} \quad \mathbf{K}_p = \mathbf{K}.$$

When multivariate polynomials are represented in this way, x_1 is called the *main* variable and the remaining variables are called *auxiliary* variables.

Example 6.1. Consider the polynomial domain $\mathbf{Q}[x, y, z]$. Then,

$$p = 3, \quad x_1 = x, \quad x_2 = y, \quad x_3 = z,$$

and

$$\begin{aligned} \mathbf{K}_3 &= \mathbf{Q}, \\ \mathbf{K}_2 &= \mathbf{K}_3[x_3] = \mathbf{Q}[z], \\ \mathbf{K}_1 &= \mathbf{K}_2[x_2] = (\mathbf{Q}[z])[y]. \end{aligned}$$

Therefore,

$$\mathbf{Q}[x, y, z] = \mathbf{K}_1[x] = (\mathbf{K}_2[y])[x] = ((\mathbf{K}_3[z])[y])[x] = ((\mathbf{Q}[z])[y])[x]. \quad \square$$

Integral Domains

The above discussion suggests that a fruitful way to approach the study of polynomials in several variables is to consider polynomials in one variable $\mathbf{K}[x]$ where $\mathbf{K}$ may not satisfy all of the properties of a field. But what axioms should define $\mathbf{K}$?

Let's take as a prototype the recursive view of polynomials in two variables with rational number coefficients $(\mathbf{Q}[y])[x]$. The coefficient domain is not a field since polynomials in $\mathbf{K} = \mathbf{Q}[y]$ with positive degree do not have inverses that are polynomials. $\mathbf{K}$ does, however, have all the properties of another algebraic system known as an integral domain.

Definition 6.2. *Let $\mathbf{K} = \{a, b, c, \ldots\}$ be a set of expressions and let $a + b$ and $a \cdot b$ be two operations defined for expressions a and b in $\mathbf{K}$. The set $\mathbf{K}$ is an **integral domain** if it satisfies the field axioms **F-1**, **F-2**,..., **F-11**, together with the axiom:*

I-1. *If $a \cdot b = 0$, then either $a = 0$ or $b = 0$.*

Axiom I-1 replaces the field axiom F-12 that is concerned with multiplicative inverses. However, since a field also satisfies axiom I-1 (Theorem 2.36, page 50), a field is also an integral domain.

The converse, of course, is not true. Both the integers **Z** and the polynomial domain **Q**[x] satisfy all the properties of an integral domain (see Example 2.33 on page 48 and Theorem 4.2 on page 112) although neither one is a field. By permitting the coefficients to come from a more general algebraic system, we extend the applicability of the polynomial concepts and algorithms described in Chapter 4 but do so, unfortunately, at the expense of a more cumbersome development.

Divisors in K

Although an integral domain may not contain inverses, the divisibility concept can still be defined.

Definition 6.3. *Let b and c be expressions in an integral domain* **K**.

1. *An expression* $b \neq 0$ *is a* **divisor** *of (or* **divides***) c if there is an expression d in* **K** *such that* $b \cdot d = c$. *We use the notation* $b \mid c$ *to indicate that b is a divisor of c and* $b \nmid c$ *if it does not. The expression d is called the* **cofactor** *of b in c and is represented by* $\mathrm{cof}(b, c)$.

2. *A* **common divisor** *of b and c is an expression* $d \neq 0$ *in* **K** *such that* $d \mid b$ *and* $d \mid c$.

Axiom I-1 states that an integral domain has no *zero divisors*. The axiom is equivalent to the following cancellation law.

Theorem 6.4. *Let* $a \neq 0$, *b, and c be expressions in an integral domain* **K**. *If* $a \cdot b = a \cdot c$, *then* $b = c$.

The proof is left to the reader (Exercise 1). □

Units and Associates

Definition 6.5. *Let b, c, and d be expressions in an integral domain* **K**.

1. *An expression b is called a* **unit** *if it has a multiplicative inverse.*

2. *Two expressions* $c \neq 0$ *and* $d \neq 0$ *are called* **associates** *if* $c = b \cdot d$, *where b is a unit.*

In $\mathbf{Q}[x]$, the unit expressions are the non-zero polynomials with zero degree (the rational numbers), and whenever one polynomial u is a (non-zero) rational multiple of another polynomial v, the two are associates. In $\mathbf{Z}$, the only units are 1 and -1. At the other extreme, in a field all non-zero expressions are units, and any two non-zero expressions are associates.

Definition 6.6. *Two expressions b and c in an integral domain $\mathbf{K}$ are* **relatively prime** *if any common divisors of b and c is a unit.*

For example, in $\mathbf{Z}$ the integers 3 and 5 are relatively prime because the only common divisors are the units 1 and -1.

In the field $\mathbf{Q}$, any two rational numbers are relatively prime because all common divisors are units.

On the other hand, in the integral domain $\mathbf{Z}[x]$, the polynomials $2x + 2$ and $2x - 2$ are not relatively prime because 2, which is not a unit, divides both polynomials.

Definition 6.7. *An expression $b \neq 0$ in an integral domain $\mathbf{K}$ is* **reducible** *if there are non-unit expressions c and d such that $b = c \cdot d$. The expression b is* **irreducible** *if it is not reducible.*

For example, in $\mathbf{Z}$ the irreducible expressions are ± 1 and $\pm p$ where p is prime number. In the field $\mathbf{Q}$, all $b \neq 0$ are irreducible because all non-zero expressions are units.

Example 6.8. Consider the expression $u = 2x + 2$ as a member of the integral domain $\mathbf{Q}[x]$. In this context, u is irreducible. However, when u is viewed as a member of the integral domain $\mathbf{Z}[x]$, u is reducible with the factorization $u = 2(x + 1)$. $\qquad\square$

Unique Factorization Domains

Our prototype coefficient domain $\mathbf{Q}[y]$ has an additional property, namely, any polynomial u in $\mathbf{Q}[y]$ with positive degree has a unique factorization in terms of irreducible polynomials (Theorem 4.38, page 138). We postulate a similar property for our coefficient domains.

Definition 6.9. *A* **unique factorization domain** $\mathbf{K}$ *is an integral domain that satisfies the following axiom.*

UFD-1. *Each $a \neq 0$ in $\mathbf{K}$ that is not a unit has a factorization in terms of non-unit, irreducible expressions in $\mathbf{K}$*

$$a = a_1 \cdot a_2 \cdots a_r.$$

The factorization is unique up to the order of the factors and associates of the factors.

The set of integers $\mathbf{Z}$ is a unique factorization domain (Theorem 2.17, page 27).

The polynomial domain $\mathbf{F}[x]$ ($\mathbf{F}$ a field) is also a unique factorization domain. By Theorem 4.38, each u in $\mathbf{F}[x]$ with positive degree has a unique factorization of the form

$$u = c\, p_1\, p_2 \cdots p_r,$$

where c is in $\mathbf{F}$ and each p_i a is a monic, irreducible polynomial in $\mathbf{F}[x]$ with $\deg(p_i) > 0$. The factorization in Definition 6.9 is obtained from this factorization by absorbing the coefficient c in one of the expressions p_i.

When $\mathbf{K}$ is a field, it is a unique factorization domain in a superficial sense because every non-zero expression is a unit.

The next theorem describes the divisor properties of a unique factorization domain.

Theorem 6.10. *Let $\mathbf{K}$ be a unique factorization domain, and let a, b, and c be in $\mathbf{K}$.*

1. *If $c \mid (a \cdot b)$ and a and c are relatively prime, then $c \mid b$.*

2. *If $c \mid (a \cdot b)$ and c is irreducible, then $c \mid a$ or $c \mid b$.*

Proof: To show (1), first if one of the expressions a, b, or c is a unit, the property follows immediately (Exercise 2(d)). Next, suppose that the expressions a, b, and c are not units and have the irreducible factorizations

$$a = a_1 \cdot a_2 \cdots a_l, \quad b = b_1 \cdot b_2 \cdots b_m, \quad c = c_1 \cdot c_2 \cdots c_n,$$

where none of the factors is a unit. Since $c \mid (a \cdot b)$, there is an expression d such that

$$c_1 \cdot c_2 \cdots c_n \cdot d = a_1 \cdot a_2 \cdots a_l \cdot b_1 \cdot b_2 \cdots b_m,$$

and since $\mathbf{K}$ is a unique factorization domain, each irreducible factor c_i is an associate of a (distinct) factor on the right side of the equation. In addition, since c and a are relatively prime, the factor c_i cannot be an associate of any of the factors a_j. Therefore, for each factor c_i, there is a (distinct) b_k so that c_i and b_k are associates. Therefore, $c \mid b$.

The proof of (2) follows directly from (1). $\qquad\qquad\qquad\qquad$ $\square$

Polynomial Domains

Let's return now to the polynomial domain $\mathbf{K}[x]$. Recall that $\mathrm{lc}(u, x)$ represents the leading coefficient of a polynomial and $\deg(u, x)$ its degree.[1] As before, the operators $\mathrm{lc}(u, x)$ and $\deg(u, x)$ satisfy the following properties.

Theorem 6.11. *Let u and v be polynomials in $\mathbf{K}[x]$.*

1. $\mathrm{lc}(u\, v) = \mathrm{lc}(u) \cdot \mathrm{lc}(v)$.

2. *If $u \neq 0$ and $v \neq 0$, then $\deg(u\, v) = \deg(u) + \deg(v)$.*

3. $\deg(u \pm v) \leq \max\left(\{\deg(u), \deg(v)\}\right)$.

Proof: The proof is similar to the one for Theorem 4.1 on page 112. $\square$

The next theorem states that the integral domain property of the coefficient domain is inherited by the polynomial domain $\mathbf{K}[x]$.

Theorem 6.12. *If $\mathbf{K}$ is an integral domain, then $\mathbf{K}[x]$ is also an integral domain.*

Proof: The theorem follows from Theorem 6.11(1), which implies that the product of two non-zero polynomials is always a non-zero polynomial. $\square$

By utilizing the recursive nature of multivariate polynomials, we extend the integral domain property to multivariate polynomials.

Theorem 6.13. *If $\mathbf{K}$ is an integral domain, then $\mathbf{K}[x_1, x_2, \ldots, x_p]$ is also an integral domain.*

Proof: Repeatedly apply Theorem 6.12. $\qquad\qquad\qquad\qquad\qquad\qquad$ $\square$

In Theorem 6.70 (page 260), we show that the unique factorization property is also inherited from $\mathbf{K}$ to $\mathbf{K}[x]$ and $\mathbf{K}[x_1, \ldots, x_p]$.

---

[1] When the variable x is evident from context, we use the simpler notations $\mathrm{lc}(u)$ and $\deg(u)$.

Exercises

1. Prove Theorem 6.4.

2. Let a, b, and c be in an integral domain $\mathbf{K}$. Prove each of the following statements.

   (a) If a and b are units, then $a \cdot b$ is a unit.

   (b) If a is a unit, then $a \mid b$.

   (c) If $a \cdot b = c$, where c is a unit, then a and b are units.

   (d) Prove Theorem 6.10(1) when at least one of the expressions a, b, or c is a unit.

3. Explain why the polynomials $u = x + x^2$ and $v = 2x + 2x^2$ are associates as polynomials in $\mathbf{Q}[x]$ but not associates as polynomials in $\mathbf{Z}[x]$.

4. Explain why $\mathbf{Z}_6$, which is described on page 53, is not an integral domain.

5. Let a and b be non-zero expressions in an integral domain $\mathbf{K}$ such that $a \mid b$ and $b \mid a$. Show that $a = c \cdot b$, where c is a unit in $\mathbf{K}$.

6. Give a procedure $Associates(u, v, x)$ that determines if polynomials u and v in $\mathbf{Q}[x]$ are associates. If the polynomials are associates, return a rational number c such that $u = c \cdot v$. If the polynomials are not associates, return the symbol **false**.

7. Show that the units in $\mathbf{K}$ and the units in $\mathbf{K}[x]$ are the same expressions.

6.2 Polynomial Division and Expansion

Recursive Polynomial Division in $\mathbf{K}[x]$

In Section 4.1, the division operation for polynomials with coefficients in a field is defined using recurrence relations which include the field computation

$$\mathrm{lc}(r_{i-1}, x)/\mathrm{lc}(v, x) \qquad (6.2)$$

(see Equations (4.1) and (4.2), page 113). This operation causes a problem when the coefficient domain $\mathbf{K}$ is not a field because $\mathrm{lc}(v, x)^{-1}$ may not exist in $\mathbf{K}$.

For u and v polynomials in $\mathbf{K}[x]$, polynomial division is often used to determine if $u \mid v$. If this is the goal, one way to define the division process is to continue the iteration as long as $\mathrm{lc}(v, x) \mid \mathrm{lc}(r_i, x)$, that is, as long as $\mathrm{cof}(\mathrm{lc}(v, x), \mathrm{lc}(r_i, x))$ exists in $\mathbf{K}$. This suggests the following division scheme.

Definition 6.14. *Let u and $v \neq 0$ be polynomials in $\mathbf{K}[x]$.* **Recursive polynomial division** *is defined by the sequence of quotients and remainders:*

$$
\begin{aligned}
q_0 &= 0, \quad r_0 = u, \\
q_i &= q_{i-1} + \mathrm{cof}(\mathrm{lc}(v,x), \mathrm{lc}(r_{i-1}, x)) \, x^{\deg(r_{i-1},x) - \deg(v,x)}, \quad (6.3) \\
r_i &= r_{i-1} - \mathrm{cof}(\mathrm{lc}(v,x), \mathrm{lc}(r_{i-1}, x)) \, x^{\deg(r_{i-1},x) - \deg(v,x)} \, v.
\end{aligned}
$$

The iteration terminates when either

$$
\deg(r_i, x) < \deg(v, x) \tag{6.4}
$$

or

$$
\mathrm{lc}(v, x) \nmid \mathrm{lc}(r_i, x). \tag{6.5}
$$

If the process stops after $i = \sigma$ iterations, then q_σ and r_σ are the **recursive quotient** *and* **recursive remainder** *of u divided by v. We also represent the quotient and the remainder by the operators* $\mathrm{recquot}(u, v, x)$ *and* $\mathrm{recrem}(u, v, x)$.

This process is called *recursive polynomial division* because it depends on a division process in the coefficient domain that determines if $\mathrm{lc}(v, x)$ divides $\mathrm{lc}(r_i, x)$. For multivariate polynomials in $\mathbf{K}[x_1, x_2, \ldots, x_p]$, this means that division in terms of the main variable x_1 depends recursively on division of polynomials in $\mathbf{K}[x_2, \ldots, x_p]$.

Recursive polynomial division satisfies the following properties.

Theorem 6.15. *Let u and $v \neq 0$ be polynomials in $\mathbf{K}[x]$.*

1. *The division process in Equation (6.3) terminates.*

2. *At termination,*

$$
u = q_\sigma v + r_\sigma, \tag{6.6}
$$

where either

$$
\deg(r_\sigma, x) < \deg(v, x) \tag{6.7}
$$

or

$$
\mathrm{lc}(v, x) \nmid \mathrm{lc}(r_\sigma, x). \tag{6.8}
$$

Proof: The proof of this theorem is similar to the proof for Theorem 4.5 on page 113 when the coefficient domain is a field. $\qquad\square$

When the coefficient domain is a field, the remainder satisfies the degree condition (6.7) which is known as the Euclidean property of polynomial division (page 114). When the coefficient domain is not a field and when the condition (6.8) terminates the division process, the remainder may not satisfy the Euclidean property.

Example 6.16. Consider $u = x^2 y^2 + x$ and $v = x y + 1$ as members of $\mathbb{Q}[x, y]$ with main variable x. Then dividing u by v, the first iteration gives

$$q_1 = x y, \qquad r_1 = x - x y.$$

Since

$$\mathrm{lc}(v, x) = y \nmid 1 - y = \mathrm{lc}(r_1, x),$$

the division terminates after one iteration with

$$u = q_1 v + r_1 = (x y)(x y + 1) + (x - x y).$$

Notice that $\deg(r_1, x) = \deg(v, x) = 1$, and so the remainder does not satisfy the Euclidean property. On the other hand, when y is the main variable, the process terminates after two iterations with

$$u = q_2 v + r_2 = (x y - 1)(x y + 1) + (x + 1).$$

Observe that $0 = \deg(r_2, y) < \deg(v, y) = 1$, and so in this case the remainder does satisfy the Euclidean property. $\qquad\square$

If the second condition (6.8) stops the iteration, the representation (6.6) may not be unique. This point is illustrated in the next example.

Example 6.17. Consider $u = x y + x + y$ and $v = x y$ as polynomials in $\mathbb{Q}[x, y]$ with main variable x. Since

$$\mathrm{lc}(v, x) = y \nmid y + 1 = \mathrm{lc}(u, x),$$

the process terminates with $\sigma = 0$, $q_0 = 0$, and $r_0 = x y + x + y$. In this case, u has another representation of the form in Equation (6.6)

$$u = 1 \cdot (x y) + (y + x),$$

where again the expression $y + x$ in the remainder position satisfies the second condition (6.8). $\qquad\square$

In the next theorem, we show that recursive polynomial division can determine if $v \mid u$.

Theorem 6.18. *Let u and $v \neq 0$ be polynomials in $\mathbf{K}[x]$, and suppose that $v \mid u$. Then, the recursive division process terminates with $q_\sigma = \mathrm{cof}(v, u)$ and $r_\sigma = 0$.*

Proof: Since $u = \mathrm{cof}(v, u)\, v = q_\sigma v + r_\sigma$, we have

$$r_\sigma = (\mathrm{cof}(v, u) - q_\sigma)\, v. \tag{6.9}$$

By Theorem 6.11(1), this relationship implies that $\mathrm{lc}(v) \mid \mathrm{lc}(r_\sigma)$, and so the termination condition (6.8) is not satisfied. In addition, if $\mathrm{cof}(v, u) - q_\sigma \neq 0$, then from Equation (6.9), $r_\sigma \neq 0$. Therefore, a degree argument shows that

$$\deg(v) \leq \deg(r_\sigma), \tag{6.10}$$

and so the termination condition (6.7) is not satisfied. However, one of the termination conditions (6.7) or (6.8) must hold, and so $q_\sigma = \mathrm{cof}(v, u)$ and $r_\sigma = 0$. □

A procedure[2] for recursive polynomial division for multivariate polynomials is given in Figure 6.1. The procedure is based on the recursive structure of multivariate polynomials where the first symbol in $L = [x_1, \ldots, x_p]$ is the main variable (line 7). Notice that division in the coefficient domain $\mathbf{K}[x_2, \ldots, x_p]$ is obtained by a recursive call on the procedure (line 15) and lines 1–5 provide a termination condition for the recursion. At lines 1–5, when L is the empty list, the division occurs in the coefficient domain which can be either $\mathbf{Z}$ or $\mathbf{Q}$.

Monomial-Based Division in $\mathbf{Q}[x_1, \ldots, x_p]$

Another approach to polynomial division is called *monomial-based division*. Like recursive division, monomial-based division can determine if $v \mid u$. However, because the process is based on the monomial structure of polynomials rather than the recursive structure, it may produce a different quotient and remainder when $v \nmid u$.

To simplify the presentation, we describe the process for the polynomial domain $\mathbf{Q}[x_1, \ldots, x_p]$. For the remainder of this section, we view polynomials in the form

$$u = u_1 + \cdots + u_n, \tag{6.11}$$

---

[2]See footnote 2 on page 115 for comments about the termination condition in line 13.

Procedure $Rec\_poly\_div(u, v, L, K)$;
Input
> u, v : multivariate polynomials in L with rational number
> coefficients and $v \neq 0$;
>
> L : list of symbols;
>
> K : the symbol **Z** or **Q**;

Output the list $[q_\sigma, r_\sigma]$;
Local Variables $x, r, q, m, n, lcv, lcr, d, c$;
Begin

1	if $L = [\ ]$ then
2	if $K = \mathbf{Z}$ then
3	if $Kind(u/v) = $ **integer then**$Return([u/v, 0])$
4	**else**$Return([0, u])$
5	else $Return([u/v, 0])$
6	else
7	$x := First(L)$;
8	$r := u$;
9	$m := Degree\_gpe(r, x)$;
10	$n := Degree\_gpe(v, x)$;
11	$q :- 0$;
12	$lcv := Leading\_Coefficient\_gpe(v, x)$;
13	while $m \geq n$ do
14	$lcr := Leading\_Coefficient\_gpe(r, x)$;
15	$d := Rec\_poly\_div(lcr, lcv, Rest(L), K)$;
16	if $Operand(d, 2) \neq 0$ then $Return([Algebraic\_expand(q), r])$
17	else
18	$c := Operand(d, 1)$;
19	$q := q + c * x^{m-n}$;
20	$r := Algebraic\_expand(r - c * v * x^{m-n})$;
21	$m := Degree\_gpe(r, x)$;
22	$Return([Algebraic\_expand(q), r])$

End

$$Rec\_quotient(u, v, L, K) \overset{\text{function}}{:=} Operand(Rec\_poly\_div(u, v, L, K), 1);$$
$$Rec\_remainder(u, v, L, K) \overset{\text{function}}{:=} Operand(Rec\_poly\_div(u, v, L, K), 2);$$

Figure 6.1. An MPL procedure for recursive polynomial division for multivariate polynomials. The first symbol in L is the main variable. (Implementation: Maple (txt), Mathematica (txt), MuPAD (txt).)

where each u_i is a monomial of the form

$$u_i = c\, x_1^{n_1} \cdots x_p^{n_p}.$$

Recall, from Section 1.4, that $x_1^{n_1} \cdots x_p^{n_p}$ is called the variable part of u_i, and the rational number c is called the coefficient part. We assume that the monomials in Equation (6.11) have distinct variable parts.[3]

Let's begin with a simple example in which the divisor v is a monomial.

Example 6.19. Consider the polynomials with x the main variable:

$$u = 2\,x^2\,y + 3\,x^2 + 4\,x\,y + 5\,x + 6\,y + 7, \qquad v = x\,y.$$

First, for recursive polynomial division of u by v, since

$$\mathrm{lc}(v, x) = y \nmid 2\,y + 3 = \mathrm{lc}(u, x),$$

the process terminates immediately with $\sigma = 0$, $q_\sigma = 0$, and $r_\sigma = u$.

On the other hand, monomial-based division recognizes that the monomial $v = x\,y$ divides the first and third monomials in u and we obtain

$$u = Q\,v + R = (2\,x + 4)\,(x\,y) + \left(3x^2 + 5x + 6y + 7\right).$$

Notice that v does not divide any of the monomials in the remainder $R.\square$

This example suggests another way to define a division process, at least when v is a monomial. In general, for $u = u_1 + \cdots + u_n$ and v a monomial, this division process gives

$$u = G(u, v)\,v + R(u, v),$$

where

$$G(u, v) = \sum_{\substack{i=1 \\ v \mid u_i}}^{n} u_i/v \qquad\qquad (6.12)$$

and

$$R(u, v) = \sum_{\substack{i=1 \\ v \nmid u_i}}^{n} u_i. \qquad\qquad (6.13)$$

In Equation (6.12), the sum for the quotient includes all monomials from u where $v \mid u_i$, while in Equation (6.13) the sum for the remainder includes all monomials where $v \nmid u_i$.

---

[3]In Maple, Mathematica, and MuPAD, in the context of automatic simplification, the monomials in a polynomial in $\mathbf{Q}[x_1, \ldots, x_p]$ have distinct variable parts.

Lexicographical Ordering of Monomials

To complete the definition of the new division process, we must show how it is defined when v is a sum of monomials. The approach we use is to repeatedly apply a process similar to that in Equations (6.12) and (6.13), but with v replaced by the leading monomial of v which is defined by the order relation in the next definition.

Definition 6.20. *Let u and v be monomials in $\mathbf{Q}[x_1, \ldots, x_p]$.*

1. *u is less than v in **lexicographical order** with respect to the list $L = [x_1, \ldots, x_p]$ if one of the following conditions is true.*

   (a) *$\deg(u, x_1) < \deg(v, x_1)$;*

   (b) *For some $1 < j \leq p$, $\deg(u, x_i) = \deg(v, x_i)$ for $i = 1, 2, \ldots, j-1$ and $\deg(u, x_j) < \deg(v, x_j)$.*

   The condition that u is less than v in lexicographical order is denoted by $u \prec v$.

2. *The monomials u and v are called **equivalent** monomials if they have the same variable part. The condition that u is equivalent to v is represented by $u \equiv v$.*

3. *The condition that $u \prec v$ or $u \equiv v$ is denoted by $u \preceq v$.*

Example 6.21. For $L = [x, y, z]$,

$$x^2 y^3 z^4 \prec x^2 y^4 z^3, \quad x y^2 z \prec x^2 y, \quad y z^5 \prec x, \quad 2xy \equiv 3xy. \quad (6.14)$$

$\square$

Lexicographical order depends on the order of the symbols in the list L. In the previous example, when the order of the symbols is changed to $L = [z, y, x]$, the lexicographical order of the first three examples in (6.14) is reversed.

The designation of this order relation as "lexicographical" suggests that it is similar to an "alphabetical" order. Indeed, the most significant factor in the order relation is the main variable x_1. It is only when u and v have the same degree in this variable that the next variable x_2 is significant. If the degrees of u and v with respect to x_2 are also the same, then x_3 is significant and so forth.

A procedure that determines if two monomials are in lexicographical order is described in Exercise 5.

The properties of lexicographical order are given in the next theorem.

Theorem 6.22. *Let u, v, and $w \neq 0$ be monomials in $\mathbf{Q}[x_1, \ldots, x_p]$.*

1. *If $u \prec v$, then $uw \prec vw$.*

2. *If v is not a rational number , then $w \prec vw$.*

The proof is left to the reader (Exercise 3).
The leading monomial of a polynomial is defined in terms of the lexicographical order of its monomials.

Definition 6.23. *Let u be in $\mathbf{Q}[x_1, \ldots, x_p]$, and let $L = [x_1, \ldots, x_p]$. If u is a sum of monomials, then the* **leading monomial** *is the monomial of u that is greatest in the lexicographical order. If u is a monomial, then the leading monomial is just u itself. The operator notation*

$$\mathrm{lm}(u, L)$$

denotes the leading monomial of u. The inclusion of the list L in this notation emphasizes that the leading monomial is defined with respect to the order of the symbols in L. If this order is implicit from the discussion, we use the simpler notation $\mathrm{lm}(u)$.

Example 6.24. If $u = 3\,x^2\,y + 4\,x\,y^2 + y^3 + x + 1$, then

$$\mathrm{lm}(u, [x, y]) = 3\,x^2\,y, \quad \mathrm{lm}(u, [y, x]) = y^3. \qquad \square$$

The leading monomial operator satisfies the following properties.

Theorem 6.25. *Let u, v, and w be polynomials in $\mathbf{Q}[x_1, \ldots, x_p]$, and let $L = [x_1, \ldots, x_p]$.*

1. $\mathrm{lm}(uv, L) = \mathrm{lm}(u, L)\,\mathrm{lm}(v, L)$.

2. *If $u \mid v$, then $\mathrm{lm}(u, L) \mid \mathrm{lm}(v, L)$.*

3. *If $\mathrm{lm}(u, L) \prec \mathrm{lm}(v, L)$ and $w \neq 0$, then $\mathrm{lm}(uw, L) \prec \mathrm{lm}(vw, L)$.*

Proof: To show (1), if either $u = 0$ or $v = 0$, then both sides of the relation in (1) are 0. If $u \neq 0$ and $v \neq 0$, let

$$u = u_1 + u_2 + \cdots + u_m,$$

$$v = v_1 + v_2 + \cdots + v_n,$$

Procedure *Leading_monomial(u, L)*;
Input
    u : multivariate polynomial in L with rational number coefficients;
    L : a list of symbols;
Output
    lm(u, L);
Local Variables
    x, m, c;
Begin
1   **if** $L = [\,]$ **then**
2   *Return(u)*
3   **else**
4        $x := First(L)$;
5        $m := Degree\_gpe(u, x)$;
6        $c := Coefficient\_gpe(u, x, m)$;
7   *Return*$(x^m * Leading\_monomial(c, Rest(L)))$
End

Figure 6.2. An MPL procedure that determines the leading monomial of a polynomial with respect to the symbol order $L = [x_1, \ldots, x_p]$. (Implementation: Maple (txt), Mathematica (txt), MuPAD (txt).)

where u_i and v_j are monomials with distinct variable parts, and lm$(u) = u_1$, lm$(v) = v_1$. We have

$$u\,v = \sum_{i=1}^{m} \left(\sum_{j=1}^{n} u_i v_j \right),$$

and by Theorem 6.22(1), $u_i v_j \prec u_1 v_1$ whenever either $i \neq 1$ or $j \neq 1$. Therefore, lm$(u\,v, L) = u_1\,v_1 = $ lm(u, L) lm(v, L).

The proofs of Parts (2) and (3) are left to the reader (Exercise 4). $\square$

A procedure that determines the leading monomial is given in Figure 6.2.

Monomial-Based Division Algorithm

Monomial-based division is defined by the following iteration scheme.

Definition 6.26. *Let u and $v \neq 0$ be polynomials in $\mathbf{Q}[x_1, \ldots, x_p]$, and let $L = [x_1, \ldots, x_p]$. Suppose that $v_l = $ lm(v, L), and define the iteration* *scheme*

$$q_0 = 0, \quad r_0 = u, \tag{6.15}$$

and for $i \geq 1$,

$$f_i = G(r_{i-1}, v_l), \tag{6.16}$$
$$q_i = q_{i-1} + f_i, \tag{6.17}$$
$$r_i = r_{i-1} - f_i v, \tag{6.18}$$

where the function G is defined in Equation (6.12). The iteration termi-nates when $G(r_i, v_l) = 0$. If $i = \tau$ is the first such index, then q_τ is the **monomial-based quotient** *of u divided by v, and r_τ is the* **monomial-based remainder**. *We also represent the quotient and the remainder with the operator notation* mbquot(u, v, L) *and* mbrem(u, v, L).

The division process repeatedly applies steps similar to Equations (6.12) and (6.13) with v replaced by v_l. Using mathematical induction, we can show that at each step

$$u = q_i v + r_i, \tag{6.19}$$

(Exercise 7). If we define

$$v_r = v - v_l, \tag{6.20}$$

then the remainder relation has the form,

$$r_i = (r_{i-1} - G(r_{i-1}, v_l) v_l) + (-G(r_{i-1}, v_l) v_r). \tag{6.21}$$

The operation $r_{i-1} - G(r_{i-1}, v_l) v_l$ eliminates some monomials from r_{i-1}, while $-G(r_{i-1}, v_l) v_r$ may contribute monomials to r_i with new variable parts that may be divisible by v_l. In fact, because of these new monomi-als, the process may continue at least one more iteration. This point is illustrated in the next example.

Example 6.27. Let $u = x^3 + 3 x^2 y + 4 x y^2$ and $v = x y + 2 y + 3 y^2$, and let $L = [x, y]$ define the variable order. Then

$$q_0 = 0, \quad r_0 = u, \quad v_l = x y, \quad v_r = 2 y + 3 y^2,$$

and the first iteration gives

$$f_1 = G(r_0, v_l) = 3 x + 4 y, \quad q_1 = 3 x + 4 y,$$

and

$$
\begin{aligned}
r_1 &= (u - f_1 v_l) + (-f_1 v_r) \\
&= (u - (3 x^2 y + 4 x y^2)) + (-9 x y^2 - 6 x y - 8 y^2 - 12 y^3) \\
&= x^3 - 9 x y^2 - 6 x y - 8 y^2 - 12 y^3.
\end{aligned}
$$

The next iteration gives

$$f_2 = -9\,y - 6$$

and

$$q_2 = 3\,x - 5\,y - 6, \quad r_2 = x^3 + 28\,y^2 + 15\,y^3 + 12\,y. \tag{6.22}$$

Since v_l does not divide any monomial of r_2, $G(r_2, v_l) = 0$ and the iteration terminates with $\tau = 2$ and the quotient and remainder in (6.22).

On the other hand, recursive polynomial division terminates after $\sigma = 0$ iterations with $q_\sigma = 0$ and $r_\sigma = u$. However, if the variable order is changed to $[y, x]$, then both monomial-based division and recursive division give the same result

$$q_\tau = q_\sigma = (4/3)\,x, \quad r_\tau = r_\sigma = x^3 + (5/3)\,x^2 - (8/3)\,x\,y. \qquad \square$$

Properties of Monomial-Based Division

In order to analyze monomial-based division, it is useful to express the iteration scheme in another form. Let $v_l = \mathrm{lm}(v, L)$ and $v_r = v - v_l$, and define the iteration scheme:

$$Q_0 = 0, \quad S_0 = 0, \quad P_0 = u, \tag{6.23}$$

and, for $i \geq 1$,

$$
\begin{aligned}
F_i &= G(P_{i-1}, v_l), & (6.24) \\
Q_i &= Q_{i-1} + F_i, & (6.25) \\
S_i &= S_{i-1} + P_{i-1} - F_i\,v_l, & (6.26) \\
P_i &= -F_i\,v_r & (6.27)
\end{aligned}
$$

where the function G is defined in Equation (6.12). The iteration terminates when $G(P_i, v_l) = 0$.

The next Theorem 6.28 describes some properties of this iteration scheme and the connection between it and the one given in Definition 6.26.

Parts (1) and (3) show that the remainder r_i is the sum of two parts, S_i where the monomials are not divisible by v_l and P_i where some of the monomials may be divisible by v_l. This point is illustrated in Example 6.29 below.

Parts (3), (4), and (5) of the theorem show that the new scheme gives the same quotient and remainder as the scheme in Definition 6.26.

Part (6) is used to show that the division process terminates (Theorem 6.30(1) below). Part (7) is similar Part (6), although it involves the full remainder and the relation $\preceq$ rather than $\prec$. In fact, this distinction between (6) and (7) is the reason the new version is easier to analyze.

Theorem 6.28. *Let u and $v \neq 0$ be polynomials in $\mathbf{Q}[x_1, \ldots, x_p]$, and let $L = [x_1, \ldots, x_p]$ define the variable order. The iteration scheme in Equations (6.23)-(6.27) satisfies the following properties:*

1. *For $i \geq 1$, v_l does not divide any monomial of S_i.*

2. *$P_i = F_{i+1}v_l + (S_{i+1} - S_i)$.*

3. *$r_i = S_i + P_i$.*

4. *$F_i = f_i$.*

5. *$Q_i = q_i$.*

6. *If $\tau \geq 1$, then $\mathrm{lm}(P_i) \prec \mathrm{lm}(P_{i-1})$, for $i = 1, \ldots, \tau$.*

7. *If $\tau \geq 1$, then $\mathrm{lm}(r_i) \preceq \mathrm{lm}(r_{i-1})$, for $i = 1, \ldots, \tau$.*

8. *Suppose that $\tau \geq 1$, and let t be a monomial in r_τ for which there is no monomial in u with the same variable part as t. Then, $t \prec \mathrm{lm}(u)$.*

Proof: We show Part (1) with mathematical induction. First, for $i = 1$, we have

$$S_1 = S_0 + P_0 + F_1\, v_l = u - G(u, v_l)\, v_l = R(u, v_l),$$

where the polynomial $R(u, v_l)$ is defined in Equation (6.13) and the monomials in this polynomial are not divisible by v_l.

Let's assume the induction hypothesis that Part (1) holds for $i - 1$ and show that it holds for i. Using Equations (6.12), (6.13), and (6.24), we have

$$
\begin{aligned}
S_i &= S_{i-1} + P_{i-1} - F_i\, v_l \\
&= S_{i-1} + (P_{i-1} - G(P_{i-1}, v_l)\, v_l) \\
&= S_{i-1} + R(P_{i-1}, v_l).
\end{aligned}
$$

Since neither term in the sum on the right is divisible by v_l, it follows that S_i is not divisible by v_l.

Part (2) is a restatement of Equation (6.26).

To show Part (3), we again use mathematical induction. First, for $i = 0$, $r_0 = u = S_0 + P_0$. Let's assume the induction hypothesis that (3) holds for $i - 1$ and show that it holds for i. From Part (1), we have

$G(P_{i-1}, v_l) = G(S_{i-1} + P_{i-1}, v_l)$, and therefore, using this relation, the induction hypothesis, Equations (6.18), (6.20), (6.24), and (6.26), we have

$$\begin{aligned}
S_i + P_i &= S_{i-1} + P_{i-1} - F_i v_l - F_i v_r \\
&= S_{i-1} + P_{i-1} - G(S_{i-1} + P_{i-1}, v_l)\, v \\
&= r_{i-1} - G(r_{i-1}, v_l)\, v \\
&= r_i.
\end{aligned}$$

Part (4) follows from Parts (1) and (3), and Part (5) follows from Part (4).

To show Part (6), we have $\text{lm}(v_r) \prec v_l$, and with $F_i \neq 0$, Theorem 6.25(3) implies that

$$\text{lm}(P_i) = \text{lm}(-F_i v_r) \prec \text{lm}(-F_i v_l). \tag{6.28}$$

In addition, since $F_i v_l = G(P_{i-1}, v_l) v_l$ includes some of the monomials of P_{i-1}, we have

$$\text{lm}(-F_i v_l) \preceq \text{lm}(P_{i-1}). \tag{6.29}$$

Therefore, the relations (6.28) and (6.29) imply that $\text{lm}(P_i) \prec \text{lm}(P_{i-1})$.

To show Part (7), consider

$$\begin{aligned}
r_i &= r_{i-1} - G(r_{i-1}, v_l)\, v \\
&= (r_{i-1} - G(r_{i-1}, v_l)\, v_l) + (-G(r_{i-1}, v_l)\, v_r). \tag{6.30}
\end{aligned}$$

For the first term in the sum, since $G(r_{i-1}, v_l)\, v_l$ cancels some of the monomials from r_{i-1}, we have

$$\text{lm}(r_{i-1} - G(r_{i-1}, v_l)\, v_l) \preceq \text{lm}(r_{i-1}). \tag{6.31}$$

For the second term in the sum in Equation (6.30), since $\text{lm}(v_r) \prec v_l$ and since for $i \leq \tau$, $G(r_{i-1}, v_l) \neq 0$, Theorem 6.25(3) implies that

$$\text{lm}(-G(r_{i-1}, v_l)\, v_r) \prec \text{lm}(-G(r_{i-1}, v_l)\, v_l). \tag{6.32}$$

In addition, since $G(r_{i-1}, v_l)\, v_l$ includes some monomials of r_{i-1}, we have

$$\text{lm}(-G(r_{i-1}, v_l)\, v_l) \prec \text{lm}(r_{i-1}). \tag{6.33}$$

Therefore, from Equation (6.30) and the relations (6.31), (6.32), and (6.33), we have $\text{lm}(r_i) \preceq \text{lm}(r_{i-1})$.

To show Part (8), there must be a monomial s in some P_i (for $i \geq 1$) with the same variable part as t. (The coefficient parts of s and t don't have to be the same since coefficients may change at a later step in the division process.) Therefore, since $P_0 = u$, Part (6) implies that $t \preceq \mathrm{lm}(P_i) \prec \mathrm{lm}(u)$. $\qquad\square$

Example 6.29. Consider again Example 6.27 where $u = x^3 + 3\,x^2\,y + 4\,x\,y^2$, $v = x\,y + 2\,y + 3\,y^2$, and x is the main variable. Then, $v_l = x\,y$, $v_r = 2\,y + 3\,y^2$, and

$$Q_0 = 0, \quad S_0 = 0, \quad P_0 = u,$$

$$F_1 = 3\,x + 4\,y, \quad Q_1 = 3\,x + 4\,y, \quad S_1 = x^3, \quad P_1 = -9\,x\,y^2 - 6\,x\,y - 8\,y^2 - 12\,y^3.$$

Observe that v_l does not divide the monomial in S_1, but does divide some of the monomials in P_1. The next iteration is

$$F_2 = -9\,y - 6, \quad Q_2 = 3\,x - 5\,y - 6,$$
$$S_2 = x^3 - 8\,y^2 - 12\,y^3, \quad P_2 = 12\,y + 36\,y^2 + 27\,y^3,$$

where now v_l does not divide any monomial of both S_2 and P_2. Therefore, the iteration terminates with $\tau = 2$, and

$$q_\tau = 3\,x - 5\,y - 6, \quad r_\tau = S_2 + P_2 = x^3 + 28\,y^2 + 15\,y^3 + 12\,y. \qquad\square$$

The next theorem shows that monomial-based division is a legitimate division process.

Theorem 6.30. *Let u and $v \neq 0$ be polynomials in $\mathbf{Q}[x_1, \ldots, x_p]$, and let $L = [x_1, \ldots, x_p]$.*

1. *The monomial-based division process terminates.*

2. *When the division process terminates,*

$$u = q_\tau\,v + r_\tau, \tag{6.34}$$

$$Q(r_\tau, v_l) = 0, \tag{6.35}$$

$$\mathrm{lm}(r_\tau, L) \preceq \mathrm{lm}(u, L), \tag{6.36}$$

$$\mathrm{lm}(u) \equiv \max(\{\mathrm{lm}(q_\tau)\,\mathrm{lm}(v), \ \mathrm{lm}(r_\tau)\}) \tag{6.37}$$

where the maximum is with respect to the monomial order relation.[4]

3. *The quotient and remainder are unique in the sense that for the order of symbols in L, they are the only polynomials that satisfy Equations (6.34) and (6.35).*

---

[4]In this context we assume that the max operator returns a monomial with coefficient part 1.

4. *If $u \neq 0$, then $\mathrm{lm}(r_\tau, L) \prec \mathrm{lm}(u, L)$ if and only if $v_l \mid \mathrm{lm}(u, L)$.*

5. *If $v \mid u$, then $q_\tau = \mathrm{cof}(v, u)$ and $r_\tau = 0$.*

Proof: To show Part (1), we use the iteration in Equations (6.23)-(6.27). First, if v is a monomial, then $v = v_l$, $v_r = 0$, and the process terminates with $\tau \leq 1$.

Suppose that v is not a monomial. If $F_i \neq 0$, from Theorem 6.28(6), $\mathrm{lm}(P_i) \prec \mathrm{lm}(P_{i-1})$, and, therefore, at some iteration $G(P_i, v_l) = 0$ which terminates the division.

To show Part (2), Equation (6.34) follows from Equation (6.19), and Equation (6.35) is the termination condition for the division. The relation (6.36) follows from Theorem 6.28(7) with $r_0 = u$. To show Equation (6.37), from Equation (6.34), we have

$$\mathrm{lm}(u) \preceq \max(\{\mathrm{lm}(q_\tau\,v), \mathrm{lm}(r_\tau)\}) = \max(\{\mathrm{lm}(q_\tau)\mathrm{lm}(v), \mathrm{lm}(r_\tau)\}). \quad (6.38)$$

In addition,

$$\mathrm{lm}(q_\tau) \cdot \mathrm{lm}(v) = \mathrm{lm}(q_\tau\,v) = \mathrm{lm}(u - r_\tau) \preceq \max(\{\mathrm{lm}(u), \mathrm{lm}(r_\tau)\}) \equiv \mathrm{lm}(u).$$

This relation and the relation (6.36) imply that

$$\max(\{\mathrm{lm}(q_\tau)\mathrm{lm}(v), \mathrm{lm}(r_\tau)\}) \preceq \mathrm{lm}(u). \quad (6.39)$$

The relation (6.37) follows from the relations (6.38) and (6.39).

To show Part (3), suppose that there were two representations,

$$u = q_{\tau_1}\,v + r_{\tau_1}, \qquad u = q_{\tau_2}\,v + r_{\tau_2},$$

where v_l does not divide any monomials in r_{τ_1} or r_{τ_2}. Then,

$$(q_{\tau_2} - q_{\tau_1})\,v = r_{\tau_1} - r_{\tau_2}.$$

If $r_{\tau_1} - r_{\tau_2} = 0$, then, since $\mathbf{Q}[x_1, \ldots, x_p]$ is an integral domain and $v \neq 0$, $q_{\tau_2} - q_{\tau_1} = 0$. In this case, the representations are identical. On the other hand, if $r_{\tau_1} - r_{\tau_2} \neq 0$, then since $v \mid r_{\tau_1} - r_{\tau_2}$, Theorem 6.25(2) implies that $v_l \mid \mathrm{lm}(r_{\tau_1} - r_{\tau_2})$. Since the monomials in $r_{\tau_1} - r_{\tau_2}$ are formed from the monomials in r_{τ_1} and r_{τ_2} and can differ from one of these monomials by at most a rational number factor (which does not affect the monomial's divisibility), v_l must divide some monomial in either r_{τ_1} or r_{τ_2} (or in both) which contradicts the definition of the remainders.

To show Part (4), first assume that $v_l \mid \mathrm{lm}(u)$. Then, in the first iteration $\mathrm{lm}(u)$ is eliminated, and since $\mathrm{lm}(P_1) \prec \mathrm{lm}(u)$, we have $\mathrm{lm}(r_1) \prec \mathrm{lm}(u)$. From this observation and Theorem 6.28(6) we have $\mathrm{lm}(r_\tau) \prec \mathrm{lm}(u)$.

Procedure $Mb\_poly\_div(u, v, L)$;
Input
    u, v : multivariate polynomials in L with rational number coefficients,
        and $v \neq 0$;
    L : list of symbols;
Output
    $[q_\tau, r_\tau]$;
Local Variables
    q, r, f, v_l;
Begin

```
1      q := 0;
2      r := u;
3      v_l := Leading_monomial(v, L);
4      f := G(r, v_l);
5      while  f ≠ 0 do
6          q := q + f;
7          r := Algebraic_expand(r − f * v);
8          f := G(r, v_l);
9      Return([q, r])
End
```

$Mb\_quotient(u, v, L) \overset{\text{function}}{:=} Operand(Mb\_poly\_div(u, v, L), 1);$

$Mb\_remaineder(u, v, L) \overset{\text{function}}{:=} Operand(Mb\_poly\_div(u, v, L), 2);$

Figure 6.3. An MPL procedure for monomial-based polynomial division and functions that extract the quotient and remainder. (Implementation: Maple (txt), Mathematica (txt), MuPAD (txt).)

To show that the implication goes in the other direction, observe that there are two ways that a monomial can be eliminated during the division process. One way is when $\text{lm}(v_l)$ divides the monomial, and the other way is when the monomial appears (with opposite sign) in some P_i. However, since $\text{lm}(P_i) \prec \text{lm}(u)$, P_i cannot eliminate $\text{lm}(u)$. Therefore, $v_l \mid \text{lm}(u)$.

The proof of (5) is left to the reader (Exercise 8).   □

A procedure for monomial-based division is given in Figure 6.3. The procedure G in lines 5 and 8 is described in Exercise 2. As a practical matter, it is only necessary to include the variables in the divisor v in the variable list L because the only use of this list is to obtain the leading monomial of the divisor (line 3).

Polynomial Expansion in $Q[x_1, \ldots, x_p]$

Let u and v be polynomials in $\mathbf{Q}[x_1, \ldots, x_p]$. A polynomial expansion for u in terms of v is a representation of u of the form

$$u = d_k v^k + d_{k-1} v^{k-1} + \cdots + d_0$$

where the d_i are also polynomials in $\mathbf{Q}[x_1, \ldots, x_p]$. One way to obtain an expansion is to modify the single-variable polynomial expansion algorithm for polynomials in $\mathbf{F}[x]$ by replacing the division operation with either recursive polynomial division or monomial-based division. The expansion is obtained with the quotient sequence:

$$
\begin{aligned}
u &= c_0 v + d_0, \\
c_0 &= c_1 v + d_1, \\
c_1 &= c_2 v + d_2, \\
&\vdots \\
c_{k-2} &= c_{k-1} v + d_{k-1}, \\
c_{k-1} &= c_k v + d_k, \\
c_k &= 0.
\end{aligned}
\tag{6.40}
$$

where c_k is the quotient of c_{i-k} divided by v and d_k is the corresponding remainder. The iteration terminates when a quotient c_k is zero.

Example 6.31. Let $u = 2x^2 + 3xy + y^2 + 3$ and $v = x + y$ be polynomials in $\mathbf{Q}[x, y]$. Using recursive polynomial division with x as the main variable, we have

$$
\begin{aligned}
c_0 &= 2x + y, & d_0 &= 3, \\
c_1 &= 2, & d_1 &= -y, \\
c_2 &= 0, & d_2 &= 2,
\end{aligned}
$$

and therefore,

$$u = d_2 v^2 + d_1 v + d_0 = 2v^2 - yv + 3.$$

In this case, expansion using monomial-based division with x as the main variable gives the same quotient sequence and same expansion. If instead y is the main variable, we obtain $u = v^2 + xv + 3$ with either form of division. $\square$

Although either form of polynomial division can be used for expansion, monomial-based division often gives more appropriate representations. This point is illustrated in the next example.

Example 6.32. Let $u = (x\,y)^2 + x\,y + x + y^2$ and $v = x\,y$ be polynomials in $\mathbf{Q}[x, y]$. Using recursive division with x the main variable, we obtain

$$c_0 = x\,y, \qquad d_0 = x\,y + x + y^2,$$

$$c_1 = 0, \qquad d_1 = 0,$$

$$c_2 = 0, \qquad d_2 = 1,$$

and therefore,

$$u = d_2\,v^2 + d_1\,v + d_0 = v^2 + x\,y + x + y^2.$$

This approach selects the $(x\,y)^2$ term but overlooks the $x\,y$ term. If instead, y is the main variable, both $(x\,y)^2$ and $x\,y$ are overlooked in the expansion.

On the other hand, using monomial-based division with either x or y as the main variable, we obtain the more reasonable representation

$$u = v^2 + v + x + y^2. \qquad \qquad \square$$

Since monomial-based division usually obtains more reasonable expansions, especially when v is a monomial, we use it for the remainder of this section. Using this approach, polynomial expansion has the following properties.

Theorem 6.33. *Let u and v be polynomials in $\mathbf{Q}[x_1, \ldots, x_p]$, and suppose that v has positive degree in some variable x_i.*

1. *The expansion process terminates and gives a polynomial*

$$u = d_k\,v^k + d_{k-1}\,v^{k-1} + \cdots + d_0. \qquad (6.41)$$

2. $Q(d_i, v_l) = 0$, *for $i = 0, \ldots, k$,*

3. *For the variable order $L = [x_1, \ldots, x_p]$, the expansion (6.41) that satisfies property (2) is unique.*

The proof of the theorem is left to the reader (Exercise 9). $\qquad \square$

A procedure for polynomial expansion that utilizes monomial-based division is given in Figure 6.4. As in the single variable case (Figure 4.2, page 122), the procedure returns a polynomial

$$P_u(t) = d_k\,t^k + d_{k-1}\,t^{k-1} + \cdots + d_1\,t + d_0$$

Procedure $Mb\_poly\_exp(u, v, L, t)$;
Input
    u : a multivariate polynomial in L with rational coefficients;
    v : a non-constant multivariate polynomial in L with
        rational coefficients;
    L : a list of symbols;
    t : an algebraic expression;
Output
    the polynomial $P_u(t)$;
Local Variables
    d, q, r;
Begin
1   **if** $u = 0$ **then**
2        $Return(0)$
3   **else**
4        $d := Mb\_poly\_div(u, v, L)$;
5        $q := Operand(d, 1)$;
6        $r := Operand(d, 2)$;
7        $Return(Algebraic\_expand(\, t * Mb\_poly\_exp(q, v, L, t) + r))$
  End

Figure 6.4. An MPL procedure for polynomial expansion that utilizes monomial-based polynomial division. (Implementation: Maple (txt), Mathematica (txt), MuPAD (txt).)

---

which is the expansion polynomial with the polynomial v replaced by the symbol t. As before, using the division relation

$$u = q_\tau\, v + r_\tau,$$

we obtain the recurrence relation

$$P_u(t) = t\, P_{q_\tau}(t) + r_\tau \qquad (6.42)$$

(Exercise 11) which is applied at line 7.

Expansion-Based Substitutions

Using polynomial expansion, we can generalize the notion of substitution for multivariate polynomials. For example, to replace $v = a + b$ in

$$u = a^2 b + 2\, ab^2 + b^3 + 2\, a + 2\, b + 3$$

by t, we use the expansion

$$Mb\_poly\_exp(u, v, [a, b], t) \to b\,t^2 + 2\,t + 3.$$

Simplification with a Side Relation

One important substitution operation is concerned with simplification with respect to a side relation. For example, to simplify the expression $u = a\,\imath^3 + b\,\imath^2 + c\,\imath + d$ with respect to the side relation $\imath^2 = -1$, we use the expansion

$$Mb\_poly\_exp(u, \imath^2, [\imath], -1) \to -a\,\imath - b + c\,\imath + d.$$

If a side relation is equated to zero, the simplification is obtained with monomial-based division. For example, to apply the side relation $\imath^2 + 1 = 0$ we use

$$\text{mbrem}\left(a\,\imath^3 + b\,\imath^2 + c\,\imath + d, \;\; \imath^2 + 1, [\imath]\right) \to -a\,\imath - b + c\,\imath + d.$$

A particularly important side relation for simplification is the trigonometric identity

$$\sin^2(x) + \cos^2(x) - 1 = 0. \tag{6.43}$$

Using division, we can apply the identity even in cases where the identity is hidden. For example, consider the simplification

$$
\begin{aligned}
u &= \sin^4(x) + \sin^3(x) + 2\sin^2(x)\cos^2(x) + \cos^4(x) \qquad (6.44)\\
&= \left(\sin^2(x) + \cos^2(x)\right)^2 + \sin^3(x),\\
&= 1 + \sin^3(x).
\end{aligned}
$$

We can simplify Equation (6.44) directly using monomial-based division[5]

$$\text{mbrem}\left(u, \sin^2(x) + \cos^2(x) - 1, [\cos(x), \sin(x)]\right) \to 1 + \sin^3(x). \quad (6.45)$$

Using polynomial division, we can formulate a procedure

$$Trig\_ident(u)$$

that applies the identity (6.43) at appropriate places in u. This procedure has the following properties.

---

[5]Although the procedures for monomial-based division and polynomial expansion have been defined when L is a list of symbols, they work as well with function forms as long as the members of L are independent.

1. The procedure should apply the identity with an appropriate variable order so that the new expression is really simpler than the old expression. For example if the division (6.45) is applied with the variable order $[\cos(x), \sin(x)]$, we obtain

$$\text{mbrem}\left(u, \sin^2(x) + \cos^2(x) - 1, [\sin(x), \cos(x)]\right)$$
$$\rightarrow -\sin(x)\cos^2(x) + \sin(x) + 1. \qquad (6.46)$$

Since the remainder in (6.45) is smaller than the remainder in (6.46), it is reasonable to consider the first remainder as the simplified form of u.

One way to define the size of an expression is with the *Tree_size* operator described in Exercise 12. Using this operator, we compare the size of the original expression to the sizes of the expressions obtained using the remainder operation with the two variable orders $[\sin(x), \cos(x)]$ and $[\cos(x), \sin(x)]$ and choose as a simplified form the one that has the smallest size. Using this criteria, the expression

$$u = \sin^3(x) + \cos^3(x)$$

is already simplified because the remainders

$$\text{mbrem}\left(u, \sin^2(x) + \cos^2(x) - 1, [\cos(x), \sin(x)]\right)$$
$$\rightarrow \sin^3(x) - \cos(x)\sin^2(x) + \cos(x),$$

$$\text{mbrem}\left(u, \sin^2(x) + \cos^2(x) - 1, [\sin(x), \cos(x)]\right)$$
$$\rightarrow \cos^3(x) - \sin(x)\cos^2(x) + \sin(x)$$

are larger in size than the original expression.

2. The procedure should apply the identity (6.43) for an arbitrary argument of the sin and cos functions. For example, if

$$u = \sin^2(z)\sin^2(x) + \sin^2(z)\cos^2(x) + \cos^2(z)\sin^2(y) + \cos^2(z)\cos^2(y),$$

then *Trig_ident*$(u) \rightarrow 1$.

We leave the formal definition of *Trig_ident* to the reader (Exercise 13).

Generalized Polynomial Operations

Let u and v be multivariate polynomials with rational number coefficients. Polynomial operators such as *Degree_gpe*(u, v) and *Coefficient_gpe*(u, v, n)

described in Section 1.4 perform in an awkward way when v is not a complete sub-expression of u. Using polynomial expansion, these operations can be obtained in a way that produces more reasonable results. For example, if

$$u = x^3 y^2 + x\, y^2 = x\, (x\, y)^2 + y\, (x\, y),$$

it is reasonable to define the degree of u with respect to $v = x\, y$ as 2. This degree operation is defined using the operations

$$w := Mb\_poly\_exp(u, v, [x, y], t);$$

$$Degree\_gpe(w, t);$$

The procedures that perform the polynomial operations in this general sense are left to the reader (Exercise 14).

Exercises

1. Let u and v be polynomials in $\mathbf{K}[x]$, and suppose that $lc(v, x)$ is a unit. Show that, in this case, the recursive polynomial division process satisfies the Euclidean property and the quotient and remainder are unique.

2. Let u be a polynomial and $v \neq 0$ be a monomial in $\mathbf{Q}[x_1, \ldots, x_p]$. Give a procedure $G(u, v)$ that computes the function in Equation (6.12).

3. Prove Theorem 6.22.

4. Prove Theorem 6.25(2),(3).

5. (a) Let u and v be monomials, and let L be a list of symbols that determines the symbol order. Give a procedure

   $$Lexicographical\_order(u, v, L)$$

   that determines if two monomials are in lexicographical order. When $u \prec v$, the procedure returns **true** and otherwise returns **false**.

   (b) Let u be a polynomial. Give a procedure, $Monomial\_list(u, L)$ that returns a list of the monomials in u in lexicographical order.

6. Show that

   $$mbquot(u_1 + u_2, v, L) = mbquot(u_1, v, L) + mbquot(u_2, v, L),$$

   $$mbrem(u_1 + u_2, v, L) = mbrem(u_1, v, L) + mbrem(u_2, v, L).$$

7. For the monomial-based division process in Definition 6.26, show that $u = q_i v + r_i$ for $i = 0, \ldots, \tau$.

8. Prove Theorem 6.30(5). *Hint:* Use the uniqueness property of monomial based division.

9. Prove Theorem 6.33.

10. Let $u = a^2 + b^2 + c^2 + ab + 2ac + 2bc$ and $v = a + b + c$. Find a polynomial expansion of u in terms of v for the variable order $L = [a, b, c]$.

11. Derive the recurrence relation in Equation (6.42) for $P_u(t)$.

12. Let u be an algebraic expression. Define the *tree-size* of u as the number of symbols, integers, algebraic operators, and function names that occur in u. For example, the expression $(x + \sin(x) + 2) * x^3$ consists of $x, +, \sin, x, 2, *, x, \wedge,$ and 3 and so has a tree-size of 9. Give a procedure *Tree_size*(u) that obtains the tree-size of u.

13. Let u be a GPE with generalized variables that are symbols, sine functions or cosine functions. In addition, assume that u is a polynomial in each generalized variable that is a sine or cosine. Give a procedure for the operator *Trig_ident*(u) that is described on page 226. The *Tree_size* operator described in Exercise 12 is useful in this exercise.

14. Let u and v be multivariate polynomials with rational number coefficients. Give procedures for the operators

$$Degree\_exp\_mv(u, v, L), \quad Coeff\_exp\_mv(u, v, n, L)$$

that return the degree and coefficient values when v is not necessarily a complete sub-expression of u. Your procedures should employ polynomial expansion as described on page 227. Make sure your procedures can determine when u is not a generalized polynomial in v. For example, for

$$u + \sin(2xy + 1)(xy)^2 + 2xy, \quad v = xy,$$

then

$$Degree\_exp\_mv(u, v, [x, y]) \rightarrow \textbf{Undefined}$$

because xy appears as part of the argument for sin.

6.3 Greatest Common Divisors

The goal of this section is the computation of greatest common divisors in a multivariate polynomial domain $\mathbf{K}[x_1, \ldots, x_p]$. Our approach is to view a multivariate polynomial in a recursive way as a polynomial in a main variable x_1 whose coefficients are polynomials in the integral domain $\mathbf{K}[x_2, \ldots, x_p]$. For this reason, the gcd computation involves computation in the original polynomial domain along with gcd computations in the coefficient domain.

We begin by describing the gcd concept in an integral domain.

Definition 6.34. *Let a and b be two expressions in an integral domain $\mathbf{K}$. The* **greatest common divisor** *of a and b (at least one of which is non-zero) is an expression d in $\mathbf{K}$ that satisfies the following three properties.*

1. *d is a common divisor of a and b.*

2. *If e is any common divisor of a and b, then e | d.*

3. *d is unit normal.* (This property is defined below.)

The operator gcd(a, b) *represents the greatest common divisor. If both of the expressions a and b are zero, the above definition does not apply. In this case, by definition,* gcd$(0, 0) = 0.$

Properties (1) and (2) are the same as the properties in the gcd definitions for integers and polynomials in $\mathbf{F}[x]$ (Definition 2.6, page 20 and Definition 4.19, page 127). In general, however, there are many expressions d that satisfy (1) and (2), although any two such expressions are associates (Exercise 5, page 207). In our previous discussions of greatest common divisors, we made the concept precise by including a third property that selects the unique gcd from among the associates. Property (3) is also included to make the gcd unique, although the definition of *unit normal* depends on the domain $\mathbf{K}$. This concept is described in the next definition.

Definition 6.35. *The* **unit normal** *expressions are a subset* $\mathbf{H}$ *of* $\mathbf{K}$ *that satisfies the following properties.*

1. *The additive identity 0 and the multiplicative identity 1 are in* $\mathbf{H}$.

2. *If a and b are in* $\mathbf{H}$, *then* $a \cdot b$ *is in* $\mathbf{H}$.

3. *For each* $b \neq 0$ *in* $\mathbf{K}$, *there is a unique unit c in* $\mathbf{K}$ *such that* $c \cdot b$ *is in* $\mathbf{H}$.

Example 6.36. Consider the integral domain $\mathbf{K} = \mathbf{Z}$ which has the units ± 1. For this domain, $\mathbf{H}$ is the set of non-negative integers. Properties (1) and (2) of Definition 6.35 follow immediately, and for property (3), let

$$c = \begin{cases} 1, & \text{if } b > 0, \\ -1, & \text{if } b < 0. \end{cases}$$ □

Example 6.37. The gcd definition is most meaningful when the two expressions are not units, although the definition includes these cases as well. Consider the domain $\mathbf{K} = \mathbf{Q}$ where all non-zero expressions are units. In this case,

$$\mathbf{H} = \{0, 1\}.$$

Properties (1) and (2) of Definition 6.35 follow immediately, and for property (3), let

$$c = \begin{cases} b^{-1}, & \text{if } b > 0, \\ -b^{-1}, & \text{if } b < 0. \end{cases}$$

In this domain, whenever a or b is non-zero, $\gcd(a,b) = 1$.   □

Observe that in property (3) we require c to be a unique unit. By doing so we restrict the size of **H** and obtain the uniqueness of the gcd. This point is the basis of the next theorem.

Theorem 6.38. *Let a and b be expressions in an integral domain* **K**.

1. *The only expression that is both a unit and unit normal is the multiplicative identity 1.*

2. *If a and b are relatively prime, then* $\gcd(a,b) = 1$.

3. *Suppose that $a \neq 0$ and $b \neq 0$ are unit normal, $a \mid b$, and $b \mid a$. Then, $a = b$.*

4. $\gcd(a,b)$ *is unique.*

Proof: To show (1), since a is a unit, for the expression c in Definition 6.35(3), we can let $c = a^{-1}$. In addition, since a is unit normal, we can also let $c = 1$. However, the multiplier c in Definition 6.35(3) is unique, and so $a^{-1} = 1$ which implies that $a = 1$.

To show (2), recall that *relatively prime* in this context means every common divisor of a and b is a unit (Definition 6.6, page 204). Therefore, $\gcd(a,b)$ is a unit, and since $\gcd(a,b)$ is unit normal, Part (1) implies that $\gcd(a,b) = 1$.

The proof of (3) is left to the reader (Exercise 1).

Since the $\gcd(a,b)$ is unit normal, Part (4) follows from Part (3).   □

Properties of the Coefficient Domain K

In order to develop our gcd algorithm, we assume that the coefficient domain **K** satisfies the following mathematical and computational properties.

1. **K** is a unique factorization domain.

2. There is a class **H** of unit normal expressions and an algorithm that determines for each $b \neq 0$ in **K**, an associate of b in **H**.

3. There is an algorithm that determines $\gcd(a, b)$ in $\mathbf{K}$.

The coefficient domains considered here are either $\mathbf{Q}$, $\mathbf{Z}$, or a polynomial domain in the auxiliary variables, where the base numerical domain is either $\mathbf{Q}$ or $\mathbf{Z}$. These domains satisfy these three properties.[6]

Greatest Common Divisors in K[x]

Since the polynomial domain $\mathbf{K}[x]$ is an integral domain, the gcd definition Definition 6.34 applies here as well. In this context, we represent the greatest common divisor by $\gcd(u, v, x)$ or by $\gcd(u, v)$ when x is evident from the discussion.

For polynomial domains, the unit normal expressions are defined in the following way.

Definition 6.39. *Let $\mathbf{K}$ be a coefficient domain. A polynomial u in $\mathbf{K}[x]$ is unit normal if $\mathrm{lc}(u, x)$ is unit normal in $\mathbf{K}$.*

It is straightforward to check that the set of unit normal polynomials satisfies the three properties in Definition 6.35 (Exercise 2).

Example 6.40. For multivariate polynomials, the unit normal property depends on which variable is the main variable. Consider

$$u = 7\,x - 2\,x\,y - 5 + y^2$$

as a polynomial with coefficients in $\mathbf{Z}$. If y is the main variable, then $\mathrm{lc}(u, y) = 1$, and so u is unit normal in $\mathbf{Z}[x, y]$.

On the other hand, if x is the main variable, then $\mathrm{lc}(u, x) = 7 - 2\,y$, which is not unit normal in $\mathbf{Z}[y]$ because -2 is not unit normal in $\mathbf{Z}$. Therefore, u is not unit normal in $\mathbf{Z}[y, x]$.   □

The next theorem gives three important properties of the greatest common divisor.

Theorem 6.41. *Let u and v be polynomials in $\mathbf{K}[x]$.*

1. $\gcd(u, v, x)$ exists.

2. $\gcd(u, v, x)$ is unique.

3. If $u \neq 0$, then $\gcd(0, u, x) = c\,u$ where c is the unique unit in $\mathbf{K}$ so that $c\,u$ is unit normal.

---

[6]For polynomial domains, this assertion follows from the recursive nature of the gcd algorithm (Figure 6.7) and Theorem 6.71 on page 260.

Proof: We show (1) by giving an algorithm to compute $\gcd(u, v, x)$ later in this section.

Part (2) follows from Theorem 6.38(4).

Part (3) follows directly from Definitions 6.34 and 6.35.   □

The recursive nature of Definition 6.39 provides a way to select a unique gcd in a multivariate polynomial domain. This point is illustrated by the next example.

Example 6.42.   Let

$$u = (2\,y + 1)\,x^2 + (10\,y + 5)\,x + (12\,y + 6),$$

$$v = (2\,y + 1)\,x^2 + (12\,y + 6)\,x + (18\,y + 9),$$

be polynomials in $\mathbf{Q}[x, y]$ with main variable x. The polynomial

$$w = (2\,y + 1)\,x + (6\,y + 3)$$

satisfies properties (1) and (2) of the gcd definition, but is not unit normal because its leading coefficient $2\,y + 1$ is not unit normal in $\mathbf{Q}[y]$. In fact, reasoning recursively, a polynomial in $\mathbf{Q}[y]$ is unit normal if its leading coefficient is unit normal in $\mathbf{Q}$. In other words, the unit normal expressions in $\mathbf{Q}[y]$ are either the monic polynomials or the 0 polynomial. Therefore,

$$\gcd(u, v) = (1/2)\,w = (y + 1/2)\,x + (3\,y + 3/2).$$

Next, let's consider u and v as polynomials in $\mathbf{Z}[x, y]$. Since the unit normal expressions in $\mathbf{Z}$ are the non-negative integers, in this setting w is unit normal and $\gcd(u, v) = w$.   □

Pseudo-Division

When the coefficient domain is a field, we find the gcd of u and v by using polynomial division to define the polynomial remainder sequence

$$
\begin{aligned}
R_{-1} &= u, \\
R_0 &= v, \\
R_{i+1} &= \operatorname{rem}(R_{i-1}, R_i, x), \quad i \geq 0,
\end{aligned}
\tag{6.47}
$$

where

$$\gcd(R_{i+1},\, R_i) = \gcd(R_i,\, R_{i-1}). \tag{6.48}$$

Since

$$\deg(R_{i+1}) < \deg(R_i), \tag{6.49}$$

some member of the remainder sequence is 0. If R_ρ is the first such remainder, then

$$\gcd(u, v) = R_{\rho-1}/\mathrm{lc}(R_{\rho-1}).$$

Let's try to extend this process to a multivariate polynomial domain by replacing the remainder operation in Equation (6.47) with either the recursive remainder or the monomial-based remainder. For both of these division operations, Equation (6.48) is valid but, since neither one satisfies the Euclidean property (6.49), there is no guarantee that the remainder sequence terminates. In fact, for

$$u = y\,x^2 + 2\,y\,x + y, \qquad v = y\,x^2 - x^2 - y + 1, \tag{6.50}$$

Equation (6.47) with recursive division gives an infinite remainder sequence (for either variable order) even though $\gcd(u, v) = x + 1$. For example, for the variable order $L = [x, y]$, we have

$$\mathrm{lc}(u, x) = y, \qquad \mathrm{lc}(v, x) = y - 1,$$

which implies

$$\mathrm{lc}(u, x) \nmid \mathrm{lc}(v, x), \qquad \mathrm{lc}(v, x) \nmid \mathrm{lc}(u, x).$$

Therefore,

$$\mathrm{recrem}(u, v, [x, y]) = u, \qquad \mathrm{recrem}(v, u, [x, y]) = v,$$

and the remainder sequence alternates between u and v. Monomial-based division (using either variable order) also gives a remainder sequence that does not terminate.

There is, however, a way to define a division-like process that gives a remainder sequence that leads to the gcd. To guarantee that the gcd remainder sequence terminates, we need a division process for u divided by v that terminates with a remainder r that satisfies the Euclidean property

$$\deg(r) < \deg(v). \tag{6.51}$$

One approach is to use recursive division but to modify u so that the division process proceeds until the inequality (6.51) is satisfied. To see how this is done, consider the polynomials in $\mathbf{Z}[x]$:

$$u = x^2 + x + 1, \qquad v = 3\,x + 1.$$

Dividing u by v using recursive division, the termination condition

$$\mathrm{lc}(v) = 3 \nmid 1 = \mathrm{lc}(u) \tag{6.52}$$

immediately stops the process with

$$\text{recrem}(u, v, x) = r_0 = x^2 + x + 1$$

which does not satisfy the degree condition (6.51).

One way to circumvent (6.52) is simply to replace u by

$$\text{lc}(v)\, u = 3\, x^2 + 3\, x + 3.$$

When this is done, the division process has one iteration, and

$$\text{recrem}(\text{lc}(v)\, u, \, v, \, x) = r_1 = 2\, x + 3$$

where the termination condition

$$\text{lc}(v) = 3 \nmid 2 = \text{lc}(r_1) \tag{6.53}$$

stops the process, and so again, the remainder r_1 does not satisfy the degree condition (6.51). We can, however, avoid the conditions in both (6.52) and (6.53) by replacing u by

$$\text{lc}(v)^2\, u = 9\, x^2 + 9\, x + 9.$$

Now, the division process has two iterations, and

$$\text{recrem}(\text{lc}(v)^2\, u, \, v, \, x) = r_2 = 7,$$

where the degree condition

$$\deg(r_2) < \deg(v)$$

terminates the process.

In general, suppose that u and $v \neq 0$ are in $\mathbf{K}[x]$ with $\deg(u) \geq \deg(v)$. To guarantee that the degree condition (6.51) stops the process, we need to find an integer $\delta \geq 0$ so that when $\text{lc}(v)^\delta\, u$ is divided by v, $\text{lc}(v) \mid \text{lc}(r_{i-1})$ at each step of the process. To find δ, suppose that the division process has a sequence of remainders

$$u = r_0, \, r_1, \ldots, r_\delta,$$

where r_δ is the first remainder with

$$\deg(r_\delta) < \deg(v). \tag{6.54}$$

As in the above example, to obtain this condition we multiply u by $\text{lc}(v)$ for each of the δ steps in the process. The maximum number of steps in the process occurs when

$$\deg(r_i) = \deg(r_{i-1}) - 1, \quad i = 1, \ldots, \delta. \tag{6.55}$$

When this happens,

$$\deg(r_\delta) = \deg(u) - \delta,$$

and so the inequality (6.54) gives

$$\deg(u) - \deg(v) < \delta.$$

Therefore, for

$$\delta = \deg(u) - \deg(v) + 1, \qquad (6.56)$$

recursive division of $\mathrm{lc}(v)^\delta u$ by v gives

$$
\begin{aligned}
\mathrm{lc}(v)^\delta u &= \mathrm{recquot}(\mathrm{lc}(v)^\delta u, v)\, v + \mathrm{recrem}(\mathrm{lc}(v)^\delta u, v) \qquad (6.57)\\
&= q\,v + r,
\end{aligned}
$$

with the termination condition

$$\deg(r, x) < \deg(v, x).$$

The division in Equation (6.57) with u replaced by $\mathrm{lc}(v, x)^\delta u$ is called the *pseudo-division* of u by v. We shall see below, when the coefficient domain **K** is not a field, that this process provides a way to compute the gcd using a remainder sequence.

Pseudo-Division Computation. From Equation (6.3), the pseudo-division in Equation (6.57) involves the iteration scheme:

$$
\begin{aligned}
q_0 &= 0, \qquad r_0 = \mathrm{lc}(v)^\delta u,\\
q_i &= q_{i-1} + \mathrm{cof}(\mathrm{lc}(v, x), \mathrm{lc}(r_{i-1}, x))\, x^{\deg(r_{i-1}, x) - \deg(v, x)}, \qquad (6.58)\\
r_i &= r_{i-1} - \mathrm{cof}(\mathrm{lc}(v, x), \mathrm{lc}(r_{i-1}, x))\, v\, x^{\deg(r_{i-1}, x) - \deg(v, x)},
\end{aligned}
$$

where the degree condition

$$\deg(r_i, x) < \deg(v, x)$$

terminates the process. In this form, the operation $\mathrm{cof}(\mathrm{lc}(v, x), \mathrm{lc}(r_{i-1}, x))$ involves division in the coefficient domain **K**.

There is another way to view pseudo-division that avoids division in the coefficient domain. Consider the iteration scheme:

$$
\begin{aligned}
p_0 &= 0, \qquad s_0 = u,\\
p_i &= \mathrm{lc}(v, x)\, p_{i-1} + \mathrm{lc}(s_{i-1}, x)\, x^{\deg(s_{i-1}, x) - \deg(v, x)}, \qquad (6.59)\\
s_i &= \mathrm{lc}(v, x)\, s_{i-1} - \mathrm{lc}(s_{i-1}, x)\, v\, x^{\deg(s_{i-1}, x) - \deg(v, x)}
\end{aligned}
$$

where the degree condition

$$\deg(s_i, x) < \deg(v, x)$$

terminates the process. Observe that this process does not involve division in the coefficient domain.

The next theorem gives the relationship between Equation (6.58) and Equation (6.59).

Theorem 6.43. *Let u and $v \neq 0$ be polynomials in $\mathbf{K}[x]$ with $\deg(u, x) \geq \deg(v, x)$. Then,*

$$q_i = \mathrm{lc}(v)^{\delta - i} p_i, \qquad (6.60)$$
$$r_i = \mathrm{lc}(v)^{\delta - i} s_i. \qquad (6.61)$$

Proof: We verify Equation (6.61) using mathematical induction. For the base case $i = 0$,

$$r_0 = \mathrm{lc}(v)^{\delta} u = \mathrm{lc}(v)^{\delta - 0} s_0.$$

Let's assume the induction hypothesis that Equation (6.61) holds for $i - 1$ and show that it also holds for i. We have

$$
\begin{aligned}
r_i &= r_{i-1} - \mathrm{cof}(\mathrm{lc}(v), \mathrm{lc}(r_{i-1})) \, v \, x^{\deg(r_{i-1}) - \deg(v)} \\
&= \mathrm{lc}(v)^{\delta - (i-1)} s_{i-1} \\
&\quad - \mathrm{cof}\left(\mathrm{lc}(v), \mathrm{lc}\left(\mathrm{lc}(v)^{\delta - (i-1)} s_{i-1}\right)\right) v \, x^{\deg(s_{i-1}) - \deg(v)} \\
&= \mathrm{lc}(v)^{\delta - (i-1)} s_{i-1} - \mathrm{lc}(v)^{\delta - (i-1) - 1} \mathrm{lc}(s_{i-1}) \, v \, x^{\deg(s_{i-1}) - \deg(v)} \\
&= \mathrm{lc}(v)^{\delta - i} \left(\mathrm{lc}(v) s_{i-1} - \mathrm{lc}(s_{i-1}) \, v \, x^{\deg(s_{i-1}) - \deg(v)}\right) \\
&= \mathrm{lc}(v)^{\delta - i} s_i. \qquad \qquad \square
\end{aligned}
$$

In the next definition we summarize the pseudo-division process that is based on Theorem 6.43.

Definition 6.44. *Let u and $v \neq 0$ be polynomials in $\mathbf{K}[x]$ and let*

$$\delta = \max(\{\deg(u, x) - \deg(v, x) + 1, \ 0\}).$$

The **pseudo-division** *of u by v is defined by the iteration*

$$
\begin{aligned}
p_0 &= 0, \quad s_0 = u, \\
p_i &= \mathrm{lc}(v, x) p_{i-1} + \mathrm{lc}(s_{i-1}, x) x^{\deg(p_{i-1}, x) - \deg(v, x)}, \qquad (6.62) \\
s_i &= \mathrm{lc}(v, x) s_{i-1} - \mathrm{lc}(s_{i-1}, x) \, v \, x^{\deg(s_{i-1}, x) - \deg(v, x)}.
\end{aligned}
$$

The process terminates when

$$\deg(s_i, x) < \deg(v, x).$$ (6.63)

Let σ be the value of i at termination. Then,

$$q = \text{lc}(v)^{\delta-\sigma} p_\sigma$$ (6.64)

is called the **pseudo-quotient** *and*

$$r = \text{lc}(v)^{\delta-\sigma} s_\sigma$$ (6.65)

is called the **pseudo-remainder** *of u divided by v. We also represent the pseudo-quotient with the operator notation* psquot(u, v, x) *and the pseudo-remainder with* psrem(u, v, x).

The definition includes the case $\deg(u, x) \geq \deg(v, x)$, for which $\delta \geq 1$, and the case $\deg(u, x) < \deg(v, x)$, for which $\delta = 0$.

When $\deg(u, x) \geq \deg(v, x)$, there is at least one iteration, and so $\sigma \geq 1$. Observe that the pseudo-quotient and pseudo-remainder include the factor $\text{lc}(v)^{\delta-\sigma}$ so that they correspond to the quotient and remainder obtained by the recursive division of $\text{lc}(v)^\delta u$ by v.

When $\deg(u, x) < \deg(v, x)$, there are no iterations, and so $\sigma = 0$. In this case, psquot$(u, v, x) = 0$ and psrem$(u, v, x) = u$.

Example 6.45. Let $u = 5\,x^4 y^3 + 3\,xy + 2$ and $v = 2\,x^3 y + 2\,x + 3$ be polynomials in $\mathbf{Q}[x, y]$ with main variable x. Then, $\delta = 2$ and

$$p_1 = 5\,y^3\,x, \qquad s_1 = -10\,x^2\,y^3 - 15x\,y^3 + 6x\,y^2 + 4y.$$

Since $\deg(s_1, x) < \deg(v, x)$, the process terminates with $\sigma = 1$, and

$$q = \text{lc}(v, x)^{\delta-\sigma} p_1 = (2\,y)^1 p_1 = 10\,y^4 x,$$

$$r = \text{lc}(v, x)^{\delta-\sigma} s_1 = -20\,x^2 y^4 - 30\,x\,y^4 + 12\,x\,y^3 + 8\,y^2. \qquad \square$$

Pseudo-division satisfies the following properties.

Theorem 6.46. *Let u and $v \neq 0$ be polynomials in $\mathbf{K}[x]$.*

1. *The pseudo-division process terminates with $\sigma \leq \delta$.*

2. *At termination,*

$$\text{lc}(v, x)^\delta u = q\,v + r$$ (6.66)

   where

$$\deg(r, x) < \deg(v, x).$$ (6.67)

3. *The polynomials q and r are the only ones that satisfy (6.66) and (6.67).*

Proof: The degree condition (6.67) is called the *Euclidean Property* for pseudo-division. The proof of the theorem is similar to the one for Theorem 4.5 on page 113. The details are left to the reader (Exercise 3).   □

A procedure[7] for pseudo-division that is based on Definition 6.44 is given in Figure 6.5.

The next example shows that the basic gcd relationship in Equation (6.48) does not always hold for pseudo-division.

Example 6.47. Consider $u = x$ and $v = xy + y$ as polynomials in $\mathbf{Q}[x, y]$ with main variable x. Then, $\delta = 1$, $\sigma = 1$, and

$$\mathrm{lc}(v, x)\, u = x\, y = q\, v + r = (1)(x\, y + y) + (-y).$$

Notice that $\gcd(u, v) = 1$ while $\gcd(v, r) = y$, and so $\gcd(u, v) \neq \gcd(v, r)$. In other words, the pseudo-division process has introduced an extraneous factor y into the greatest common divisor of v and r.   □

Since pseudo-division is such a simple process (it doesn't utilize the gcd properties of the coefficient domain), it is not surprising that it cannot be used directly to find the gcd. However, we show below that by first preprocessing the polynomials and modifying the gcd remainder sequence, pseudo-division can be used as the basis for a gcd algorithm. Before we do this, however, we need a few more mathematical tools.

Content and Primitive Part

Definition 6.48. *Let u be a polynomial in $\mathbf{K}[x]$. The **content** of u is defined by one of the following rules.*

1. *If $u = \displaystyle\sum_{i=0}^{n} a_i x^i$ is a sum with two or more (non-zero) terms, then the content is the greatest common divisor in $\mathbf{K}$ of the non-zero coefficients of u.*

---

[7]See footnote 2 on page 115 for comments about the termination condition in line 8.

Procedure $Pseudo\_division(u, v, x)$;
Input
    u, v : multivariate polynomials with x the main variable;
    x : a symbol;
Output
    the list $[q, r]$;
Local Variables
    $p, s, m, n, \delta, lcv, \sigma, lcs$;
Begin

1    $p := 0$;
2    $s := u$;
3    $m := Degree\_gpe(s, x)$;
4    $n := Degree\_gpe(v, x)$;
5    $\delta := Max(\{m - n + 1,\ 0\})$;
6    $lcv := Coefficient\_gpe(v, x, n)$;
7    $\sigma := 0$;
8   **while** $m \geq n$ **do**
9      $lcs := Coefficient\_gpe(s, x, m)$;
10     $p := lcv * p + lcr * x^{m-n}$;
11     $s := Algebraic\_expand\left(lcv * s - lcs * v * x^{m-n}\right)$;
12     $\sigma := \sigma + 1$;
13     $m := Degree\_gpe(s, x)$;
14    $Return\left(\left[Algebraic\_expand\left(lcv^{\delta-\sigma} * p\right), Algebraic\_expand\left(lcv^{\delta-\sigma} * s\right)\right]\right)$
End

$$Pseudo\_quotient(u, v, x) \overset{\text{function}}{:=} Operand(Pseudo\_division(u, v, x), 1);$$

$$Pseudo\_remainder(u, v, x) \overset{\text{function}}{:=} Operand(Pseudo\_division(u, v, x), 2);$$

Figure 6.5. An MPL procedure for pseudo-division of u by v and functions that extract the pseudo-quotient and pseudo-remainder. (Implementation: Maple (txt), Mathematica (txt), MuPAD (txt).)

2. If $u = a_i x^i$ is a non-zero monomial, then the content is $d\,a_i$ where d is the unique unit in **K** so that the content is unit normal.

3. If $u = 0$, then the content is 0.

The content is denoted by $\text{cont}(u, x)$ or by $\text{cont}(u)$ if x is understood from the context.

Observe that $\text{cont}(u, x)$ is always unit normal.

Example 6.49. In $\mathbf{Z}[x]$, we have

$$\mathrm{cont}(4\,x^2 - 6\,x,\ x) = 2, \quad \mathrm{cont}(2\,x,\ x) = 2, \quad \mathrm{cont}(-x,\ x) = 1.$$

In $\mathbf{Q}[x]$, all non-zero polynomials have content 1.

In $\mathbf{Q}[x, y]$ with main variable x, we have

$$\mathrm{cont}((1/2)\,x\,y + 6\,y,\ x) = y, \quad \mathrm{cont}(2\,x,\ x) = 1, \quad \mathrm{cont}(-x,\ x) = 1. \quad \square$$

Definition 6.50. *A polynomial $u \neq 0$ in $\mathbf{K}[x]$ is a* **primitive** *polynomial if* $\mathrm{cont}(u, x) = 1$.

Example 6.51. In $\mathbf{Z}[x]$, the polynomials $2\,x^2 + 3\,x$ and $-x$ are primitive, while $4x^2 + 6x$ and $2x$ are not.

In $\mathbf{Q}[x]$, all non-zero polynomials are primitive.

In $\mathbf{Q}[x, y]$ with main variable x, the polynomials $x\,y + y + 1$ and $2x$ are primitive, while $(1/2)x\,y + 6y$ and $x^2 y^2$ are not. $\quad\square$

Theorem 6.52. *A polynomial $u \neq 0$ in $\mathbf{K}[x]$ can be factored uniquely as* $u = c\,v$, *where c is unit normal in $\mathbf{K}$ and v is primitive in $\mathbf{K}[x]$.*

Proof: To obtain the factorization, let

$$u = \sum_{i=0}^{n} a_i x^i, \quad c = \mathrm{cont}(u, x).$$

Therefore, $u = c\,v$ where

$$v = \sum_{i=0}^{n} \mathrm{cof}(c, a_i)x^i.$$

To show that v is primitive, we show that an expression r that divides each of the non-zero coefficients of v must be a unit. Since $r\,c$ divides each coefficient of u, and since c is the greatest common divisor of the coefficients of u, $r\,c \mid c$. Therefore, $c = r\,c\,\mathrm{cof}(r\,c, c)$, and by Theorem 6.4, $1 = r\,\mathrm{cof}(r\,c, c)$. This implies r is a unit, and so v is primitive.

To show the uniqueness property, suppose that there were two such factorizations

$$u = c_1\,v_1 = c_2\,v_2$$

where

$$v_1 = \sum_{i=0}^{n} e_i x^i, \quad v_2 = \sum_{i=0}^{n} f_i x^i.$$

Since $c_1 e_i = c_2 f_i$ for $i = 0, \ldots, n$, and since v_1 and v_2 are primitive, we have

$$c_1 = \gcd(c_1 e_0, \ldots, c_1 e_n) = \gcd(c_2 f_0, \ldots, c_2 f_n) = c_2.$$

This implies $v_1 = v_2$, and so the representation is unique. $\qquad \square$

Definition 6.53. *Let $u \neq 0$ be a polynomial in $\mathbf{K}[x]$. The polynomial v in Theorem 6.52 is called the* **primitive part** *of u with respect to x and is denoted by $\mathrm{pp}(u, x)$. If $u = 0$, the primitive part of u is, by definition, 0. If x is understood from the context, we use the simpler notation $\mathrm{pp}(u)$.*

Example 6.54. Let $u = (y^2 + 2y + 1) x^2 + (2y^2 - 2)x + (3y + 3)$ be a polynomial in $\mathbf{Q}[x, y]$ with main variable x. Then,

$$\mathrm{cont}(u, x) = y + 1, \quad \mathrm{pp}(u, x) = (y + 1) x^2 + (2y - 2) x + 3. \qquad \square$$

Gauss's Theorem

To develop our algorithm, we need the fact that the product of primitive polynomials is primitive (Theorem 6.56 below). The next theorem, which is a fundamental one for the analysis of polynomials in $\mathbf{K}[x]$, is used to prove this property.

Theorem 6.55. *Let u and v be in $\mathbf{K}[x]$, and let c be an irreducible expression in $\mathbf{K}$ that divides each coefficient of the product $u\,v$. Then, c divides each coefficient of u or each coefficient of v.*

Proof: Let

$$
\begin{aligned}
u &= u_m x^m + u_{m-1} x^{m-1} + \cdots + u_1 x + u_0, \\
v &= v_n x^n + v_{n-1} x^{n-1} + \cdots + v_1 x + v_0.
\end{aligned}
$$

Then, the coefficient c_k of x^k in $u\,v$ is given by

$$
\begin{aligned}
c_0 &= u_0 v_0, \\
c_1 &= u_0 v_1 + u_1 v_0, \\
c_2 &= u_0 v_2 + u_1 v_1 + u_2 v_0, \\
&\;\;\vdots
\end{aligned}
$$

$$c_k = \sum_{i=0}^{k} u_i v_{k-i} \tag{6.68}$$

$$\vdots$$

where the coefficients of u and v in the sum (6.68) that have subscripts larger than their degrees are assumed to be zero. Suppose that c divides all the coefficients c_k. If c does not divide all the coefficients of u or v, then there are indices i_0 and j_0 such that

$$c \mid u_0, u_1, \ldots, u_{i_0-1}, \quad c \nmid u_{i_0}, \tag{6.69}$$

$$c \mid v_0, v_1, \ldots, v_{j_0-1}, \quad c \nmid v_{j_0}. \tag{6.70}$$

Let

$$R = \sum_{i=0}^{i_0-1} u_i v_{i_0+j_0-i}, \qquad S = \sum_{i=i_0+1}^{i_0+j_0} u_i v_{i_0+j_0-i}.$$

Then, by Equation (6.68),

$$c_{i_0+j_0} = R + u_{i_0} v_{j_0} + S$$

where $c \mid R$, $c \mid S$, and $c \mid c_{i_0+j_0}$. Therefore, $c \mid u_{i_0} v_{j_0}$, and since c is irreducible, Theorem 6.10(2) implies $c \mid u_{i_0}$ or $c \mid v_{j_0}$, which contradicts the conditions (6.69) or (6.70). Therefore, c divides all the coefficients of u or v. ☐

Theorem 6.56. [Gauss's Theorem] *If u and v are primitive polynomials in $\mathbf{K}[x]$, then uv is also primitive.*

Proof: If uv were not primitive, then there would be a (non-unit) irreducible expression c in $\mathbf{K}$ that divides each of the coefficients of uv. Then, by Theorem 6.55, c divides all the coefficients of either u or v which implies that one of these polynomials is not primitive. ☐

The next theorem is a restatement of Gauss's theorem in terms of contents and primitive parts.

Theorem 6.57. *Let u and v be polynomials in $\mathbf{K}[x]$. Then,*

$$\mathrm{cont}(uv) = \mathrm{cont}(u)\,\mathrm{cont}(v), \tag{6.71}$$

$$\mathrm{pp}(uv) = \mathrm{pp}(u)\,\mathrm{pp}(v). \tag{6.72}$$

Proof: By representing $u\,v$, u, and v in terms of contents and primitive parts, we have

$$u\,v = \operatorname{cont}(u\,v)\,\operatorname{pp}(u\,v) \tag{6.73}$$

and

$$\begin{aligned}
u\,v &= (\operatorname{cont}(u)\,\operatorname{pp}(u)) \cdot (\operatorname{cont}(v)\,\operatorname{pp}(v)) \\
&= (\operatorname{cont}(u)\,\operatorname{cont}(v)) \cdot (\operatorname{pp}(u)\,\operatorname{pp}(v)). \tag{6.74}
\end{aligned}$$

However, since the contents are unit normal, $\operatorname{cont}(u)\,\operatorname{cont}(v)$ is also unit normal, and by Theorem 6.56, $\operatorname{pp}(u)\,\operatorname{pp}(v)$ is primitive. With these observations, the relations in Equations (6.71) and (6.72) follow from Equations (6.73) and (6.74) and the uniqueness property in Theorem 6.52. $\qquad\square$

The next theorem, which is the basis for the preprocessing part of the gcd algorithm, expresses the gcd in terms of the greatest common divisors of the contents and primitive parts.

Theorem 6.58. *Let u and v be polynomials in $\mathbf{K}[x]$, and let*

$$\begin{aligned}
d &= \gcd(\operatorname{cont}(u), \operatorname{cont}(v)) \quad (\text{in } \mathbf{K}), \\
w &= \gcd(\operatorname{pp}(u), \operatorname{pp}(v)) \quad (\text{in } \mathbf{K}[x]).
\end{aligned}$$

Then, $\gcd(u, v) = d\,w$.

Proof: Since d divides both $\operatorname{cont}(u)$ and $\operatorname{cont}(v)$, and w divides both $\operatorname{pp}(u)$ and $\operatorname{pp}(v)$, the product $d\,w$ divides both u and v. Now let p be any divisor of u and v. Then, by Theorem 6.57

$$\begin{aligned}
\operatorname{cont}(p) \mid \operatorname{cont}(u), &\quad \operatorname{pp}(p) \mid \operatorname{pp}(u), \\
\operatorname{cont}(p) \mid \operatorname{cont}(v), &\quad \operatorname{pp}(p) \mid \operatorname{pp}(v),
\end{aligned}$$

and so $\operatorname{cont}(p) \mid d$ and $\operatorname{pp}(p) \mid w$. Therefore, $p \mid d\,w$. Finally, since d and w are greatest common divisors, they are unit normal, and so $d\,w$ is unit normal as well. Therefore, $\gcd(u, v) = d\,w$. $\qquad\square$

The Primitive Gcd Algorithm for K[x]

Although pseudo-division cannot be used directly to compute the gcd, it can be used for primitive polynomials. In this case, the algorithm is based on the relationship in the next theorem.

Theorem 6.59. *Let u and $v \neq 0$ be primitive polynomials in $\mathbf{K}[x]$. Then,*

$$\gcd(u, v) = \gcd(v, \operatorname{pp}(\operatorname{psrem}(u, v, x), x)). \tag{6.75}$$

Proof: Let $r = \text{psrem}(u, v, x)$. To prove the theorem, we show that the two pairs of polynomials u, v and v, $\text{pp}(r, x)$ have the same common divisors.

Let p be a polynomial that divides u and v. Observe that p is primitive because it divides a primitive polynomial (Exercise 8). In addition, since

$$\text{lc}(v, x)^\delta u = \text{psquot}(u, v, x) v + r, \tag{6.76}$$

$p \mid r$, and there is a polynomial s such that $r = p s$. By Theorem 6.57,

$$\text{pp}(r, x) = \text{pp}(p s, x) = \text{pp}(p, x) \text{pp}(s, x) = p \cdot \text{pp}(s, x),$$

and so $p \mid \text{pp}(r, x)$.

Next, let's suppose that p divides v and $\text{pp}(r, x)$ where, by Exercise 8, p is primitive. Therefore, by Equation (6.76), $p \mid \text{lc}(v, x)^\delta u$, and there is a polynomial s such that

$$
\begin{aligned}
\text{lc}(v, x)^\delta u &= s p \\
&= \text{cont}(s, x) \text{pp}(s, x) \text{cont}(p, x) \text{pp}(p, x) \\
&= \text{cont}(s, x) \text{pp}(s, x) p
\end{aligned}
$$

where, by Theorem 6.56, the product $\text{pp}(s, x) p$ is primitive. Since u is primitive and since $\text{cont}(s, x)$ is unit normal, Theorem 6.52 implies

$$c \, \text{lc}(v, x)^\delta = \text{cont}(s, x)$$

where c is the unique unit in $\mathbf{K}$ so that $c \, \text{lc}(v, x)^\delta$ is unit normal. Therefore,

$$u = \text{pp}(u) = c \, \text{pp}(s, x) \, p,$$

and $p \mid u$.

This analysis shows that the polynomials u and v and the polynomials v and $\text{pp}(r, x)$ have the same common divisors, and therefore $\gcd(u, v) = \gcd(v, \text{pp}(r, x))$. $\qquad\square$

The next example shows that the pp operator in Equation (6.75) is required since the pseudo-division of a primitive polynomial u by a primitive polynomial v may give a pseudo-remainder that is not primitive.

Example 6.60. The polynomials

$$u = x^2 + x + (y + 1), \qquad v = x^2 y^2 + x$$

are primitive in $\mathbf{Z}[x, y]$ with main variable x. Then,

$$\mathrm{psrem}(u, v, x) = (y^2 - 1)\, x + y^2\,(y + 1)$$

which is not primitive. $\square$

We describe now a gcd algorithm for polynomials that is based on Theorems 6.58 and 6.59. Let u and v be two non-zero polynomials in $\mathbf{K}[x]$. First, compute

$$d = \gcd(\mathrm{cont}(u, x), \mathrm{cont}(v, x)). \tag{6.77}$$

This step, which constitutes the preprocessing step of the algorithm, uses a gcd algorithm in the coefficient domain $\mathbf{K}$ to compute the contents and the gcd in Equation (6.77). Next, when $v \neq 0$, consider the remainder sequence:

$$
\begin{aligned}
R_{-1} &= \mathrm{pp}(u), \\
R_0 &= \mathrm{pp}(v), \\
R_1 &= \mathrm{pp}(\mathrm{psrem}(R_{-1}, R_0, x)), \\
&\ \vdots \\
R_i &= \mathrm{pp}(\mathrm{psrem}(R_{i-2}, R_{i-1}, x)), \\
&\ \vdots
\end{aligned}
\tag{6.78}
$$

This sequence is called the *primitive* remainder sequence because each of the remainders is primitive. Since pseudo-division satisfies the Euclidean property, there is some remainder in this sequence which is zero. Let $R_\rho = 0$ be the first such remainder. By Theorem 6.59,

$$
\begin{aligned}
\gcd(\mathrm{pp}(u), \mathrm{pp}(v)) &= \gcd(R_{-1}, R_0) \\
&= \gcd(R_0, R_1) \\
&\ \vdots \\
&= \gcd(R_{\rho-1}, R_\rho) = \gcd(R_{\rho-1}, 0) \\
&= R_{\rho-1}
\end{aligned}
\tag{6.79}
$$

where $R_{\rho-1}$ is the non-normalized form of the gcd. Therefore, by Theorem 6.58 and Equation (6.77),

$$\gcd(u, v) = c \cdot d \cdot R_{\rho-1}$$

where the unit c in $\mathbf{K}$ transforms the gcd to unit normal form.

This discussion is summarized in the following theorem.

Theorem 6.61. *Let u and v be polynomials in $\mathbf{K}[x]$ at least one of which is non-zero, and let*

$$d = \gcd(\operatorname{cont}(u, x), \operatorname{cont}(v, x)). \tag{6.80}$$

Then,

$$\gcd(u, v) = c \cdot d \cdot R_{\rho-1} \tag{6.81}$$

where c is the unique unit so that transforms the gcd to unit normal form.

Proof: Although the theorem was derived assuming $v \neq 0$, it is true when $u \neq 0$ and $v = 0$ as well. In this case, $\rho = 0$ and

$$R_{\rho-1} = \operatorname{pp}(u), \quad d = \operatorname{cont}(u).$$

Therefore,

$$\gcd(u, 0) = c \cdot d \cdot R_{\rho-1} = c \cdot \operatorname{cont}(u) \operatorname{pp}(u) = c\,u$$

where c transforms u to unit normal form.   □

The algorithm based on Equations (6.78), (6.79), (6.80), and (6.81) is called the *primitive* gcd algorithm because the remainders in Equation (6.78) are primitive.

Example 6.62. Let

$$u = -yx^2 + y^3, \qquad v = yx^2 + 2y^2x + y^3$$

be polynomials in $\mathbf{Z}[x, y]$ with main variable x. Then,

$$\operatorname{cont}(u, x) = y, \quad \operatorname{cont}(v, x) = y,$$
$$d = \gcd(\operatorname{cont}(u, x), \operatorname{cont}(v, x)) = y,$$
$$R_{-1} = \operatorname{pp}(u, x) = -x^2 + y^2,$$
$$R_0 = \operatorname{pp}(v, x) = x^2 + 2yx + y^2,$$

and

$$
\begin{aligned}
R_1 &= \operatorname{pp}(\operatorname{psrem}(R_{-1}, R_0, x), x) \\
&= \operatorname{pp}(2\,y^2 + 2\,y\,x, x) \\
&= y + x,
\end{aligned}
$$

$$
\begin{aligned}
R_2 &= \operatorname{pp}(\operatorname{psrem}(R_0, R_1, x), x) \\
&= \operatorname{pp}(0, x) \\
&= 0.
\end{aligned}
$$

At this point, the process terminates, and using Equation (6.81),

$$\gcd(u, v) = d\, R_1 = y\, x + y^2.$$

□

The *Mv_poly_gcd* and *Mv_poly_gcd_rec* Procedures. Procedures that obtain the gcd in the multivariate polynomial domain $\mathbf{K}[x_1, \ldots, x_p]$ when $\mathbf{K}$ is either $\mathbf{Z}$ or $\mathbf{Q}$ are given in Figures 6.6 and 6.7. The input polynomials are multivariate polynomials (in expanded form) in the symbols in the list L where the first symbol in L is the main variable. The main procedure *Mv_poly_gcd* handles some simple cases in lines 1–4, and, at line 6, calls on the recursive procedure *Mv_poly_gcd_rec* which performs the calculations. The *Normalize* operator in lines 2, 4, and 6, which transforms the gcd to unit normal form, is described in Exercise 12. Since the unit normal form depends on the base coefficient domain, the domain variable K is included in the parameter list for *Normalize*.

The procedure *Mv_poly_gcd_rec*, which returns the greatest common divisor in non-normalized form, is highly recursive because the content computations in lines 7, 8, and 17 and the gcd computation in line 9 involve recursive calls on the procedure. (The *Polynomial_content* operator is described in Exercise 7.)

Procedure *Mv_poly_gcd*(u, v, L, K);
Input
    u, v : multivariate polynomials with variables in L and coefficients
        in $\mathbf{Z}$ or $\mathbf{Q}$;
    L : list of symbols;
    K : the symbol $\mathbf{Z}$ or $\mathbf{Q}$;
Output
    $\gcd(u, v)$;
Begin
1   **if** $u = 0$ **then**
2       $Return(Normalize(v, L, K))$
3   **elseif** $v = 0$ **then**
4       $Return(Normalize(u, L, K))$
5   **else**
6       $Return(Normalize(Mv\_poly\_gcd\_rec(u, v, L, K), L, K))$
End

Figure 6.6. The main MPL procedure that obtains the gcd of two multivariate polynomials with coefficients in $\mathbf{Z}$ or $\mathbf{Q}$. (Implementation: Maple (txt), Mathematica (txt), MuPAD (txt).)

Procedure $Mv\_poly\_gcd\_rec(u, v, L, K)$;
Input
  u, v : non-zero multivariate polynomials with variables in L and
      coefficients in **Z** or **Q**;
  L : a list of symbols;
  K : the symbol **Z** or **Q**;
Output
  $\gcd(u, v)$ (not normalized);
Local Variables
  $x, R, cont\_u, cont\_v, d, pp\_u, pp\_v, r, pp\_r, cont\_r$;
Begin
1   **if** $L = [\,]$ **then**
2   **if** $K = \mathbf{Z}$ **then** $Return(Integer\_gcd(u, v))$
3   **elseif** $K = \mathbf{Q}$ **then** $Return(1)$
4   **else**
5       $x := First(L)$;
6       $R := Rest(L)$;
7       $cont\_u := Polynomial\_content(u, x, R, K)$;
8       $cont\_v := Polynomial\_content(v, x, R, K)$;
9       $d := Mv\_poly\_gcd\_rec(cont\_u, cont\_v, R, K)$;
10      $pp\_u := Rec\_quotient(u, cont\_u, L, K)$;
11      $pp\_v := Rec\_quotient(u, cont\_v, L, K)$;
12   **while** $pp\_v \neq 0$ **do**
13        $r := Pseudo\_remainder(pp\_u, pp\_v, x)$;
14   **if** $r = 0$ **then**
15          $pp\_r := 0$
16   **else**
17          $cont\_r := Polynomial\_content(r, x, R, K)$;
18          $pp\_r := Rec\_quotient(u, cont\_r, L, K)$;
19        $pp\_u := pp\_v$;
20        $pp\_v := pp\_r$;
21      $Return(Algebraic\_expand(d * pp\_u))$
    End

Figure 6.7. The MPL procedure that obtains a (non-normalized) gcd of two
multivariate polynomials with coefficients in **Z** or **Q**. (Implementation: Maple
(txt), Mathematica (txt), MuPAD (txt).)

Lines 1–3 provide a termination condition for the recursion. When
the input list L is empty, the input expressions are in the base coefficient
domain. When this domain is **Z**, we return the gcd of the two integers
(line 2), and when the domain is **Q**, we simply return the integer 1 (line 3).

In lines 5–9, we obtain the expression d in Equation (6.80), and, in lines 10–11, we obtain the primitive parts of the input polynomials. In lines 12–20, we form the polynomial remainder sequence (6.78), and, in line 21, we obtain the gcd in non-normalized form using Equation (6.81). (The constant c that obtains the normalized form is obtained in *Normalize* at line 6 of *Mv_poly_gcd*.)

Appraisal of the Algorithm

In Example 4.26 on page 130 we showed that the gcd algorithm in $\mathbf{Q}[x]$ exhibits a phenomenon called coefficient explosion where succeeding polynomials in the remainder sequence have coefficients that require more and more digits. The next example shows that this problem can be even more pronounced for multivariate polynomials.

Example 6.63. Consider the polynomials

$$
\begin{aligned}
u &= x^3y^2 + 6x^4y + 9x^5 + 4x^2y^2 + 24x^3y + 36x^4 + 5xy^2 + 30yx^2 \\
  &\quad + 45x^3 + 2y^2 + 12yx + 18x^2, \\
v &= x^5y^2 + 8x^4y + 16x^3 + 12x^4y^2 + 96x^3y + 192x^2 + 45x^3y^2 \\
  &\quad + 360yx^2 + 720x + 50x^2y^2 + 400yx + 800
\end{aligned}
$$

as polynomials in $\mathbf{Z}[x, y]$ with x the main variable. Observe that

$$
\deg(u, x) = 5, \quad \deg(u, y) = 2, \quad \deg(v, x) = 5, \quad \deg(v, y) = 2,
$$

and the maximum number of digits in the integer coefficients in u and v is 3.

By collecting coefficients in powers of x, we obtain

$$
\begin{aligned}
u &= 9x^5 + (6y + 36)x^4 + \left(y^2 + 24y + 45\right)x^3 + \left(4y^2 + 30y + 18\right)x^2 \\
  &\quad + \left(5y^2 + 12y\right)x + 2y^2, \\
v &= y^2x^5 + \left(8y + 12y^2\right)x^4 + \left(96y + 16 + 45y^2\right)x^3 \\
  &\quad + \left(192 + 50y^2 + 360y\right)x^2 + (720 + 400y)x + 800.
\end{aligned}
$$

In this form, we see that

$$
\mathrm{cont}(u, x) = 1, \quad \mathrm{cont}(v, x) = 1,
$$

and so,

$$
d = \gcd(\mathrm{cont}(u, x), \mathrm{cont}(v, x)) = 1.
$$

Therefore, $R_{-1} = u$ and $R_0 = v$, and at the first iteration, we have (in collected form)

$$
\begin{aligned}
r &= \text{psrem}(R_{-1}, R_0, x) \\
&= \left(-72\,y - 72\,y^2 + 6\,y^3\right) x^4 \\
&\quad + \left(y^4 - 360\,y^2 - 144 - 864\,y + 24\,y^3\right) x^3 \\
&\quad + \left(4\,y^4 - 3240\,y - 1728 + 30\,y^3 - 432\,y^2\right) x^2 \\
&\quad + \left(5\,y^4 - 6480 - 3600\,y + 12\,y^3\right) x \\
&\quad + 2\,y^4 - 7200.
\end{aligned}
$$

Although pseudo-division has decreased the degree in the main variable x by 1, it has increased the degree in y to 4 and increased the size of the integer coefficients. Because of this, the content computation at line 17 of *Mv_poly_gcd_rec* involves coefficient explosion of up to 7 digits. Since the content of this polynomial is 1, we have $R_1 = r$.

At the next iteration, we obtain

$$
\begin{aligned}
r &= \text{psrem}(R_0, R_1, x) \\
&= \left(-6048\,y^6 + y^{10} + 672\,y^8 - 2448\,y^7 + 20736\,y^2\right. \\
&\quad \left. - 48\,y^9 + 82944\,y^3 + 221184\,y^4 + 29376\,y^5\right) x^3 \\
&\quad + \left(4\,y^{10} + 57024\,y^5 - 89856\,y^6 + 2040\,y^8 + 4464\,y^7\right. \\
&\quad \left. + 995328\,y^3 + 248832\,y^2 - 192\,y^9 + 1254528\,y^4\right) x^2 \\
&\quad + \left(-144288\,y^6 + 1848\,y^8 + 5\,y^{10} - 176256\,y^5\right. \\
&\quad + 1710720\,y^4 - 240\,y^9 + 19296\,y^7 + 3732480\,y^3 \\
&\quad \left. + 933120\,y^2\right) x \\
&\quad + 1036800\,y^2 - 345600\,y^5 + 22464\,y^6 + 172800\,y^4 \\
&\quad + 912\,y^8 - 96\,y^9 + 2\,y^{10} + 1152\,y^7 + 4147200\,y^3.
\end{aligned}
$$

Again, pseudo-division has decreased the degree in x by 1, but has increased the degree in y to 10, and has increased the size of the integer coefficients. At this step, the content computation at line 17 involves coefficients with up to 41 digits. Since the content of this polynomial is y^2, we have

$$
R_2 = \text{Rec\_quotient}(r, y^2, [x, y]).
$$

The iteration continues until $R_5 = 0$. The coefficient and degree explosion for the remaining iterations is summarized in Figure 6.8. Observe that, at iterations 3 and 4, the remainder r involves more coefficient explosion and degree explosion in y (columns 2 and 4). In addition, the degree of

i	Max. Digits r (line 13)	$\deg(r, x)$ (line 13)	$\deg(r, y)$ (line 13)	$\deg(\text{cont}(r)), y)$ (line 17)	Max. Digits in computation of cont_r (line 17)
1	4	4	4	0	7
2	7	3	10	2	41
3	14	2	18	10	44
4	17	1	24	24	17
5	1	$-\infty$	$-\infty$	$-\infty$	1

Figure 6.8. The degree and coefficient explosion in the procedure *Mv_poly_gcd_rec* for the polynomials in Example 6.63.

the content increases (column 5), and the most significant coefficient explosion occurs during the content computation (column 6). Although, at the fourth iteration $\deg(r, y) = 24$, all of this involves content, and once this content is removed, the algorithm obtains

$$\gcd(u, v) = x + 2. \qquad \qquad \square$$

The Subresultant gcd Algorithm in K[x]

There are two competing actions at work in the gcd algorithm that affect the coefficient and degree explosion. First, the content computation and division at lines 17–18 of *Mv_poly_gcd_rec* tend to reduce the size of the remainders. On the other hand, the content computation for these remainders involves significant coefficient explosion and degree explosion in the auxiliary variables.

We describe now another approach that avoids the content computation of the pseudo-remainders but still tends to reduce the size of the remainders. The algorithm is known as the *subresultant gcd algorithm* because its mathematical justification is based on a mathematical concept with this name. Since the mathematics behind this algorithm is quite involved, we only describe the algorithm and refer the reader to the references cited at the end of the chapter for the mathematical details.

Let u and v be in $\mathbf{K}[x]$. As before, we first compute

$$d = \gcd(\text{cont}(u, x), \text{cont}(v, x)), \qquad (6.82)$$

and, for this algorithm, compute

$$g = \gcd(\text{lc}(u, x), \text{lc}(v, x)) \qquad (6.83)$$

where the gcd and content computations are done in $\mathbf{K}$. For $v \neq 0$ and $\deg(u, x) \geq \deg(v, x)$, this approach replaces the sequence in Equation (6.78) by the sequence

$$
\begin{aligned}
S_{-1} &= \mathrm{pp}(u), \\
S_0 &= \mathrm{pp}(v), \\
S_1 &= \frac{\mathrm{psrem}(S_{-1}, S_0, x)}{\beta_1},
\end{aligned}
$$

$$\vdots$$

$$
S_i = \frac{\mathrm{psrem}(S_{i-2}, S_{i-1}, x)}{\beta_i}, \tag{6.84}
$$

$$\vdots$$

where the sequence terminates when $S_\rho = 0$. In this sequence, the content removal in Equation (6.78) is replaced with the division by the expression $\beta_i \neq 0$ that is defined in the following way. Let

$$
\delta_i = \deg(S_{i-2}, x) - \deg(S_{i-1}, x) + 1, \qquad i \geq 1, \tag{6.85}
$$

and consider the recurrence relations

$$
\psi_1 = -1, \qquad \beta_1 = (-1)^{\delta_1}, \tag{6.86}
$$

$$
\psi_i = \frac{(-\mathrm{lc}(S_{i-2}, x))^{\delta_{(i-1)}-1}}{\psi_{i-1}^{\delta_{(i-1)}-2}}, \qquad i > 1, \tag{6.87}
$$

$$
\beta_i = -\mathrm{lc}(S_{i-2}, x)\, \psi_i^{\delta_i - 1}, \qquad i > 1. \tag{6.88}
$$

Once the remainder sequence (6.84) terminates, the gcd is obtained with

$$
s = \frac{\mathrm{lc}(S_{\rho-1})}{g}, \tag{6.89}
$$

$$
W = \frac{S_{\rho-1}}{s}, \tag{6.90}
$$

$$
\gcd(u, v) = c \cdot d \cdot \mathrm{pp}(W, x) \tag{6.91}
$$

where c is the unit that transforms the gcd to unit normal form.

The divisor sequence β_i is defined in terms of an auxiliary sequence ψ_i, and both sequences involve computation in the coefficient domain $\mathbf{K}$. Using some involved mathematics, it can be shown that

$$
\beta_i \mid \mathrm{cont}(S_i, x),
$$

and so, the division in Equation (6.84) removes part of the content from S_i. For this reason, S_i tends to have somewhat larger coefficients and higher degree in the auxiliary variables than R_i. On the other hand, β_i does not involve a gcd computation which significantly reduces the coefficient and degree explosion in $\mathbf{K}$. These points are illustrated in Example 6.64 below.

Finally, this approach requires that $\deg(u, x) \geq \deg(v, x)$, and when this is not so, it is necessary to interchange u and v.

The *Sub_resultant_gcd_rec* Procedure. In Figures 6.9 and 6.10, we give a procedure that uses this algorithm to obtain a gcd in the multivariate polynomial domain $\mathbf{K}[x_1, \ldots, x_p]$ when $\mathbf{K}$ is either $\mathbf{Z}$ or $\mathbf{Q}$. The procedure, which is similar to the procedure for *Mv_poly_gcd_rec* in Figure 6.7, returns the gcd in non-normalized form. The normalized form is obtained using the procedure *Mv_poly_gcd* (Figure 6.6) with *Mv_poly_gcd_rec* at line 6 replaced by *Sub_resultant_gcd_rec*.

In lines 6–11, we select U and V so that $\deg(U, x) \geq \deg(V, x)$. In lines 13–15 we obtain d in Equation (6.82) using the *Polynomial_content_sr* operator described in Exercise 7 and a recursive call on the procedure. The remainder sequence is initialized in lines 16–17, and g in Equation (6.83) is obtained in line 18. In lines 19–38, we obtain the remainder sequence (6.84) where β_i and ψ_i are obtained in lines 23–32. Finally, a non-normalized form of the gcd is obtained in lines 39–43 using Equations (6.89), (6.90), and (6.91). (The constant c that obtains the normalized form is obtained in *Normalize* at line 6 of *Mv_poly_gcd*.)

Example 6.64. Consider the gcd calculation for the polynomials in Example 6.63. In Figure 6.11 we summarize the degree and coefficient explosion obtained by *Sub_resultant_gcd_rec* in the pseudo-remainder calculation (line 21) and β calculation (lines 26 and 32).

Comparing these results with those in Figure 6.8, we see, for this procedure, the remainder r involves coefficient explosion (column 2) of up to 24 digits and degree explosion (column 4) up to degree 32. This is somewhat larger than for r in *Mv_poly_gcd_rec* which involves up to 17 digits and degree 24.

On the other hand, β involves coefficient explosion (column 6) of up to 11 digits and degree explosion (column 5) up to degree 16, which is significantly smaller than the computation of *cont_r* in *Mv_poly_gcd_rec*, which involves up to 44 digits and degree 24. Because of this, the computation time for this example for *Sub_resultant_gcd_rec* is about one third the time for *Mv_poly_gcd_rec*. □

Procedure *Sub_resultant_gcd_rec*(u, v, L, K);
Input
    u, v : non-zero multivariate polynomials with variables in L with
            coefficients in **Z** or **Q**;
    L : a list of symbols;
    K : the symbol **Z** or **Q**;
Output $\gcd(u, v)$ (not normalized);
Local Variables
    $x, U, V, R, cont\_U, cont\_V, d, g, i, r, \delta, \psi, \beta, \delta p, f, s, W, cont\_W, pp\_W$;
Begin
1   **if** $L = [\,]$ **then**
2   **if** $K = \mathbf{Z}$ **then** *Return*$(Integer\_gcd(u, v))$
3   **elseif** $K = \mathbf{Q}$ **then** *Return*(1)
4   **else**
5      $x := First(L)$;
6   **if** $Degree\_gpe(u, x) \geq Degree\_gpe(v, x)$ **then**
7        $U := u$;
8        $V := v$
9   **else**
10       $U := v$;
11       $V := u$
12      $R := Rest(L)$;
13      $cont\_U := Polynomial\_content\_sr(U, x, R, K)$;
14      $cont\_V := Polynomial\_content\_sr(V, x, R, K)$;
15      $d := Sub\_resultant\_gcd\_rec(cont\_U, cont\_V, R, K)$;
16      $U := Rec\_quotient(U, cont\_U, L, K)$;
17      $V := Rec\_quotient(V, cont\_V, L, K)$;
18      $g := Sub\_resultant\_gcd\_rec\Big($
           $Leading\_Coefficient\_gpe(U, x), Leading\_Coefficient\_gpe(V, x), R, K\Big)$;

Continued in Figure 6.10.

Figure 6.9. An MPL procedure that obtains a gcd of two multivariate polynomials using the subresultant algorithm. (Implementation: Maple (txt), Mathematica (txt), MuPAD (txt).)

```
19        i := 1;
20        while  V ≠ 0 do
21            r := Pseudo_remainder(U, V, x);
22            if  r ≠ 0 then
23                if  i = 1 then
24                    δ = Degree_gpe(U, x) − Degree_gpe(V, x) + 1;
25                    ψ := −1;
26                    β = (−1)^δ
27                elseif  i > 1 then
28                    δp = δ;
29                    δ = Degree_gpe(U, x) − Degree_gpe(V, x) + 1;
30                    f := Leading_Coefficient_gpe(U, x);
31                    ψ := Rec_quotient (Algebraic_expand ((−f)^{δp−1}),
                                        Algebraic_expand (ψ^{δp−2}),  R,  K);
32                    β := Algebraic_expand (−f * ψ^{δ−1});
33                U := V;
34                V := Rec_quotient(r, β, L, K);
35                i := i + 1
36            elseif  r = 0 then
37                U := V;
38                V := r;
39        s := Rec_quotient(Leading_Coefficient_gpe(U, x), g, R, K);
40        W := Rec_quotient(U, s, L, K);
41        cont_W := Polynomial_content_sr(W, x, R, K);
42        pp_W := Rec_quotient(W, cont_W, L, K);
43        Return(Algebraic_expand(d * pp_W))
      End
```

Figure 6.10. Continuation of Figure 6.9.

i	Max. digits r (line 21)	$\deg(r, x)$ (line 21)	$\deg(r, y)$ (line 21)	$\deg(\beta, y)$ (lines 26 and 32)	Max. digits in computation of β (lines 26 and 32)
1	4	4	3	0	1
2	7	3	10	2	1
3	15	2	18	6	5
4	24	1	32	16	11
5	1	$-\infty$	$-\infty$	−	−

Figure 6.11.   The degree and coefficient explosion in the procedure *Sub_resultant_gcd_rec* for the polynomials in Example 6.63.

Because of coefficient and degree explosion, gcd computation for polynomials with large coefficients and many variables of high degree is a difficult problem. There are, however, more sophisticated algorithms that obtain the gcd more efficiently than those given here. The references cited at the end of the chapter survey some of the recent developments for this problem.

Rational Simplification

One application of multivariate polynomial gcd computation is to obtain the rationally simplified form of an algebraic expression. Consider the class of expressions RS-IN defined by the following rules.

RS-IN-1. u is an integer.

RS-IN-2. u is a fraction.

RS-IN-3. u is a symbol.

RS-IN-4. u is sum with operands that are in RS-IN.

RS-IN-5. u is a product with operands that are in RS-IN.

RS-IN-6. u is a power with a base that is in RS-IN and exponent that is an integer.

In the next definition we describe a standard form for expressions in RS-IN.

Definition 6.65. *Let u be an expression in RS-IN. The expression u is in* **rationally simplified form** *if it satisfies one of the following rules.*

RS-1. *u is a multivariate polynomial with integer coefficients.*

RS-2. *Let*
$$n = Numerator(u), \quad d = Denominator(u).$$

Then, n and d satisfy the following:

1. *n and d are multivariate polynomials in a list of symbols L with integer coefficients.*
2. *$d \neq 0$ and $d \neq 1$.*
3. *n and d are relatively prime.*
4. *n and d are in unit normal form.*

RS-3. *$u = (-1)\,v$ where v satisfies RS-2.*

The process of transforming an expression in the class RS-IN to a rationally simplified form in the class RS is called *rational simplification*.

Example 6.66. The following transformations illustrate the rational simplification process:

$$\frac{-4\,a^2 + 4\,b^2}{8\,a^2 - 16\,ab + 8\,b^2} \quad \rightarrow \quad -\frac{a+b}{2\,a - 2\,b}, \quad L = [a, b], \qquad (6.92)$$

$$\rightarrow \quad \frac{a+b}{2\,b - 2\,a}, \quad L = [b, a], \qquad (6.93)$$

$$\frac{1}{\left(\dfrac{1}{a} + \dfrac{c}{a\,b}\right)} + \frac{a\,b\,c + a\,c^2}{(b+c)^2} \rightarrow a, \quad L = [a, b, c],$$

$$\frac{2\,a^3 + 22\,ab + 6\,a^2 + 7\,a + 6\,a^2b + 12\,b^2 + 21\,b}{7\,a^2 - 2\,a^2b - 5\,a - 5\,ab^2 + 21\,ab - 15\,b + 3\,b^3} \qquad (6.94)$$
$$\rightarrow \frac{2\,a^2 + 4\,b + 6\,a + 7}{7\,a - 2\,ab + b^2 - 5}, \quad L = [b, a].$$

The expression to the right of the arrow is in rationally simplified form. In each case the variable order is indicated by a list L where the main variable is the first symbol in L. Observe that in Equations (6.92) and (6.93), the rational simplified form depends on the variable order. In Equation (6.94), there is a common factor of $a + 3\,b$ in the numerator and denominator. □

In Exercise 13 we describe a procedure that obtains the rationally simplified form of an expression.

The Extended Euclidean Algorithm in K[x]

When the coefficient domain is a field, using the extended Euclidean algorithm, we obtain polynomials A and B such that $A\,u + B\,v = \gcd(u, v)$. When the coefficient domain is not a field, this relationship takes the following form.

Theorem 6.67. *Let u and v be polynomials in $\mathbf{K}[x]$. Then, there are polynomials A and B in $\mathbf{K}[x]$ and a unit normal expression s in $\mathbf{K}$ such that*

$$A\,u + B\,v = s\,\gcd(u, v). \qquad (6.95)$$

Proof: A constructive proof of the theorem is similar to the proofs for Theorem 2.12 (page 24) and Theorem 4.27 (page 132) with modifications to accommodate the properties of the coefficient domain $\mathbf{K}$. The details of the proof are left to the reader (Exercise 14).   □

Example 6.68. Consider the polynomials

$$u = -4\,x^3 y + 4\,xy^3 + 4\,xy, \qquad v = 6\,x^4 y + 12\,x^3 y^2 + 6\,y^3 x^2$$

in $\mathbf{Z}[x, y]$ with main variable x. Then,

$$s = 24\,y^4 + 24\,y^2, \qquad \gcd(u, v) = 2\,xy,$$

$$A = 12\,y^2 + 24\,y^4 x^2 + 24\,y^5 x + 12\,y^2 x^2, \quad B = 16\,y^4 x + 8\,xy^2 - 16\,y^5 - 16\,y^3.$$

                                                                                □

The next theorem extends the divisor properties described in Theorem 6.10 (page 205) from the coefficient domain $\mathbf{K}$ to $\mathbf{K}[x]$.

Theorem 6.69. *Suppose that u, v, and w are in $\mathbf{K}[x]$ with $w \mid u\,v$.*

1. If w and u are relatively prime, then $w \mid v$.

2. If w is irreducible, then $w \mid u$ or $w \mid v$.

Proof: To show (1), by Theorem 6.67, there are polynomials A and B in $\mathbf{K}[x]$ and an s in $\mathbf{K}$ such that

$$A\,w + B\,u = s\,\gcd(w, u)$$

where, since w and u are relatively prime, $s\,\gcd(w, u)$ is in $\mathbf{K}$. Therefore,

$$A\,w\,v + B\,u\,v = s\,\gcd(w, u)\,v,$$

and, since w divides each term in the sum, $w \mid s\,\gcd(w, u)\,v$. Furthermore, Theorem 6.57 implies that

$$\mathrm{pp}(w, x) \mid \mathrm{pp}(s\,\gcd(w, u), x)\,\mathrm{pp}(v, x),$$

and, since $\mathrm{pp}(s\,\gcd(w, u), x) = 1$,

$$\mathrm{pp}(w, x) \mid \mathrm{pp}(v, x). \tag{6.96}$$

In addition, since $w \mid u\,v$, Theorem 6.57 implies

$$\mathrm{cont}(w, x) \mid \mathrm{cont}(u, x)\,\mathrm{cont}(v, x).$$

However, w and u are relatively prime, and so $\text{cont}(w, x)$ and $\text{cont}(u, x)$ are relatively prime as well. Therefore, Theorem 6.10(1) implies

$$\text{cont}(w, x) \mid \text{cont}(v, x). \tag{6.97}$$

By combining the relations (6.96) and (6.97), we have $w \mid v$.

Part (2) is a direct consequence of Part (1). □

Theorem 6.70. *If* $\mathbf{K}$ *is a unique factorization domain, then* $\mathbf{K}[x]$ *is also a unique factorization domain.*

Proof: A proof, which is based on Theorem 6.69, is similar to the proof of Theorem 4.38. The details are left to the reader (Exercise 15). □

By repeatedly applying this theorem, we obtain the following.

Theorem 6.71. *If* $\mathbf{K}$ *is a unique factorization domain, then* $\mathbf{K}[x_1, \ldots, x_p]$ *is also a unique factorization domain.*

Exercises

1. Prove Theorem 6.38(3).

2. Show that the set of unit normal polynomials defined in Definition 6.39 satisfies the properties of Definition 6.35.

3. Prove Theorem 6.46.

4. Let $u = 2yx^3 + 3yx + 4y$ and $v = y\,x + 2x + 5$ be polynomials in $\mathbf{Z}[x, y]$ with main variable x. Find the pseudo-quotient and pseudo-remainder for u divided by v.

5. Consider the polynomial $u = -2y^2x^2 + 8x^2 - 2y\,x - 4x - 2y^2 + 8$ with main variable x.

    (a) Find $\text{cont}(u, x)$ and $\text{pp}(u, x)$ in $\mathbf{Z}[x, y]$.

    (b) Find $\text{cont}(u, x)$ and $\text{pp}(u, x)$ in $\mathbf{Q}[x, y]$.

6. Prove or disprove: the sum of primitive polynomials is primitive.

7. Let $L = [x, y, \ldots, z]$ be a list of symbols, and let u be a polynomial in $\mathbf{K}[x, y, \ldots, z]$ where the base coefficient domain $\mathbf{K}$ is either $\mathbf{Z}$ or $\mathbf{Q}$. In addition, let $R = Rest[L]$.

    (a) Give a procedure

    $$Polynomial\_content(u, x, R, K)$$

    that finds the content of u with respect to the main variable x and auxiliary variables R using Mv_poly_gcd to find the gcd for the coefficients in the auxiliary variables R.

(b) Give a procedure

$$Polynomial\_content\_sr(u, x, R, K)$$

that finds the content that uses the subresultant algorithm to find the gcd for the coefficients in the auxiliary variables R.

8. Let p, u, and v be in $\mathbf{K}[x]$.

    (a) Suppose that u is primitive and $v \mid u$. Show that v is also primitive.

    (b) For p a primitive polynomial and $p \mid \gcd(u, v, x)$, show that

    $$p \mid \gcd(\mathrm{pp}(u, x), \mathrm{pp}(v, x), x).$$

9. Suppose that u and v are polynomials in $\mathbf{K}[x]$, and suppose that

    $$\deg(\gcd(u, v), x) = 0.$$

    Show that $\mathrm{pp}(u, x)$ and $\mathrm{pp}(v, x)$ are relatively prime.

10. Suppose that u and v are polynomials in $\mathbf{K}[x]$ with positive degree, and suppose that $\mathrm{psrem}(u, v, x) = 0$. Show that u and v have a common factor with positive degree.

11. Let u and v be polynomials in $\mathbf{K}[x]$. Show that

    (a) $\gcd(\mathrm{cont}(u, x), \mathrm{cont}(v, x), x) = \mathrm{cont}(\gcd(u, v, x), x)$.

    (b) $\gcd(\mathrm{pp}(u, x), \mathrm{pp}(v, x), x) = \mathrm{pp}(\gcd(u, v, x), x)$.

12. Let u be a polynomial in $\mathbf{K}[x, y, \ldots, z]$ with x the main variable where the coefficient domain $\mathbf{K}$ is either $\mathbf{Z}$ or $\mathbf{Q}$, and let $L = [x, y, \ldots, z]$. Give a procedure

    $$Normalize(u, L, K)$$

    that finds the unit normal form of u. If $u = 0$, return 0. *Note:* The normalized form depends on the base coefficient domain.

13. Let u be an expression that satisfies the RS-IN rules on page 257, and let L be the list of the symbols in u. Give a procedure $Rational\_simplify(u, L)$ that transforms u to rationally simplified form. If a division by zero is encountered, return the global symbol **Undefined**.

14. Prove Theorem 6.67 using the following approach. Assume that

    $$\deg(u, x) \geq \deg(v, x),$$

    and consider the remainder sequence

    $$
    \begin{aligned}
    T_{-1} &= u, \\
    T_0 &= v, \\
    T_1 &= \mathrm{psrem}(\mathrm{pp}(T_{-1}, x), \mathrm{pp}(T_0, x), x),
    \end{aligned}
    $$

$$\vdots$$
$$T_i \;\; = \;\; \mathrm{psrem}(\mathrm{pp}(T_{i-2}, x), \mathrm{pp}(T_{i-1}, x), x),$$
$$\vdots$$

Observe that $R_i = \mathrm{pp}(T_i, x)$, and so the iteration terminates when $T_\rho = 0$. Let
$$l_i = \mathrm{lc}(\mathrm{pp}(T_{i-1}, x))^{\delta_i}, \quad i \geq 1,$$
where $\delta_i = \deg(T_{i-2}, x) - \deg(T_{i-1}, x) + 1$, and let
$$k_i = \begin{cases} 1, \text{ if } i = -1, \\ k_{i-1}\,\mathrm{cont}(T_{i-1}, x), & \text{if } i \geq 0. \end{cases}$$

Define sequences of polynomials A_i and B_i with the recurrence relations:
$$A_{-1} = 1, \quad A_0 = 0,$$
$$B_{-1} = 0, \quad B_0 = \mathrm{cont}(u, x),$$
and for $i \geq 1$,
$$\begin{aligned} A_i &= \mathrm{cont}(T_{i-1}, x)\,l_i\,A_{i-2} - Q_i A_{i-1}, \\ B_i &= \mathrm{cont}(T_{i-1}, x)\,l_i\,B_{i-2} - Q_i B_{i-1} \end{aligned}$$
where $Q_i = \mathrm{psquot}(\mathrm{pp}(T_{i-2}, x), \mathrm{pp}(T_{i-1}, x), x)$.

  (a) Show that for $i \geq -1$,
$$k_i\,T_i = A_i\,u + B_i\,v.$$

  (b) Let $d = \gcd(\mathrm{cont}(u, x), \mathrm{cont}(v, x))$, and let c be the unit in $\mathbf{K}$ that transforms the greatest common divisor to unit normal form. Show that
$$A = c\,d\,A_{\rho-1}, \qquad B = c\,d\,B_{\rho-1}, \qquad k = k_\rho.$$

15. Prove Theorem 6.70.

Further Reading

6.1 Integral Domains. Integral domains are described in greater detail in Akritas [2], Geddes, Czapor, and Labahn [39], Herstein [46], and Dean [31]. Artin [3], page 349 gives an example of an integral domain that is not a unique factorization domain.

6.2 Polynomial Division and Expansion. The *Trigident* operator is similar to the **trigsimp** command in Macsyma. See also the **algsubs** command in Maple.

6.3 Greatest Common Divisors. The approach here is similar to the approach in Knuth [55]. The subresultant algorithm presented here is given in Brown [14], [15], Brown and Traub [16], and Collins [25], [27]. For modern developments in gcd calculations, see Akritas [2], Geddes, Czapor, and Labahn [39], von zur Gathen and Gerhard [96], Winkler [101], Yap [105], and Zippel [108].

7

The Resultant

In this chapter we introduce the concept of the resultant of two polynomials. The resultant has applications in calculations with algebraic numbers, the factorization of polynomials with coefficients in algebraic number fields, the solution of systems of polynomial equations, and the integration of rational functions.

The resultant of polynomials u and v is formally defined in Section 7.1 as the determinant of a matrix whose entries depend on the coefficients of the polynomials. In addition, we describe a Euclidean type algorithm that obtains the resultant without a determinant calculation. In Section 7.2, we use resultants to find polynomial relations for explicit algebraic numbers.

7.1 The Resultant Concept

Suppose that we are given two polynomials

$$u = x^2 + t\,x + 2, \qquad v = x^2 + 5\,x + 6,$$

and want to know for which values of t the polynomials have a common factor with positive degree. Since the polynomials do not have a common factor for all values of t, a greatest common divisor algorithm cannot be used for this purpose. In this section, we introduce the resultant of two polynomials which provides a way to determine t.

Let's generalize things a bit. Let u and v be polynomials in $\mathbf{F}[x]$ with degree 2:

$$u = u_2\,x^2 + u_1\,x + u_0, \qquad v = v_2\,x^2 + v_1\,x + v_0.$$

Our goal is to find a function of the coefficients of the two polynomials that determines if there is a common factor. Suppose, for the moment, that $\gcd(u, v) = 1$. Then, by Theorem 4.36 on page 137, there are unique polynomials $A(x)$ and $B(x)$ with $\deg(A) < \deg(v)$ and $\deg(B) < \deg(u)$ such that

$$A\,u + B\,v = 1. \tag{7.1}$$

With $A = a_1\,x + a_0$ and $B = b_1\,x + b_0$, Equation (7.1) becomes

$$(a_1\,x + a_0)\,(u_2\,x^2 + u_1\,x + u_0) + (b_1\,x + b_0)\,(v_2\,x^2 + v_1\,x + v_0) = 1.$$

Expanding the left side of this equation and comparing the coefficients of powers of x, we obtain a system of linear equations:

$$u_2\,a_1 + v_2\,b_1 = 0, \quad \text{(coefficient of } x^3),$$
$$u_1\,a_1 + u_2\,a_0 + v_1\,b_1 + v_2\,b_0 = 0, \quad \text{(coefficient of } x^2),$$
$$u_0\,a_1 + u_1\,a_0 + v_0\,b_1 + v_1\,b_0 = 0, \quad \text{(coefficient of } x^1),$$
$$u_0\,a_0 + v_0\,b_0 = 1, \quad \text{(coefficient of } x^0).$$

In matrix form these equations become:

$$M \begin{bmatrix} a_1 \\ a_0 \\ b_1 \\ b_0 \end{bmatrix} = \begin{bmatrix} 0 \\ 0 \\ 0 \\ 1 \end{bmatrix}, \tag{7.2}$$

where M is the matrix of coefficients

$$M = \begin{bmatrix} u_2 & 0 & v_2 & 0 \\ u_1 & u_2 & v_1 & v_2 \\ u_0 & u_1 & v_0 & v_1 \\ 0 & u_0 & 0 & v_0 \end{bmatrix}.$$

Since the polynomials A and B are the unique solution to Equation (7.1), the system (7.2) also has a unique solution. Therefore, when $\gcd(u, v) = 1$, we have $\det(M) \neq 0$.

Conversely, if we assume that $\det(M) \neq 0$, then Equation (7.2) has a solution, which implies that Equation (7.1) has a solution, and so by Theorem 4.30 on page 133, u and v are relatively prime.

The value $\det(M)$ is called the *resultant* of the polynomials u and v. In practice, we usually work with the transpose of the matrix M, which is obtained by interchanging the rows and columns

$$M^T = \begin{bmatrix} u_2 & u_1 & u_0 & 0 \\ 0 & u_2 & u_1 & u_0 \\ v_2 & v_1 & v_0 & 0 \\ 0 & v_2 & v_1 & v_0 \end{bmatrix}.$$

Since $\det(M^T) = \det(M)$, the determinant of the transpose gives an identical result.

Example 7.1. Let's return to the problem proposed at the beginning of the section with $u = x^2 + tx + 2$ and $v = x^2 + 5x + 6$. Since we want u and v to have a common factor, their resultant must be 0. Therefore,

$$\det(M^T) = \det \begin{bmatrix} 1 & t & 2 & 0 \\ 0 & 1 & t & 2 \\ 1 & 5 & 6 & 0 \\ 0 & 1 & 5 & 6 \end{bmatrix} = 6t^2 - 40t + 66 = 0.$$

This equation has the roots $t = 3, 11/3$, which are the values of t for which u and v have a common factor. For $t = 3$, the common factor is $x + 2$, and for $t = 11/3$, the common factor is $x + 3$. $\square$

The resultant concept is formally described in the next two definitions.

Definition 7.2. *Let* **K** *be a coefficient domain,*[1] *and let*

$$u(x) = u_m x^m + u_{m-1} x^{m-1} + \cdots + u_0,$$

$$v(x) = v_n x^n + v_{n-1} x^{n-1} + \cdots + v_0$$

be polynomials in **K**$[x]$ *with* $m = \deg(u) > 0$ *and* $n = \deg(v) > 0$. *Define the matrix*

$$S(u, v, x) = \begin{bmatrix} u_m & u_{m-1} & \cdots & u_0 & & & \\ & u_m & u_{m-1} & \cdots & u_0 & & \\ & & \ddots & & & \ddots & \\ & & & u_m & u_{m-1} & \cdots & u_0 \\ v_n & v_{n-1} & \cdots & v_0 & & & \\ & v_n & v_{n-1} & \cdots & v_0 & & \\ & & \ddots & & & \ddots & \\ & & & v_n & v_{n-1} & \cdots & v_0 \end{bmatrix} \tag{7.3}$$

which has n *rows of* u's *coefficients and* m *rows of* v's *coefficients. The blank locations in the matrix contain zeroes. The matrix* $S(u, v, x)$ *is called the* **Sylvester** *matrix*[2] *of the two polynomials* u *and* v.

---

[1] In this section, the coefficient domain **K** satisfies the properties on page 231.

[2] The matrix is named after the mathematician James Joseph Sylvester (1814- 1897) who first studied the matrix in relation to the elimination of variables from systems of polynomial equations.

The Sylvester matrix has $n + m$ rows and columns and is only defined when u and v have positive degree. For example, if $\deg(u) = 2$ and $\deg(v) = 3$,

$$S(u, v, x) = \begin{bmatrix} u_2 & u_1 & u_0 & 0 & 0 \\ 0 & u_2 & u_1 & u_0 & 0 \\ 0 & 0 & u_2 & u_1 & u_0 \\ v_3 & v_2 & v_1 & v_0 & 0 \\ 0 & v_3 & v_2 & v_1 & v_0 \end{bmatrix}.$$

Definition 7.3. *Let u and v be non-zero polynomials in $\mathbf{K}[x]$, with $m = \deg(u)$ and $n = \deg(v)$. The **resultant**[3] of u and v is defined as*

$$\operatorname{res}(u, v, x) = \begin{cases} 1, & \text{if } \deg(u, x) = 0, \quad \deg(v, x) = 0, \\ u_0^n, & \text{if } \deg(u, x) = 0, \quad \deg(v, x) > 0, \\ v_0^m, & \text{if } \deg(u, x) > 0, \quad \deg(v, x) = 0, \\ \det(S(u, v, x)), & \text{if } \deg(u, x) > 0, \quad \deg(v, x) > 0. \end{cases} \quad (7.4)$$

The resultant is not defined when either u or v is the zero polynomial.[4] When there is no chance for confusion, we omit the variable x and use the notation $\operatorname{res}(u, v)$.

The most important case of the definition is when both polynomials have positive degree, and the other cases are natural extensions of this case. For example, if $\deg(u, x) = 0$ and $\deg(v, x) > 0$, then a natural extension of the Sylvester matrix is one with $n = \deg(v, x)$ rows of u's coefficients and $m = \deg(u, x) = 0$ rows of v's coefficients. Since $u = u_0$, this matrix has the form

$$S = \begin{bmatrix} u_0 & \cdots & & 0 \\ 0 & u_0 & \cdots & 0 \\ & & \cdots & \\ 0 & 0 & \cdots & u_0 \end{bmatrix},$$

where there are n rows and columns. Therefore, $\det(S) = u_0^n$, and so it is reasonable for this case to define $\operatorname{res}(u, v) = u_0^n$.

Example 7.4. Let

$$u = 2\,x + 3, \qquad v = 4\,x^2 + 5\,x + 6. \quad (7.5)$$

---

[3]The resultant is obtained in Maple with the **resultant** operator, in Mathematica with the **Resultant** operator, and in MuPAD with the **polylib :: resultant** operator. (Implementation: Maple (mws), Mathematica (nb), MuPAD (mnb).)

[4]In Maple, Mathematica, and MuPAD, the resultant operator returns 0 when either u or v is 0.

The resultant is

$$\text{res}(u, v) = \det \begin{bmatrix} 2 & 3 & 0 \\ 0 & 2 & 3 \\ 4 & 5 & 6 \end{bmatrix} = 30. \tag{7.6}$$

In the definition of the Sylvester matrix, we emphasized that the leading coefficients of u and v cannot be 0. Indeed, if the resultant is computed with a Sylvester matrix that is too large, the value may change. For example, suppose that we redefine u in Equation (7.5) as

$$u = c x^2 + 2 x + 3$$

where c is an undefined symbol. Then,

$$\text{res}(u, v) = \det \begin{bmatrix} c & 2 & 3 & 0 \\ 0 & c & 2 & 3 \\ 4 & 5 & 6 & 0 \\ 0 & 4 & 5 & 6 \end{bmatrix} = 36\,c^2 - 129\,c + 120. \tag{7.7}$$

To return to the situation in Equation (7.5), we let $c = 0$, and so $\text{res}(u, v) = 120$ rather than the value 30 obtained in Equation (7.6). $\square$

Properties of the Resultant

The resultant satisfies the following properties.

Theorem 7.5. *Let b and c be in the coefficient domain $\mathbf{K}$, and let u and v be non-zero polynomials in $\mathbf{K}[x]$ with $m = \deg(u)$ and $n = \deg(v)$.*

1. $\text{res}(u, \ v) = (-1)^{m\,n}\,\text{res}(v, \ u).$

2. $\text{res}(c\,u, \ v) = c^n\,\text{res}(u, \ v).$

3. $\text{res}(u, \ c\,v) = c^m\,\text{res}(u, \ v).$

4. $\text{res}(x, \ v) = v(0).$

5. *If $u = b\,(x - c)$, then $\text{res}(u, \ v) = b^n v(c).$*

6. *If the coefficients of u are linear in the symbol c, then*

$$\deg(\text{res}(u, \ v), \ c) \leq n.$$

7. $\text{res}((x - b)\,u, \ v) = v(b)\,\text{res}(u, \ v).$

Proof: To prove (1), we assume first that both u and v have positive degree and show that $S(v, u, x)$ is obtained from $S(u, v, x)$ by $m \cdot n$ row interchanges. Let S_i be the ith row of $S(u, v, x)$ where $S_1, S_2, \ldots S_n$ are the rows with coefficients of u and $S_{n+1}, S_{n+2}, \ldots S_{n+m}$ are the rows with coefficients of v. Consider the following row interchanges:

$$
\begin{bmatrix} S_1 \\ S_2 \\ \vdots \\ S_n \\ S_{n+1} \\ S_{n+2} \\ \vdots \\ S_{n+m} \end{bmatrix}
\rightarrow
\begin{bmatrix} S_{n+1} \\ S_1 \\ \vdots \\ S_n \\ S_{n+2} \\ S_{n+3} \\ \vdots \\ S_{n+m} \end{bmatrix}
\rightarrow
\begin{bmatrix} S_{n+1} \\ S_{n+2} \\ S_1 \\ \vdots \\ S_n \\ S_{n+3} \\ S_{n+4} \\ \vdots \\ S_{n+m} \end{bmatrix}
\rightarrow \cdots \rightarrow
\begin{bmatrix} S_{n+1} \\ S_{n+2} \\ \vdots \\ S_{n+m} \\ S_1 \\ S_2 \\ \vdots \\ S_n \end{bmatrix} .
$$

To move each S_i $(n + 1 \le i \le n + m)$ to its new place requires n row interchanges, and since there are m such rows to move, the row interchange property for determinants implies that $\mathrm{res}(u,\ v) = (-1)^{m\,n}\mathrm{res}(v,\ u)$. If $\deg(u) = 0$ or $\deg(v) = 0$, (1) can also be easily verified from the definition.

The proofs of (2) and (3), which also follow from the properties for determinants, are left to the reader (Exercise 5).

To show (4), we assume first that $\deg(v) > 0$. Then, the Sylvester matrix is the $n + 1$ by $n + 1$ matrix

$$
S(x,\ v,\ x) = \begin{bmatrix} 1 & 0 & \cdots & & & \\ 0 & 1 & 0 & \cdots & & \\ & & \vdots & & & \\ & & & & 1 & 0 \\ v_n & v_{n-1} & \cdots & & & v_0 \end{bmatrix},
$$

and, therefore,

$$\mathrm{res}(x\,v) = \det(S(x,\ v,\ x)) = v_0 = v(0).$$

When $\deg(v) = 0$, the relationship follows directly from the definition of the resultant.

To show (5), first assume that $\deg(v) > 0$. Then,

$$
\mathrm{res}(b\,(x - c),\ v) = \det \begin{bmatrix} b & -b\,c & 0 & \cdots & & & \\ 0 & b & -b\,c & 0 & \cdots & & \\ & & \vdots & & & & \\ & & & 0 & \cdots & b & -b\,c \\ v_n & v_{n-1} & \cdots & & & & v_0 \end{bmatrix} .
$$

Let C_j be the jth column of this $n+1$ by $n+1$ matrix and form a new matrix with C_{n+1} replaced by

$$c^n\, C_1 + c^{n-1}\, C_2 + \cdots c\, C_n + C_{n+1} =$$

$$
\begin{bmatrix}
c^n\, b - c^{n-1}\, bc \\
c^{n-1}\, b - c^{n-2}\, bc \\
\vdots \\
cb - bc \\
v_n\, c^n + v_{n-1}\, c^{n-1} + \cdots + v_0
\end{bmatrix}
=
\begin{bmatrix}
0 \\
0 \\
\vdots \\
0 \\
v(c)
\end{bmatrix}.
$$

Since this column operation does not change the value of the determinant, we have

$$
\operatorname{res}(b\,(x-c),\ v) = \det
\begin{bmatrix}
b & -bc & 0 & \cdots & & & 0 \\
0 & b & -bc & 0 & \cdots & & 0 \\
& & \vdots & & & & \\
& & & 0 & \cdots & b & 0 \\
v_n & v_{n-1} & \cdots & & & v_1 & v(c)
\end{bmatrix}.
$$

Evaluating the determinant by expanding with the last column, we have

$$
\operatorname{res}(b\,(x-c),\ v) =
$$

$$
v(c) \cdot \det
\begin{bmatrix}
b & -bc & 0 & \cdots & & & \\
0 & b & -bc & 0 & \cdots & & \\
& & \vdots & & & & \\
& & & 0 & \cdots & b & -bc \\
& & & 0 & \cdots & 0 & b
\end{bmatrix}
$$

$$
= \quad v(c)\, b^n.
$$

The case where $\deg(v) = 0$ is left to the reader.

We show (6) when $\deg(u) > 0$ and $\deg(v) > 0$ and leave the other cases to the reader. Recall that the evaluation of a determinant involves the sum of terms which are products with one element from each row and column of the matrix. Since the matrix S has n rows with u's coefficients, and since each coefficient has degree at most 1 in c, the resultant is at most degree n in c.

We show (7) when $\deg(u) > 0$ and $\deg(v) > 0$ and leave the other cases to the reader. Let

$$
w = (x - b)\, u = w_{m+1} x^{m+1} + \cdots + w_0
$$

where

$$w_i = \begin{cases} u_m, & i = m+1, \\ u_{i-1} - b\,u_i, & i = 1, \ldots, m, \\ -b\,u_i, & i = 0. \end{cases} \tag{7.8}$$

We have $\mathrm{res}(w,\ v) = \det(S(w,v,x))$ where $S(w,v,x)$ is the $m+1+n$ by $m+1+n$ matrix

$$S(w,v,x) =$$

$$\begin{bmatrix} w_{m+1} & w_m & \cdots & w_0 & & & & \\ & w_{m+1} & w_m & \cdots & & w_0 & & \\ & & \ddots & & & & \ddots & \\ & & & w_{m+1} & w_m & \cdots & w_0 \\ v_n & v_{n-1} & \cdots & v_0 & & & \\ & v_n & v_{n-1} & \cdots & & v_0 & \\ & & \ddots & & & & \ddots \\ & & & v_n & v_{n-1} & \cdots & v_0 \end{bmatrix}. \tag{7.9}$$

Multiplying column C_i $(1 \le i < m+1+n)$ by $b^{m+1+n-i}$ and adding these columns to the last column, we obtain a new matrix S_2 where the last column has the entries

$$b^{m+n}C_1 + b^{m+n-1}C_2 + \cdots + b\,C_{m+n}$$

$$= \begin{bmatrix} b^{m+n}\,w_{m+1} + \cdots + b^{n-1}\,w_0 \\ \vdots \\ b^{m+1}\,w_{m+1} + \cdots + w_0 \\ b^{m+n}\,v_n + \cdots + b^m\,v_0 \\ \vdots \\ b^n\,v_n + \cdots + v_0 \end{bmatrix} = \begin{bmatrix} b^{n-1}w(b) \\ \vdots \\ w(b) \\ b^m\,v(b) \\ \vdots \\ v(b) \end{bmatrix} = \begin{bmatrix} 0 \\ \vdots \\ 0 \\ b^m\,v(b) \\ \vdots \\ v(b) \end{bmatrix}.$$

We can factor $v(b)$ out of the last column and obtain a new matrix S_3 where the last column has the entries

$$0, \ldots, 0, b^m, \cdots, 1,$$

and

$$\mathrm{res}(w,v) = \det(S(w,v,x)) = \det(S_2) = v(b) \cdot \det(S_3). \tag{7.10}$$

Observe that $\det(S(w,v,x))$ is a polynomial in b, and by Part (6),

$$\deg(\det(S(w,v,x)), b) \le n.$$

However, since $\deg(v(b), b) = n$, Equation (7.10) implies

$$
\begin{aligned}
\deg(v(b) \cdot \det(S_3), b) &= \deg(v(b), b) + \deg(\det(S_3), b,) \\
&= n + \deg(\det(S_3), b) \\
&\leq n,
\end{aligned}
$$

and so

$$
\deg(\det(S_3), b) = 0.
$$

In other words, the symbol b is eliminated in the evaluation of $\det(S_3)$. Therefore, we substitute 0 for b in S_3 and obtain a new matrix S_4 with the $m + 1 + n$ entries

$$
0, \ldots, 0, 1
$$

in the last column. From Equation (7.10), evaluating $\det(S_4)$ by expanding about the last column, we obtain

$$
\mathrm{res}(w, v) =
$$

$$
v(b) \det
\begin{bmatrix}
w_{m+1} & w_m & \cdots & w_0 & & & \\
 & w_{m+1} & w_m & \cdots & & w_0 & \\
 & & \ddots & & & & \ddots \\
 & & & w_{m+1} & w_m & \cdots & w_1 \\
v_n & v_{n-1} & \cdots & v_0 & & & \\
 & v_n & v_{n-1} & \cdots & v_0 & & \\
 & & \ddots & & & & \ddots \\
 & & & v_n & v_{n-1} & \cdots & v_0
\end{bmatrix}
\tag{7.11}
$$

where the matrix has n rows of w coefficients, m rows of v coefficients, and the last entry in row n is w_1. However, when $b = 0$, Equation (7.8) gives

$$
w_i =
\begin{cases}
u_{i-1}, & i = 1, \cdots, m, \\
0, & i = 0,
\end{cases}
$$

and therefore, Equation (7.11) gives

$$
\mathrm{res}(w, v) = v(b) \det
\begin{bmatrix}
u_m & u_{m-1} & \cdots & u_0 & & & \\
 & u_m & u_{m-1} & \cdots & u_0 & & \\
 & & \ddots & & & & \ddots \\
 & & & u_m & u_{m-1} & \cdots & u_1 \\
v_n & v_{n-1} & \cdots & v_0 & & & \\
 & v_n & v_{n-1} & \cdots & v_0 & & \\
 & & \ddots & & & & \ddots \\
 & & & v_n & v_{n-1} & \cdots & v_0
\end{bmatrix}
$$

$$
= v(b)\,\mathrm{res}(u, v). \qquad \square
$$

Fundamental Theorems

The next three theorems are fundamental to the application of resultants.

Theorem 7.6. *Let u and v be polynomials of positive degree in $\mathbf{K}[x]$. Then, u and v have a common factor of positive degree if and only if there are non-zero polynomials C and D in $\mathbf{K}[x]$ with*

$$\deg(C) < \deg(v), \quad \deg(D) < \deg(u) \tag{7.12}$$

such that

$$C\,u + D\,v = 0.$$

Proof: First, assume that u and v have a common factor w of positive degree. Then, $u = p\,w$, $v = q\,w$, and so let $C = -q$ and $D = p$. Therefore, C and D are non-zero and satisfy the degree conditions (7.12). In addition,

$$C\,u + D\,v = -q\,(p\,w) + p\,(q\,w) = 0.$$

Conversely, assume that C and D exist with the properties stated in the theorem, and suppose that u and v do not have a common factor of positive degree. Then, $\mathrm{pp}(u)$ and $\mathrm{pp}(v)$ are relatively prime (Exercise 9, page 261) and

$$\mathrm{pp}(C)\,\mathrm{pp}(u) = -\mathrm{pp}(D)\,\mathrm{pp}(v).$$

Using Theorem 6.69(1), we have

$$\mathrm{pp}(u) \mid \mathrm{pp}(D). \tag{7.13}$$

But since $\mathrm{pp}(D) \neq 0$ and

$$\deg(\mathrm{pp}(D)) = \deg(D) < \deg(u) = \deg(\mathrm{pp}(u)),$$

the condition (7.13) can't be true. Therefore, u and v have a common factor of positive degree. □

The next theorem is the resultant version of the relationship obtained with the extended Euclidean algorithm.

Theorem 7.7. *Let u and v be polynomials in $\mathbf{K}[x]$ with $\deg(u) = m > 0$ and $\deg(v) = n > 0$. Then, there are polynomials A and B in $\mathbf{K}[x]$ with $\deg(A) < n$ and $\deg(B) < m$ such that*

$$A\,u + B\,v = \mathrm{res}(u, v). \tag{7.14}$$

Proof: Let C_i be the ith column of the Sylvester matrix, and form a new matrix S_2 by replacing column C_{m+n} by

$$x^{m+n-1}\,C_1 + x^{m+n-2}\,C_2 + \cdots + x\,C_{m+n-1} + C_{m+n} = \begin{bmatrix} x^{n-1}\,u \\ x^{n-2}\,u \\ \vdots \\ u \\ x^{m-1}\,v \\ x^{m-2}\,v \\ \vdots \\ v \end{bmatrix}.$$

Therefore,

$$S_2 = \begin{vmatrix} u_m & u_{m-1} & \cdots & u_0 & & & \cdots & x^{n-1}u \\ & u_m & u_{m-1} & \cdots & u_0 & & \cdots & x^{n-2}u \\ & & \ddots & & & \ddots & & \vdots \\ & & & u_m & u_{m-1} & \cdots & & u \\ v_n & v_{n-1} & \cdots & v_0 & & & \cdots & x^{m-1}v \\ & v_n & v_{n-1} & \cdots & v_0 & & \cdots & x^{m-2}v \\ & & \ddots & & & \ddots & & \vdots \\ & & & v_n & v_{n-1} & & \cdots & v \end{vmatrix}, \quad (7.15)$$

and since this column operation does not change the value of the determinant we have

$$\mathrm{res}(u, v) = \det(S) = \det(S_2).$$

If we evaluate $\det(S_2)$ by expanding about the last column, we obtain Equation (7.14), where the coefficients of A and B are determinant cofactors of the elements in the last column of S_2. The degree conditions for A and B are obtained from the powers of x that multiply u and v in S_2. $\qquad\square$

Example 7.8. We consider pairs of polynomials in $\mathbf{Q}[x]$ and the polynomials A and B obtained by evaluating $\det(S_2)$.

For $u = x^2 + 5x + 6$ and $v = x^2 - 1$, then

$$\gcd(u, v) = 1, \quad \mathrm{res}(u, v) = 24, \quad A = -5x + 7, \quad B = 5x + 18.$$

For $u = x^3 + 2x^2 + x + 2$ and $v = x^3 - 2x^2 + x - 2$, then

$$\gcd(u, v) = x^2 + 1, \quad \mathrm{res}(u, v) = 0, \quad A = 0, \quad B = 0.$$

In this case, the polynomials A and B are not unique. Indeed

$$A\,u + B\,v = (-x + 2)\,u + (x + 2)\,v = 0. \qquad \qquad \square$$

The next theorem describes the fundamental relationship between the resultant and the common structure (in x) of u and v.

Theorem 7.9. *Let u and v be polynomials in $\mathbf{K}[x]$ with positive degree. Then, u and v have a common factor w with $\deg(w, x) > 0$ if and only if $\mathrm{res}(u, v, x) = 0$.*

Proof: To prove the theorem first suppose that w is a common factor of u and v with $\deg(w, x) > 0$. Then, by Equation (7.14), $w \mid \mathrm{res}(u, v)$. However, since w has positive degree and since $\mathrm{res}(u, v)$ is in $\mathbf{K}$, we have $\mathrm{res}(u, v) = 0$.

Conversely, suppose that $\mathrm{res}(u, v) = 0$. By Theorem 7.7 there are polynomials A and B with $\deg(A) < n$ and $\deg(B) < m$ such that

$$A\,u + B\,v = 0. \tag{7.16}$$

Observe that if $A \neq 0$ and $B \neq 0$, then Theorem 7.6 implies that there is a common factor of u and v with positive degree. Unfortunately, the polynomials A and B obtained in Theorem 7.7 may both be zero (see Example 7.8 above). When this occurs, we obtain a proof as follows. If

$$A(x) = a_{n-1}x^{n-1} + a_{n-2}x^{n-2} + \cdots + a_0,$$

$$B(x) = b_{m-1}x^{m-1} + b_{m-2}x^{m-2} + \cdots + b_0,$$

the matrix form of Equation (7.16) is given by

$$S^T \cdot \begin{vmatrix} a_{n-1} \\ a_{n-2} \\ \vdots \\ a_0 \\ b_{m-1} \\ b_{m-2} \\ \vdots \\ b_0 \end{vmatrix} = \begin{bmatrix} 0 \\ 0 \\ \vdots \\ \\ 0 \\ 0 \end{bmatrix} \tag{7.17}$$

where S^T is the transpose of the Sylvester matrix S. Since $\det(S^T) = \det(S) = \mathrm{res}(u, v) = 0$, the homogeneous system (7.17) has a solution with some a_i or b_i not zero. This implies that there are polynomials A and

B, not both zero, which satisfy Equation (7.16). In fact, Equation (7.16) implies that $Au = -Bv$ which implies that both A and B are not zero. Therefore, Theorem 7.6 implies that u and v have a common factor of positive degree.   □

Euclidean Algorithms for Resultant Calculation

Although the resultant can be computed directly from the determinant definition, it can also be obtained with an algorithm similar to the Euclidean algorithm. Since the development is simpler when the coefficient domain is a field $\mathbf{F}$, we consider this case first.

Resultant Calculation in F[x].   In this case, the algorithm is based on the next theorem which is the resultant version of Theorem 4.22 on page 128.

Theorem 7.10. *Suppose that u and v are polynomials in $\mathbf{F}[x]$ with positive degree, where $m = \deg(u)$ and $n = \deg(v)$. Let r_σ be defined by the polynomial division*

$$u = q_\sigma v + r_\sigma,$$

and let $s = \deg(r_\sigma)$. Then,

$$\operatorname{res}(u, v, x) = \begin{cases} 0, & \text{if } r_\sigma = 0, \\ (-1)^{mn} v_n^{m-s} \operatorname{res}(v, r_\sigma, x), & \text{if } (r_\sigma) \neq 0. \end{cases} \tag{7.18}$$

Proof: To verify the first case, if $r_\sigma = 0$, then $\deg(\gcd(u, v)) > 0$, and by Theorem 7.9, $\operatorname{res}(u, v) = 0$.

Assume next that $r_\sigma \neq 0$. First observe that when $m < n$, we have $\operatorname{rem}(u, v) = u$, and the second case follows from Theorem 7.5(1).

For the remainder of the proof we assume that $m \geq n$. Let's look at the interplay between the division process and the matrix S. Recall that the remainder in polynomial division is defined by the iteration scheme

$$r_i = r_{i-1} - \frac{\operatorname{lc}(r_{i-1})}{\operatorname{lc}(v)} v\, x^{\deg(r_{i-1}) - \deg(v)}, \qquad r_0 = u. \tag{7.19}$$

Since $m \geq n$, there is a least one iteration step and

$$r_1 = u - \frac{u_m}{v_n} v\, x^{m-n} \tag{7.20}$$

where $\deg(r_1) \leq m - 1$. Suppose that

$$r_1 = u_{m-1}^{(1)} x^{m-1} + u_{m-2}^{(1)} x^{m-2} + \cdots + u_0^{(1)}$$

where the notation $u_j^{(1)}$ represents the coefficient of x^j in r_1. Observe that the iteration (7.19) gives

$$
\begin{aligned}
0 &= u_m - \frac{u_m}{v_n} v_n, \\
u_{m-1}^{(j)} &= u_{m-1} - \frac{u_m}{v_n} v_{n-1}, \\
u_{m-2}^{(j)} &= u_{m-2} - \frac{u_m}{v_n} v_{n-2},
\end{aligned}
\tag{7.21}
$$

$$\vdots$$

Let S_j be the jth row of the Sylvester matrix S. In matrix terms, the operation (7.21) corresponds to replacing row S_j, by $S_j - (u_m/v_n) S_{j+n}$, for $1 \le j \le n$, and this row operation does not change the value of the determinant. Therefore,

$$
\mathrm{res}(u,v) = \det
\begin{bmatrix}
0 & u_{m-1}^{(1)} & \cdots & u_0^{(1)} & & & & \\
0 & 0 & u_{m-1}^{(1)} & \cdots & u_0^{(1)} & & & \\
& & \ddots & & & \ddots & & \\
& & & 0 & u_{m-1}^{(1)} & \cdots & u_0^{(1)} & \\
v_n & v_{n-1} & \cdots & v_0 & & & & \\
& v_n & v_{n-1} & \cdots & v_0 & & & \\
& & \ddots & & & \ddots & & \\
& & & v_n & v_{n-1} & \cdots & v_0 &
\end{bmatrix},
$$

where the matrix has n rows of r_1's coefficients and m rows of v's coefficients.

The division process continues in this manner obtaining the remainders r_2, r_3, ... with the iteration (7.19). At each step we modify the matrix with the corresponding row operations as was done for r_1, and, as before, the value of the determinant is not changed. The process terminates with the remainder

$$
r_\sigma = u_s^{(\sigma)} x^s + u_{s-1}^{(\sigma)} x^{s-1} + \cdots + u_0^{(\sigma)}
$$

with $s = \deg(r_\sigma) < \deg(v)$ and

$$\operatorname{res}(u, v) =$$

$$\det \begin{bmatrix} 0 & \cdots & 0 & u_s^{(\sigma)} & \cdots & u_0^{(\sigma)} & & & \\ 0 & 0 & \cdots & 0 & u_s^{(\sigma)} & \cdots & u_0^{(\sigma)} & & \\ & & \ddots & & & \ddots & & & \\ 0 & 0 & \cdots & & & 0 & u_s^{(\sigma)} & \cdots & u_0^{(\sigma)} \\ v_n & v_{n-1} & \cdots & v_0 & & & & & \\ 0 & v_n & v_{n-1} & \cdots & v_0 & & & & \\ & \ddots & & & & \ddots & & & \\ 0 & & v_n & v_{n-1} & \cdots & & & & v_0 \end{bmatrix}.$$

$$(7.22)$$

This matrix has n rows of r_σ coefficients, m rows of v coefficients, and the first row has 0 in the first $m - s$ columns. Observe that v_n is the only member of column 1 that is not zero. Therefore, evaluating the determinant by expanding about column 1 we obtain

$$\operatorname{res}(u, v) = v_n \cdot (-1)^n$$

$$\cdot \det \begin{bmatrix} 0 & \cdots & 0 & u_s^{(\sigma)} & \cdots & u_0^{(\sigma)} & & & \\ 0 & 0 & \cdots & 0 & u_s^{(\sigma)} & \cdots & u_0^{(\sigma)} & & \\ & & \ddots & & & \ddots & & & \\ 0 & 0 & \cdots & & & 0 & u_s^{(\sigma)} & \cdots & u_0^{(\sigma)} \\ v_n & v_{n-1} & \cdots & v_0 & & & & & \\ & v_n & v_{n-1} & \cdots & v_0 & & & & \\ & \ddots & & & & \ddots & & & \\ & v_n & v_{n-1} & \cdots & & & & & v_0 \end{bmatrix}.$$

$$(7.23)$$

Although the matrices (7.22) and (7.23) look the same, the one in (7.23) has n rows of r_σ's coefficients, $m - 1$ rows of v's coefficients, and the first $m - s - 1$ coefficients in row 1 are 0. Again, we evaluate this determinant by expanding about column 1. By repeating this process $m - s - 1$ times, we obtain

$$\mathrm{res}(u,v) =$$

$$(v_n(-1)^n)^{m-s} \cdot \det
\begin{bmatrix}
u_s^{(\sigma)} & \cdots & u_0^{(\sigma)} & & & & \\
0 & u_s^{(\sigma)} & \cdots & u_0^{(\sigma)} & & & \\
& & \ddots & & \ddots & & \\
& & \cdots & & u_s^{(\sigma)} & \cdots & u_0^{(\sigma)} \\
v_n & v_{n-1} & \cdots & v_0 & & & \\
& v_n & v_{n-1} & \cdots & v_0 & & \\
& & \ddots & & & \ddots & \\
& & v_n & v_{n-1} & \cdots & & v_0
\end{bmatrix},$$

$$(7.24)$$

where there are n rows of r_σ's coefficients and $s = \deg(r_\sigma)$ rows of v's coefficients.

Let's assume first that $\deg(r_\sigma) \geq 1$. Since the determinant in Equation (7.24) is $\mathrm{res}(r_\sigma, v)$, it follows from Theorem 7.5(1) that

$$\mathrm{res}(u,v) = (v_n(-1)^n)^{m-s}\,(-1)^{n\,s}\,\mathrm{res}(v, r_\sigma) = (-1)^{n\,m}\,v_n^{m-s}\,\mathrm{res}(v, r_\sigma),$$

which proves the second case of the theorem when $\deg(r_\sigma) \geq 1$. Finally, suppose that $\deg(r_\sigma) = 0$. Evaluating the determinant and using Definition 7.3, we obtain

$$\mathrm{res}(u,v) = ((-1)^n v_n)^m \left(u_0^{(\sigma)}\right)^n = (-1)^{n\,m}\,v_n^m\,r_\sigma^n = (-1)^{n\,m}v_n^m\mathrm{res}(v, r_\sigma),$$

which proves the second case of the theorem. $\square$

We can compute $\mathrm{res}(u, v)$ by repeatedly applying Theorem 7.10 to members of the polynomial remainder sequence R_i (see Equation (4.24), page 128).

Example 7.11. Let $u = 2\,x^3 - 3\,x + 1$ and $v = 3\,x^2 - 4\,x + 3$. The remainder sequence is given by

$$R_{-1} = u, \qquad R_0 = v, \qquad R_1 = -13/9\,x - 5/3, \qquad R_2 = 1962/169.$$

Therefore, applying Theorem 7.10, we have

$$\begin{aligned}
\mathrm{res}(u,v) &= \mathrm{res}(R_{-1}, R_0) = (-1)^{3\cdot 2}\,3^{3-1}\,\mathrm{res}(R_0, R_1) \\
&= 9 \cdot \mathrm{res}(R_0, R_1).
\end{aligned}$$

$$(7.25)$$

Since $\deg(R_2) = 0$, using Equation (7.4), we have $\mathrm{res}(R_1, R_2) = (1962/169)^1$, and therefore

$$\mathrm{res}(R_0, R_1) = (-1)^{1\cdot 2}\,(-13/9)^2\,\mathrm{res}(R_1, R_2) = 218/9. \qquad (7.26)$$

Procedure *Polynomial_resultant(u, v, x)*;
Input
    u, v : non-zero polynomials in $\mathbf{F}[x]$ where all field
        in $\mathbf{F}$ are obtained with automatic simplification;
    x : a symbol;
Output
    $res(u, v, x)$;
Local Variables
    m, n, r, s, l;
Begin

```
1     m :=  Degree_gpe(u, x);
2     n :=  Degree_gpe(v, x);
3     if  n = 0 then
4         Return(vᵐ)
5     else
6         r :=  Remainder(u, v, x);
7         if  r = 0 then
8             Return(0)
9         else
10            s := Degree_gpe(r, x);
11            l := Coefficient_gpe(v, x, n);
12            Return( Algebraic_expand(
                     (−1)^(m∗n) ∗ l^(m−s) ∗ Polynomial_resultant(v, r, x)))
```

End

Figure 7.1. An MPL procedure for a Euclidean algorithm that obtains the resultant for polynomials in $\mathbf{F}[x]$. (Implementation: Maple (txt), Mathematica (txt), MuPAD (txt).)

Therefore, using Equation (7.25), we have $res(u, v) = 218$. $\square$

A procedure for the resultant that is based on Theorem 7.10 is given in Figure 7.1. Lines 3 and 4, which are based on Equation (7.4), and Lines 7 and 8, which are based on Equation (7.18), provide termination conditions for the recursion. Lines 7 through 12 are based on the recursion relation in the theorem.

Resultant Calculation in K[x]. When the coefficient domain $\mathbf{K}$ is not a field, resultant calculation is based on pseudo-division rather than ordinary polynomial division. Let u and v be polynomials in $\mathbf{K}[x]$ with positive degree where $m = \deg(u, x)$, $n = \deg(v, x)$, $\delta = m - n + 1$, and $m \geq n$. In addition, assume that

$$r = \text{psrem}(u, v, x) \neq 0.$$

To motivate the relation between $\text{res}(u, v, x)$ and $\text{res}(v, r, x)$, recall that we can view pseudo-division as the recursive division

$$\text{lc}(v)^\delta u = q v + r, \tag{7.27}$$

and, in this case, the division terminates when $\deg(r, x) < \deg(v, x)$. If we assume, for the moment, that the recurrence relation (7.18) holds in this setting for Equation (7.27), we have, using Theorem 7.5(2),

$$\text{res}\left(\text{lc}(v)^\delta u, v, x\right) = \text{lc}(v)^{\delta n} \text{res}(u, v, x) = (-1)^{mn} \text{lc}(v)^{m-s} \text{res}(v, r, x)$$

where $s = \deg(r)$. Therefore,

$$\text{lc}(v)^{\delta n - m + s} \text{res}(u, v, x) = (-1)^{mn} \text{res}(v, r, x).$$

However, since $n \geq 1$, $m \geq n$, and $s \geq 0$,

$$
\begin{aligned}
\delta n - m + s &= (m - n + 1) n - m + s \\
&\geq m n - n^2 + n - m \\
&= (n - 1)(m - n) \\
&\geq 0,
\end{aligned}
$$

we have

$$\text{res}(u, v, x) = \frac{(-1)^{mn} \text{res}(v, r, x)}{\text{lc}(v)^{\delta n - m + s}}.$$

This discussion suggests that in this setting, Theorem 7.10 has the following form.

Theorem 7.12. *Suppose u and v are polynomials in $\mathbf{K}[x]$ with positive degree where $m = \deg(u)$, $n = \deg(v)$, and $m \geq n$. For the pseudo-division $\text{lc}(v)^\delta u = q v + r$ with $s = \deg(r, x)$, we have*

$$
\text{res}(u, v, x) = \begin{cases} 0, & \text{if } r = 0, \\ \dfrac{(-1)^{mn} \text{res}(v, r, x)}{\text{lc}(v)^{\delta n - m + n}}, & \text{if } r \neq 0. \end{cases} \tag{7.28}
$$

Proof: A proof of the theorem that is based on the definition of pseudo-division (Definition 6.44, page 237) is similar to the one for Theorem 7.10. The details are left to the reader (Exercise 8). $\square$

Procedure *Polynomial_resultant_mv*(u, v, L, K);
Input
   u, v : non-zero multivariate polynomials in symbols in L
            with coefficients in **Z** or **Q**;
   L : a list of symbols with main variable $x = Operand(x, 1)$;
   K : the symbol **Z** or **Q**;
Output
   res(u, v, x);
Local Variables
   $x, m, n, \delta, r, s, l, w, f, k$;
Begin

```
1     x = Operand(L, 1);
2     m := Degree_gpe(u, x);
3     n := Degree_gpe(v, x);
4     if  m < n then
5         Return ((−1)^{m*n} * Polynomial_resultant_mv(v, u, L, K))
6     elseif  n = 0 then
7         Return(v^m)
8     else
9         δ := m − n + 1;
10        r := Pseudo_remainder(u, v, x);
11        if  r = 0 then
12            Return(0)
13        else
14            s := Degree_gpe(r, x);
15            w := Algebraic_expand(
                    (−1)^{m*n} Polynomial_resultant_mv(v, r, L, K));
16            l := Leading_Coefficient_gpe(v, x);
17            k := δ * n − m + s;
18            f := Algebraic_expand(l^k);
19            Return(Rec_quotient(w, f, L, K))
      End
```

Figure 7.2. An MPL procedure for a Euclidean algorithm that obtains the resultant for multivariate polynomials. (Implementation: Maple (txt), Mathematica (txt), MuPAD (txt).)

A procedure for resultant calculation in this setting is given in Figure 7.1. The input polynomials u and v are multivariate polynomials in the symbols in L where the first operand x of L is the main variable. The procedure returns the resultant with respect to x. Observe that the coefficient domain K is included as a formal parameter because it is required by *Rec_quotient* in line 19.

Recursive call	$\deg(r, x)$	$\deg(r, y)$	Max. digits r
1	4	4	4
2	3	10	6
3	2	22	14
4	1	54	33

Figure 7.3. Coefficient and degree explosion for the monomials in r obtained at line 10 in the procedure *Polynomial_resultant_mv* for the polynomials in Example 7.13.

Since *Polynomial_resultant* and *Polynomial_resultant_mv* are based on polynomial remainder sequences, they produce coefficient and degree explosion similar to the gcd procedures in Figure 4.3 and Figure 6.6. This point is illustrated in the next example.

Example 7.13. Consider the polynomials

$$
\begin{aligned}
u &= x^3y^2 + 6x^4y + 9x^5 + 4x^2y^2 + 24x^3y + 36x^4 + 5xy^2 + 30yx^2 \\
&\quad +45x^3 + 2y^2 + 12yx + 18x^2, \\
v &= x^5y^2 + 8x^4y + 16x^3 + 12x^4y^2 + 96x^3y + 192x^2 + 45x^3y^2 \\
&\quad +360yx^2 + 720x + 50x^2y^2 + 400yx + 800
\end{aligned}
$$

as polynomials in $\mathbf{Z}[x, y]$ with x the main variable. In Example 6.63 (page 250), we obtained $\gcd(u, v) = x + 2$, and, therefore, by Theorem 7.9,

$$\mathrm{res}(u, v, x) = 0.$$

In Figure 7.3, we give a summary of the coefficient and degree explosion obtained in the monomials in the pseudo-remainder r at line 10 in Figure 7.1. Observe that there is a dramatic increase in the coefficient and degree explosion for subsequent procedure calls. Eventually, the pseudo-remainder $r = 0$ terminates the recursion (lines 11–12), and so the recursion unwinds giving a resultant of 0. □

Resultant Calculation: The Subresultant Remainder Sequence. There are a number of ways to reduce the coefficient and degree explosion in resultant calculation. We briefly describe an approach based on the subresultant remainder sequence that was used in Section 6.3 to reduce coefficient and degree explosion in multivariate gcd calculations. This approach involves the remainder sequence

$$S_{-1} = \mathrm{pp}(u),$$

Recursive call	$\deg(r, x)$	$\deg(r, y)$	Max. digits r
1	4	4	4
2	3	8	7
3	2	12	9
4	1	16	14

Figure 7.4. Coefficient and degree explosion for the monomials in r obtained at line 9 in the procedure *Sr_polynomial_resultant_rec* for the polynomials in Examples 7.13 and 7.14.

$$\begin{aligned}
S_0 &= \mathrm{pp}(v), \\
S_1 &= \frac{\mathrm{psrem}(S_{-1}, S_0, x)}{\beta_1}, \\
&\vdots \\
S_i &= \frac{\mathrm{psrem}(S_{i-2}, S_{i-1}, x)}{\beta_i}, \\
&\vdots
\end{aligned} \tag{7.29}$$

which reduces the coefficient and degree explosion by dividing the pseudo-remainders by the expression β_i given by the recurrence relations

$$\psi_1 = -1, \qquad \beta_1 = (-1)^{\delta_1}, \tag{7.30}$$

$$\psi_i = \frac{(-\mathrm{lc}(S_{i-2}, x))^{\delta_{(i-1)}-1}}{\psi_{i-1}^{\delta_{(i-1)}-2}}, \quad i > 1, \tag{7.31}$$

$$\beta_i = -\mathrm{lc}(S_{i-2}, x)\, \psi_i^{\delta_i - 1}, \quad i > 1 \tag{7.32}$$

where

$$\delta_i = \deg(S_{i-2}, x) - \deg(S_{i-1}, x) + 1, \qquad i \geq 1. \tag{7.33}$$

Procedures to compute the resultant using this approach are given in Figures 7.5, 7.6, and 7.7. The main procedure *Sr_polynomial_resultant* removes the content of the polynomials and then obtains the resultant of the primitive parts using the recursive procedure *Sr_polynomial_resultant_rec*. At line 10, the resultant is obtained using Theorem 7.5(2),(3).

The recursive procedure *Sr_polynomial_resultant_rec* performs the calculations using the subresultant remainder sequence. Observe that the formal parameter list includes δp and ψp which correspond to the previous

---

Procedure $Sr\_polynomial\_resultant(u, v, L, K)$;
Input
   u, v : non-zero multivariate polynomials in symbols in L
        with coefficients in **Z** or **Q**;
   L : a list of symbols with main variable $x = Operand(x, 1)$;
   K : the symbol **Z** or **Q**;
Output
   $res(u, v, x)$;
Local Variables
   $x, R, m, n, cont\_u, pp\_u, cont\_v, pp\_v, s$;
Begin
1    $x = Operand(L, 1)$;
2    $R = Rest(L)$;
3    $m := Degree\_gpe(u, x)$;
4    $n := Degree\_gpe(v, x)$;
5    $cont\_u := Polynomial\_content\_sr(u, x, R, K)$;
6    $pp\_u := Rec\_quotient(u, cont\_u, L, K)$;
7    $cont\_v := Polynomial\_content\_sr(v, x, R, K)$;
8    $pp\_v := Rec\_quotient(v, cont\_v, L, K)$;
9    $s := Sr\_polynomial\_resultant\_rec(pp\_u, pp\_v, L, K, 1, 0, 0)$;
10   $Return((cont\_u)^n * (cont\_v)^m * s)$
End

Figure 7.5. The main MPL procedure that obtains the resultant for multivariate polynomials using the subresultant remainder sequence. (Implementation: Maple (txt), Mathematica (txt), MuPAD (txt).)

values δ_{i-1} and ψ_{i-1} in Equations (7.31) and (7.32). The initial call on the procedure at line 9 of Figure 7.5 uses the actual parameters $\delta_p = 0$ and $\psi_p = 0$ because δ_p and ψ_p are not used when $i = 1$ (see lines 15–17).

Since this approach requires $\deg(u, x) \geq \deg(v, x)$, at lines 4–5 we check this condition and, if necessary, apply Theorem 7.5(1). Lines 6–7 use a case in Definition 7.4 as one of the termination conditions for the recursion. In lines 9–11, we compute the pseudo-remainder and then check the termination condition in Equation (7.28). We compute the current values of ψ_i and β_i in lines 15–21, and obtain the division in Equation (7.29) at line 22. At line 23, we obtain (recursively) $res(v, r)$ and then apply Theorem 7.5(1),(2). Finally, in lines 24–28 we perform the division in Equation (7.28).

Example 7.14. Consider again the resultant calculation for the polynomials in Example 7.13. In Figure 7.4 (page 285), we give a summary of the

Procedure $Sr\_polynomial\_resultant\_rec(u, v, L, K, i, \delta p, \psi p)$;
Input
    u, v : non-zero multivariate polynomials in symbols in L
           with coefficients in **Z** or **Q**;
    L : a list of symbols with main variable $x = Operand(x, 1)$;
    K : the symbol **Z** or **Q**;
    $i, \delta p, \psi p$: non-negative integers;
Output
    $res(u, v, x)$;
Local Variables
    $x, m, n, r, \delta, R, \psi, \beta, f, s, w, l, k$;
Begin
1    $x = Operand(L, 1)$;
2    $m := Degree\_gpe(u, x)$;
3    $n := Degree\_gpe(v, x)$;
4   **if** $m < n$ **then**
5        $Return((-1)^{m*n} * Sr\_polynomial\_resultant\_rec(v, u, L, K, i, \delta p, \psi p))$
6   **elseif** $n = 0$ **then**
7        $Return(v^m)$
8   **else**
9        $r := Pseudo\_remainder(u, v, x)$;
10   **if** $r = 0$ **then**
11         $Return(0)$

Continued in Figure 7.7

Figure 7.6. A recursive MPL procedure that obtains the resultant for multivariate polynomials using the sub-resultant remainder sequence. (Implementation: Maple (txt), Mathematica (txt), MuPAD (txt).)

coefficient and degree explosion obtained in the monomials in the pseudo-remainder r at line 9 in the procedure $Sr\_polynomial\_resultant\_rec$ in Figure 7.6. Comparing these results with those in Figure 7.3, we see that sub-resultant remainder sequence significantly reduces degree and coefficient explosion. $\square$

Exercises

1. Let $u = x^3 - 1$ and $v = x^2 + 2$.

   (a) Give the Sylvester matrix $S(u, v, x)$.

```
12        elseif  r ≠ 0 then
13             δ := m − n + 1;
14             R := Rest(L);
15             if  i = 1 then
16                  ψ = −1;
17                  β = (−1)^δ
18             elseif  i > 1 then
19                  f := Leading_Coefficient_gpe(u, x);
20                  ψ := Rec_quotient (Algebraic_expand ((−f)^{δp−1}),
                                      Algebraic_expand (ψ^{δp−2}),  R,  K);
21                  β := Algebraic_expand (−f * ψ^{δ−1});
22             r := Rec_quotient(r, β, L, K);
23             w := Algebraic_expand((−1)^{m*n} * β^n
                              *Sr_polynomial_resultant(v, r, L, K, i + 1, δ, ψ));
24             l := Coefficient_gpe(v, x, n);
25             s := Degree_gpe(r, x);
26             k = n * δ − m + s;
27             f := Algebraic_expand (l^k);
28             Return( Rec_quotient(w, f, L, K))
    End
```

Figure 7.7. Continuation of Figure 7.6.

(b) Compute res(u, v) using the determinant definition.

(c) Compute res(u, v) using the algorithm in Figure 7.1.

2. For what values of t do the two polynomials $u = x^2 + 5x + t$ and $v = x^2 + tx + 9$ have a common factor of positive degree?

3. Consider the two equations $u = ax^2 + bx + a = 0$ and $v = x^3 - 2x^2 + 2x - 1 = 0$. Show that the two equations have a common root if $a = -b$ or $a = -b/2$.

4. Suppose u and v in $\mathbf{K}[y]$ and res(u, v) $\neq 0$. Show that A and B in Equation (7.14) are unique.

5. Prove Theorem 7.5, Parts (2) and (3).

6. Let m be a positive integer, and let v be in $\mathbf{F}[x]$. Show that res(x^m, v) = v_0^m.

7. Let $u = x^2 + 1$ and $v = x^2 + 2x + 1$. Find res(u, v) and the polynomials A and B described in Theorem 7.7.

8. Prove Theorem 7.12.

7.2 Polynomial Relations for Explicit Algebraic Numbers

Let α be an explicit algebraic number. The algorithm described in this section uses the resultant to find a non-zero polynomial $u(x)$ in $\mathbf{Q}[x]$ that has α as a root. The algorithm is based on the following theorem.

Theorem 7.15. *Let $v(x)$ and $w(x)$ be polynomials in $\mathbf{Q}[x]$ with positive degree, and suppose that $v(\alpha) = 0$ and $w(\beta) = 0$. Then:*

1. *$\alpha + \beta$ is a root of*

$$u(x) = \operatorname{res}(v(x - y), w(y), y) = 0.$$

2. *$\alpha\beta$ is a root of*

$$u(x) = \operatorname{res}(y^m v(x/y), w(y), y) = 0$$

   where $m = \deg(v(x), x)$.

3. *If $\alpha \neq 0$, then α^{-1} is a root of*

$$u(x) = \operatorname{res}(x\, y - 1, v(y), y) = 0.$$

4. *If a/b is a positive rational number in standard form, then $\alpha^{a/b}$ is a root of*

$$u(x) = \operatorname{res}(v(y), x^b - y^a, y) = 0.$$

Proof: To show (1), define the polynomial

$$f(y) = v(\alpha + \beta - y).$$

Since $f(\beta) = v(\alpha) = 0$ and $w(\beta) = 0$, the polynomials f and w have a common root β, and Theorem 7.9 implies that $\operatorname{res}(f(y), w(y), y) = 0$. Therefore,

$$u(\alpha + \beta) = \operatorname{res}(v(\alpha + \beta - y), w(y), y) = \operatorname{res}(f(y), w(y), y) = 0.$$

The other statements in the theorem are proved in a similar way (Exercise 2). □

As a consequence of Theorem 7.15, we have the following theorem.

Theorem 7.16. *The algebraic numbers $\mathbf{A}$ are a field.*

Proof: Theorem 7.15 shows that the algebraic numbers are closed under addition (Axiom F-1) and multiplication (Axiom F-2), and that the multiplicative inverse of a non-zero algebraic number is also an algebraic number (Axiom F-12). The remaining field properties follow because the algebraic numbers are included in the field of complex numbers. $\qquad\square$

In the next few examples, we show that by repeatedly applying Theorem 7.15 we obtain a polynomial that has a given explicit algebraic number as a root.

Example 7.17. Consider the explicit algebraic number

$$\alpha = 2^{1/2} + 3^{1/2} + 5^{1/2}.$$

The three radicals in the sum have minimal polynomials

$$p_2(x) = x^2 - 2, \qquad p_3(x) = x^2 - 3, \qquad p_5(x) = x^2 - 5.$$

Using Theorem 7.15(1), $2^{1/2} + 3^{1/2}$ is a root of

$$u(x) = \text{res}(p_2(x - y), p_3(y), y) = x^4 - 10\,x^2 + 1.$$

Applying the theorem again, we have α is a root of

$$\text{res}(u(x - y), p_5(y), y) = x^8 - 40\,x^6 + 352\,x^4 - 960\,x^2 + 576.$$

Since this polynomial is monic and irreducible, it is the minimal polynomial for α. $\qquad\square$

Example 7.18. Consider the explicit algebraic number

$$\alpha = \frac{1}{\sqrt{2} + \sqrt{3}} + \frac{1}{\sqrt{5} + \sqrt{7}}.$$

From the last example, the polynomial $u_1(x) = x^4 - 10\,x^2 + 1$ has $2^{1/2} + 3^{1/2}$ as a root, and applying Theorem 7.15(3), we obtain

$$u_2(x) = \text{res}(x\,y - 1, u_1(y), y) = x^4 - 10\,x^2 + 1$$

which has $1/(2^{1/2} + 3^{1/2})$ as a root. (Although $u_1 = u_2$ in this case, this is not generally true.) In a similar way, $1/(5^{1/2} + 7^{1/2})$ is a root of $u_3(x) = 4\,x^4 - 24\,x^2 + 1$. Applying Theorem 7.15(1), α is a root of

$$
\begin{aligned}
u_4(x) \;=\;& \text{res}(u_2(x - y), u_3(y), y) \\
=\;& 256\,x^{16} - 16384\,x^{14} + 363776\,x^{12} - 3729408\,x^{10} \\
& + 18468704\,x^8 - 39450624\,x^6 + 23165264\,x^4 \\
& - 3909760\,x^2 + 46225.
\end{aligned}
$$

Since this polynomial is irreducible in $\mathbf{Q}[x]$, $(1/\mathrm{lc}(u_4)) \cdot u_4$ is the minimal polynomial for α. $\qquad\square$

Example 7.19. Let

$$\alpha = \sqrt{2}\sqrt{3} + \sqrt{6}.$$

Repeated application of Theorem 7.15 gives

$$u(x) = x^8 - 48\,x^6 + 576\,x^4 = x^4\,(x^2 - 24)^2 \qquad (7.34)$$

which has α as a root. In this case, $u(x)$ is reducible and the factors of u give additional information about α. In a numerical sense, α can have up to eight different values that correspond to the sign interpretations of its three square root symbols

$$(\pm\sqrt{2})(\pm\sqrt{3}) \pm \sqrt{6}. \qquad (7.35)$$

These numerical values are indicated by the factors of the polynomial. For example, since x^4 is a factor, there are four choices of signs where the expression (7.35) simplifies to zero:

$$(+\sqrt{2})(+\sqrt{3}) - \sqrt{6}, \qquad (+\sqrt{2})(-\sqrt{3}) + \sqrt{6},$$

$$(-\sqrt{2})(+\sqrt{3}) + \sqrt{6}, \qquad (-\sqrt{2})(-\sqrt{3}) - \sqrt{6}.$$

In addition, since $(x^2 - 24)^2$ is a factor, there are two choices of signs where the expression (7.35) simplifies to $+\sqrt{24} = 2(+\sqrt{6})$, and two choices of signs where it simplifies to $-\sqrt{24} = 2(-\sqrt{6})$.

It is interesting to see what happens when Theorem 7.15(3) is applied to $u(x)$ in Equation (7.34) to find the polynomial associated with

$$\frac{1}{(\pm\sqrt{2})(\pm\sqrt{3}) \pm \sqrt{6}}, \qquad (7.36)$$

because, for some choices of signs, the denominator is 0. In this case we obtain

$$
\begin{aligned}
\mathrm{res}(x\,y - 1, u(y), y) &= \mathrm{res}(x\,y - 1, y^8 - 48\,y^6 + 576\,y^4, y) \\
&= 576\,x^4 - 48\,x^2 + 1 \\
&= (24\,x^2 - 1)^2
\end{aligned}
$$

which has as its solutions the four expressions in (7.36) where the signs are chosen so that division by 0 does not occur. In Exercise 3, we give a resultant relation that shows why this is so. $\qquad\square$

Example 7.20. Let

$$\alpha = \sqrt{2} + \sqrt{3} + \sqrt{5 + 2\sqrt{6}}.$$

Repeated application of Theorem 7.15 gives the reducible polynomial

$$
\begin{aligned}
u(x) &= x^{16} - 80\,x^{14} + 2208\,x^{12} - 28160\,x^{10} \\
&\quad + 172288\,x^8 - 430080\,x^6 + 147456\,x^4 \\
&= x^4\,(x^2 - 8)^2(x^2 - 12)^2(x^4 - 40\,x^2 + 16) \qquad (7.37)
\end{aligned}
$$

which has α for a root. In a numerical sense, α can have up to 16 different values that correspond to the sign interpretations of its four square root symbols

$$\pm\sqrt{2} \pm \sqrt{3} \pm \sqrt{5 \pm 2\sqrt{6}}. \qquad (7.38)$$

Since x^4 is a factor of $u(x)$, there are four possible choices of signs where the expression (7.38) simplifies to zero:

$$+\sqrt{2} + \sqrt{3} - \sqrt{5 + 2\sqrt{6}}, \qquad +\sqrt{2} - \sqrt{3} + \sqrt{5 + 2\sqrt{6}},$$

$$-\sqrt{2} + \sqrt{3} - \sqrt{5 - 2\sqrt{6}}, \qquad -\sqrt{2} - \sqrt{3} + \sqrt{5 + 2\sqrt{6}}.$$

In addition, since $(x^2 - 8)^2$ is a factor, there are four choices of signs where it simplifies to $\pm\sqrt{8} = \pm 2\sqrt{2}$. In a similar way, there are four choices of signs where it simplifies to $\pm\sqrt{12} = \pm 2\sqrt{3}$, and four choices where it simplifies to one of the roots $\pm 2\sqrt{2} \pm 2\sqrt{3}$ of $x^4 - 40\,x^2 + 16 = 0$.   □

The procedure *Find_polynomial* in Figures 7.8 and 7.9 is based on Theorem 7.15. The procedure returns a polynomial that has the explicit algebraic number α as a root. Notice that the procedure uses a global mathematical symbol y to avoid using a local variable that would not be assigned before it was used (see page 3).

Degree and Coefficient Explosion.   Although *Find_polynomial* returns a polynomial relation for any explicit algebraic number, both the degree of the polynomial and the number of digits in its coefficients increase rapidly with the complexity of the number (Exercise 4). For example, for $\alpha = 2^{1/2} + 3^{1/2} + 5^{1/2} + 7^{1/2} + 11^{1/2} + 13^{1/2}$ the procedure returns a monic, irreducible polynomial in $\mathbf{Z}[x]$ of degree 64 with some coefficients that have more than 35 digits.

Procedure $Find\_polynomial(\alpha, x)$;
Input
    α : an explicit algebraic number;
    x : a symbol;
Output
    a polynomial in $\mathbf{Q}[x]$;
Local Variables $base, exponent, n, d, j, p, q, w, m$;
Global y;
Begin

```
1       if  Kind(α) ∈ {integer, fraction} then  Return(x − α)
2       elseif  Kind(α) = ” ∧ ” then
3           base := Operand(α, 1);
4           exponent := Operand(α, 2);
5           n := Numerator(exponent);
6           d := Denominator(exponent);
7           p := Substitute(Find_polynomial(b, x), x = y);
8           if exponent > 0 then
```
9             $Return\Big(Sr\_polynomial\_resultant\Big(p, x^d - y^n, [y, x], \mathbf{Q}\Big)\Big)$
```
10          else
```
11           $w := Substitute\Big($
                     $Sr\_polynomial\_resultant\Big(p, x^d - y^{|n|}, [y, x], \mathbf{Q}\Big), x = y\Big)$;

12           $Return(Sr\_polynomial\_resultant(x * y - 1, w, [y, x], \mathbf{Q}))$

Continued in Figure 7.9.

Figure 7.8. The MPL *Find_polynomial* procedure. (Implementation: Maple (txt), Mathematica (txt), MuPAD (txt).)

Exercises

1. For each of the following explicit algebraic numbers, find a non-zero polynomial in $\mathbf{Q}[x]$ which has α for a root.

    (a) $\alpha = 1 + \sqrt{2} + \sqrt{3}$.
    (b) $\alpha = (\sqrt{2} + \sqrt{3})^2$.
    (c) $\alpha = 1 + 3^{1/2} + 4^{1/3}$.
    (d) $\alpha = 1/(2\sqrt{2} + 3\sqrt{3})$.

2. Prove Parts (2), (3), and (4) of Theorem 7.15.

3. Let v be a polynomial in $\mathbf{Q}[y]$ with positive degree. Prove each of the following statements.

```
13      elseif  Kind(α) = ” + ” then
14         w := Find_polynomial(Operand(α, 1), x);
15         for  j := 2 to  Number_of_operands(α) do
16            p := Find_polynomial(Operand(α, j), x);
17            q := Algebraic_expand(Substitute(p, x = x − y));
18            w := Sr_polynomial_resultant(q, Substitute(w, x = y), [y, x], Q);
19         Return(w)
20      elseif  Kind(α) = ” ∗ ” then
21         w := Find_polynomial(Operand(α, 1), x);
22         for  j := 2 to  Number_of_operands(α) do
23            p := Find_polynomial(Operand(α, j), x);
24            m := Degree_gpe(p, x);
```
$$q := Algebraic\_expand\Big(y^m * Substitute(p, x = x/y)\Big);$$
```
26            w := Sr_polynomial_resultant(q, Substitute(w, x = y), [y, x], Q);
27         Return(w)
      End
```

Figure 7.9. Continuation of Figure 7.8.

(a) If $y \mid v(y)$, then $\text{res}(x\,y - 1, v(y), y) = \text{res}(x\,y - 1, v(y)/y, y)$.

(b) Suppose that $v(y) = 0$ has zero as a root of multiplicity r. Show that $\text{res}(x\,y - 1, v(y), y) = \text{res}(x\,y - 1, v(y)/y^r, y)$.

Notice that this exercise implies that if $\alpha \neq 0$ is a root of $v(y) = 0$, which also has zero as a root of multiplicity r, we can still apply the relation in Theorem 7.15 because the resultant is computed as if the factor y^r (which corresponds to the root $y = 0$ of multiplicity r) has been removed.

4. In this exercise, we describe a way to find the degrees of the polynomials $u(x)$ in Theorem 7.15. Let v and w be polynomials in $\mathbf{F}[x]$ with $m = \deg(v, x) > 0$ and $n = \deg(w, x) > 0$. Prove each of the following statements.

(a) $\deg(\text{res}(v(x - y), w(y), y), x) = m\,n$.

(b) $\deg(\text{res}(y^m\, v(x/y), w(y), y), x) = m\,n$.

(c) If r is the maximum positive integer such that $y^r \mid v(y)$, then

$$\deg(\text{res}(x\,y - 1, v(y), y), x) = m - r.$$

(d) If a/b is a positive rational number in standard form, then

$$\deg(\text{res}(v(y), x^b - y^a, y), x) = b\,m.$$

5. Suppose that α is a root of $x^m - 1 = 0$ and β is a root of $x^n - 1 = 0$. Show that $\alpha\beta$ is a root of $x^{mn} - 1$.

6. Let α be an algebraic number which is a root of $ax^2 + bx + c = 0$, and let β be an algebraic number which is a root of $dx^2 + ex + f = 0$.

   (a) Show that if $c \neq 0$, then $1/\alpha$ is a root of $cx^2 + bx + a = 0$. (The condition $c \neq 0$ implies that $\alpha \neq 0$.)

   (b) Show that $\alpha\beta$ is a root of

   $$a^2 d^2 x^4 - abde x^3 + \left(-2acdf + ae^2c + db^2f\right)x^2$$
   $$-bcef x + c^2 f^2 = 0.$$

   (c) Show that $\alpha + \beta$ is a root of

   $$a^2 d^2 x^4 + \left(2abd^2 + 2dea^2\right)x^3$$
   $$+ \left(b^2 d^2 + 2d^2ac + 3abde + a^2e^2 + 2a^2df\right)x^2$$
   $$+ \left(2a^2ef + deb^2 + 2abdf + ae^2b + 2d^2bc + 2acde\right)x$$
   $$+ c^2 d^2 + abef - 2acdf + a^2f^2 + fdb^2 + ae^2c + bcde$$
   $$= 0.$$

Further Reading

7.1 The Resultant Concept. See Akritas [2], Buchberger et al. [17] ("Computing in Algebraic Extensions"), Cohen [22], Geddes et al. [39], von zur Gathen and Gerhard [96], and Zippel [108] for a discussion of resultants. See Collins [26] for a discussion of the computation of multivariate polynomial resultants.

7.2 Polynomial Relations For Explicit Algebraic Numbers. The material in this section is based on the work of R. Loos (see Buchberger et al. [17], "Computing in Algebraic Extensions").

8

Polynomial Simplification with Side Relations

In Section 4.3 we considered the simplification of polynomials with symbols for algebraic numbers that are defined as the solutions of polynomial equations. In Section 6.2 we considered the simplification of multivariate polynomials with respect to a single polynomial side relation. In this chapter we again consider the polynomial simplification problem but this time allow for several multivariate polynomial side relations that may have symbols in common. Because of this, the supporting mathematical theory is more involved than the material in Sections 4.3 and 6.2. The algorithm in this chapter is a generalization of the algorithms in these earlier sections.

In Section 8.1 we describe the division process that is used in the simplification algorithm. In Section 8.2 we give a precise definition of the simplification problem and introduce the concept of a Gröbner basis which plays a key role in our algorithm. Finally, in Section 8.3, we give an algorithm that finds a Gröbner basis and the polynomial simplification algorithm.

8.1 Multiple Division and Reduction

To simplify the presentation, we assume that all polynomials are in the domain $\mathbf{Q}[x_1, \ldots, x_p]$. We begin with some examples.

Example 8.1. Suppose that w, x, y, and z, satisfy the two side relations

$$f_1 = w^2 + x^2 - 1 = 0, \tag{8.1}$$

$$f_2 = y\,z - 1 = 0, \tag{8.2}$$

and consider the simplification of the polynomial

$$
\begin{aligned}
u &= \left(y\,z^2 + x\,w\right) f_1 + (w - z\,x)\,f_2 \tag{8.3}\\
&= yz^2\,w^2 + y\,z^2\,x^2 - y\,z^2 + x\,w^3 + x^3\,w - x\,w + w\,y\,z \\
&\quad -w - z^2\,x\,y + z\,x \tag{8.4}
\end{aligned}
$$

with respect to the two side relations. Although it is evident from the unexpanded form in Equation (8.3) that u simplifies to 0 at all points that satisfy the side relations, this is not apparent from the expanded form in Equation (8.4). We can simplify the expanded form, however, using monomial-based division. First, dividing u by f_1 and then using the side relation $f_1 = 0$, we obtain

$$
\begin{aligned}
u &= \mathrm{mbquot}(u, f_1, [w,\,x])\,f_1 + \mathrm{mbrem}(u, f_1, [w,\,x]) \\
&= \mathrm{mbrem}(u, f_1, [w,\,x]) \\
&= w\,y\,z - w - z^2\,x\,y + z\,x
\end{aligned}
$$

In a similar way, dividing the new u by f_2 and then using the side relation $f_2 = 0$, we obtain the simplified form for u

$$u = \mathrm{mbrem}(u,\,f_2, [y,\,z]) = 0.$$

In this example, since the two side relations are expressed in terms of different variables, we obtain the simplification with two division steps. □

The next example shows that the simplification problem is more involved when the two side relations have variables in common.

Example 8.2. Let x and y be solutions of the two side relations:

$$
\begin{aligned}
f_1 &= 3\,x^2 + 5x\,y - 5\,x - 2\,y^2 + 11\,y - 12 = 0, \tag{8.5}\\
f_2 &= 2\,x^2 + x\,y - 5\,x - 3\,y^2 + 10\,y - 7 = 0, \tag{8.6}
\end{aligned}
$$

and consider the simplification of the polynomial

$$
\begin{aligned}
u &= x\,f_1 + y\,f_2 \tag{8.7}\\
&= 3\,x^3 + 7\,x^2 y - 5\,x^2 - x\,y^2 + 6\,x\,y - 12\,x \tag{8.8}\\
&\quad -3\,y^3 + 10\,y^2 - 7y
\end{aligned}
$$

with respect to these side relations. Again, it is evident from the unexpanded form in Equation (8.7) that u simplifies to 0, but not from the expanded form in Equation (8.8).

Let's try to obtain this simplification using monomial-based division. Dividing u by f_1 with x as the main variable and then using the side relation $f_1 = 0$, we obtain

$$\begin{aligned} u &= \mathrm{mbquot}(u, f_1, [x, y])\, f_1 + \mathrm{mbrem}(u, f_1, [x, y]) \\ &= \mathrm{mbrem}(u, f_1, [x, y]) \\ &= -\frac{7}{3}\, x\, y^2 - \frac{5}{3}\, x\, y - \frac{5}{3}\, y^3 + \frac{8}{3}\, y^2 + y \end{aligned} \qquad (8.9)$$

However, since $\mathrm{lm}(f_2, [x, y]) = 2\, x^2$ does not divide any monomial of u in Equation (8.9), division by f_2 does not obtain the simplified form 0. Although we can continue the division process on u with y as the main variable, this again does not lead to the simplified form 0. In fact, no matter how we order the divisions or the variables we cannot obtain the simplified form 0 using monomial-based division.

The problem here is not so much with the division process but with the form of the side relations. We shall see (in Section 8.3) that the first step in our simplification algorithm is to obtain an equivalent set of side relations (called the Gröbner basis), and then to apply the division process with these new side relations. In this case, the algorithm in Section 8.3 gives the Gröbner basis relations

$$\begin{aligned} g_1 &= 258\, x + 462\, y^3 - 1471\, y^2 + 339\, y = 0, & (8.10) \\ g_2 &= 66\, y^4 - 163\, y^3 - 128\, y^2 + 217\, y + 116 = 0. & (8.11) \end{aligned}$$

Dividing u by g_1 with x as the main variable, we obtain the remainder

$$\begin{aligned} r &= -\frac{1369599}{79507}\, y^9 + \frac{26164677}{159014}\, y^8 - \frac{171536827}{318028}\, y^7 + \frac{3026492827}{5724504}\, y^6 \\ &\quad + \frac{3730217111}{5724504}\, y^5 - \frac{8003771203}{5724504}\, y^4 + \frac{240501299}{5724504}\, y^3 \\ &\quad + \frac{1288231829}{1431126}\, y^2 - \frac{191757295}{1431126}\, y - \frac{137520262}{715563}, \end{aligned}$$

and then division of r by g_2 gives remainder 0. $\qquad\square$

The goal in this chapter is the description of an algorithm that simplifies a multivariate polynomial u with respect to the relations $f_1 = 0, \ldots, f_n = 0$. A precise description of this problem is given in Section 8.2, and the simplification algorithm is given in Section 8.3.

Multiple Division

Let's begin by examining in greater detail the division of a polynomial by a list of polynomial divisors. The condition in the next definition is used to terminate the division process.

Definition 8.3. *Let F be a set or list of non-zero distinct polynomials $f_1, \ldots, f_n$, and let L be a list of symbols . The polynomial u is* **reduced**[1] *with respect to F if either $u = 0$ or if no monomial in u is divisible by $\mathrm{lm}(f_i, L)$, $i = 1, \ldots, n$.*

This condition is similar to the one that terminates monomial-based division with a single divisor (Definition 6.26, page 215). In Example 8.2 above, Equation (8.9) is reduced with respect to the side relations in Equations (8.5) and (8.6).

Definition 8.4. *Let $F = [f_1, \ldots, f_n]$ be a list of non-zero polynomials, and let L be a list of symbols that defines the variable order.* **Multiple division** *of a polynomial u by the divisor list F is defined by the iteration scheme:*

$$c_1 := c_2 := \cdots := c_n := 0; \quad r := u;$$

> **while** r is not reduced with respect to F **do**
>   **for** $i := 1$ **to** n **do**
> $q := \mathrm{mbquot}(r, f_i, L)$;
> $c_i := c_i + q$;
> $r := \mathrm{mbrem}(r, f_i, L)$;

The process terminates when r is reduced with respect to F. Upon termination, the polynomials $c_1, \ldots, c_n$ are the **multiple quotients** *with respect to F, and r is the* **multiple remainder**.

Observe that the relations for q, c_i, and r are defined (by assignments) in terms of the previous values of r and c_i. The algorithm includes a **for** loop which is nested within a **while** loop. We have done this because division by f_j may introduce new opportunities for division by some other f_i with $i < j$, which means the process may require several passes through the **while** loop. (See Example 8.5 below and Exercise 2.) In addition, at each division step in the **for** loop,

$$r_{old} = q_{new}\, f_i + r_{new},$$

where the r_{old} is the old remainder value, and r_{new} and q_{new} are the new values. Therefore, since $r = u$ at the start, at each iteration step we have the representation

$$u = c_1\, f_1 + \cdots + c_n\, f_n + r \qquad (8.12)$$

---

[1]Similar terms *reducible* and *irreducible* were introduced to describe the factorization of polynomials (Definition 4.11, page 118) and the factorization of expressions in an integral domain (Definition 6.7, page 204). In this chapter the term *reduced* refers to the concept in Definition 8.4.

where the polynomials c_i and r are the most recently computed values. At termination, the remainder r is reduced with respect to F.

The division process depends on both the order of the polynomials in F and the order of the symbols in L. This point is illustrated in the next example.

Example 8.5. Let

$$u = x^3 + 2\,x^2 y - 5\,x + y^3 - 3\,y, \quad f_1 = x\,y - 1, \quad f_2 = x^2 + y^2 - 4,$$

$F = [f_1,\, f_2]$, and $L = [x, y]$. Monomial-based division of u by f_1 gives

$$c_1 = 0 + q = \mathrm{mbquot}(u,\, f_1,\, [x, y]) = 2\,x,$$

$$r = \mathrm{mbrem}(u,\, f_1,\, [x, y]) = x^3 - 3\,x + y^3 - 3\,y.$$

Next, division of r by f_2 gives

$$c_2 = 0 + q = \mathrm{mbquot}(r,\, f_2,\, [x, y]) = x,$$

$$r = \mathrm{mbrem}(r,\, f_2,\, [x, y]) = -x\,y^2 + x + y^3 - 3\,y.$$

Observe that division by f_2 has created a new opportunity for division by f_1. Division of r by f_1 for a second time gives

$$c_1 = 2\,x + q = 2\,x + \mathrm{mbquot}(r,\, f_1,\, [x, y]) = 2\,x - y,$$

$$r = \mathrm{mbrem}(r,\, f_1,\, [x, y]) = x + y^3 - 4\,y.$$

Since r is reduced with respect to F, the process terminates, and we obtain the representation for u in the form of Equation (8.12)

$$u = (2\,x - y)\,f_1 + x\,f_2 + (x + y^3 - 4\,y). \tag{8.13}$$

The representation in Equation (8.12) is not unique. To obtain another representation, we reverse the order of the divisions. Starting the process by dividing u by f_2, we obtain

$$c_2 = 0 + q = \mathrm{mbquot}(u,\, f_2, [x, y]) = x + 2\,y,$$

$$r = \mathrm{mbrem}(u,\, f_2,\, [x, y]) = -x\,y^2 - x - y^3 + 5\,y,$$

and division of r by f_1 gives

$$c_1 = 0 + q = \mathrm{mbquot}(r,\, f_1,\, [x, y]) = -y,$$

$$r = \mathrm{mbrem}(r,\, f_1,\, [x, y]) = -x - y^3 + 4\,y.$$

Since r is reduced with respect to F, the division process terminates with another representation

$$u = (-y)\, f_1 + (x + 2\, y)\, f_2 + (-x - y^3 + 4\, y). \qquad (8.14)$$

A third representation of the form in Equation (8.12) is obtained by using y as the main variable and dividing u by $[f_1, f_2]$

$$u = x\, f_1 + y\, f_2 + (y + x^3 - 4\, x).$$

In this case, division of u by $[f_2, f_1]$ with y as the main variable gives the same representation. In fact, there are infinitely many representations of the form in Equation (8.12), although only three of them are obtained by multiple division (Exercise 1). $\qquad\qquad\square$

The properties of multiple division are given in the next theorem.

Theorem 8.6. *Let u be a polynomial, $F = [f_1, \ldots, f_n]$ a list of non-zero polynomials, and L a list of symbols that defines the variable order.*

1. *The multiple division algorithm terminates.*

2. *When the algorithm terminates,*

$$u = c_1 f_1 + \cdots + c_n f_n + r \qquad (8.15)$$

where r is reduced with respect to F. In addition,

$$\mathrm{lm}(r) \preceq \mathrm{lm}(u), \qquad (8.16)$$

and

$$\mathrm{lm}(u) \equiv \max\left(\{\mathrm{lm}(c_1)\,\mathrm{lm}(f_1), \ldots, \mathrm{lm}(c_n)\,\mathrm{lm}(f_n),\ \mathrm{lm}(r)\}\right) \qquad (8.17)$$

where the maximum is with respect to the monomial order relation.

Proof: Consider the algorithm in Definition 8.4. First, if $u = 0$, the algorithm terminates immediately, and so we assume that $u \neq 0$. After one pass through the **while** loop, the remainder $r = R_1$ has the form

$$R_1 = M_1 + T_1,$$

where M_1 is reduced with respect to F, and each monomial in T_1 is divisible by some $\mathrm{lm}(f_i)$, $i = 1, \ldots, n-1$. (We have not included $\mathrm{lm}(f_n)$ in this list

because after completing the **for** loop, R_1 is reduced with respect to f_n.)
We show that
$$\mathrm{lm}(T_1) \prec \mathrm{lm}(u).$$
To see why, let t be a monomial of T_1, and suppose that $\mathrm{lm}(f_i) \mid t$. However, after division by f_i, all monomials in the current remainder are not divisible by $\mathrm{lm}(f_i)$. Therefore, for some $j > i$, the operation $\mathrm{mbrem}(r, f_j)$ has created a monomial s with the same variable part as t. By Theorem 6.30(2), at each step in the **for** loop, $\mathrm{lm}(r) \preceq \mathrm{lm}(u)$. In addition, by Theorem 6.28(8), $s \prec \mathrm{lm}(r)$, and therefore $\mathrm{lm}(T_1) \prec \mathrm{lm}(u)$.

Consider now the second pass through the **while** loop. Since M_1 is reduced with respect to F, the termination of the process depends only on T_1. Therefore, for the sake of analysis, let's apply the division operations to M_1 and T_1 separately.[2] First, since M_1 is reduced with respect to F, the execution of the **for** loop beginning with $r = M_1$ simply gives M_1 for a remainder. Next, beginning with $r = T_1$, the execution of the **for** loop gives a remainder $M_2 + T_2$ where M_2 is reduced with respect to F and each monomial in T_2 is divisible by some $\mathrm{lm}(f_i)$, $i = 1, \ldots, n - 1$. Therefore, starting with $r = M_1 + T_1$, the execution of the **for** loop gives a remainder
$$R_2 = M_1 + M_2 + T_2.$$
In addition, using an argument similar to the one in the previous paragraph, we have
$$\mathrm{lm}(T_2) \prec \mathrm{lm}(T_1).$$
In general, after s passes through the **while** loop, the remainder has the form
$$R_s = M_1 + \cdots + M_s + T_s$$
where each M_i is reduced with respect to F, each monomial in T_s is divisible by some $\mathrm{lm}(f_i)$, $i = 1, \ldots, n - 1$, and
$$\mathrm{lm}(T_s) \prec \mathrm{lm}(T_{s-1}) \prec \cdots \prec \mathrm{lm}(T_1) \prec \mathrm{lm}(u).$$
Therefore, after some pass through the **while** loop, $T_i = 0$ and the process terminates.

To show (2), Equation (8.15) is given in Equation (8.12), and the reduction (8.16) follows by repeatedly applying the relation (6.36) in Theorem 6.30 (page 220) at each monomial-based division step in the **for** loop.

For now, we omit the proof of Equation (8.17), because we prove it later for a more general division process called reduction. □

A procedure that performs multiple division is described in Exercise 3.

---

[2]See Exercise 6 on page 228 for a justification for this approach.

Reduction

From an algorithmic perspective, multiple division is adequate for our simplification process as long as the side relations have the proper form. From a theoretical (and practical) perspective, however, we need a more flexible division process. The next two examples illustrate some limitations of multiple division.

Example 8.7. Let $f_1 = x\,y - 1$, $f_2 = x^2 + y^2 - 4$, $L = [x, y]$, and

$$u = 2\,x^3 + 3\,x^2 y + x^2 + 2\,y^3 + 4\,y + 1.$$

Using multiple division of u by $F_1 = [f_1, f_2]$, we obtain the representation

$$u = (3\,x - 2\,y)\,f_1 + (2\,x + 1)\,f_2 + r \tag{8.18}$$

with remainder

$$r = 11\,x + 2\,y^3 - y^2 + 2\,y + 5.$$

Next, let $f_3 = r$, and consider the division of u by

$$F_2 = [f_1, f_2, f_3]. \tag{8.19}$$

Since we have included the remainder r in F_2, we might expect the remainder in this division to be 0. Instead, we obtain the representation

$$
\begin{aligned}
u \;=\; & 3\,x\,f_1 + (2\,x + 1)\,f_2 + \left(-\frac{2}{11}\,y^2 + 1\right) f_3 \\
& + \left(\frac{4}{11}\,y^5 - \frac{2}{11}\,y^4 + \frac{4}{11}\,y^3 + \frac{10}{11}\,y^2 + 2\,y\right)
\end{aligned}
\tag{8.20}
$$

where the sum in parentheses on the last line is the non-zero remainder. The problem here is the division of u by F_1 requires two passes through the **while** loop using the divisor sequence f_1, f_2, f_1, f_2. By including f_3 in F_2, multiple division uses the divisor sequence f_1, f_2, f_3, f_1, f_2, f_3 which gives a non-zero remainder. In fact, no matter how we order the divisors in Equation (8.19), multiple division obtains a representation with a non-zero remainder.

There is, however, a way to divide u by the set $\{f_1, f_2, f_3\}$ that obtains the remainder 0, although it doesn't use the divisor sequence specified by the **for** loop in Definition 8.4. In fact, if we perform division using the divisor sequence

$$f_1, \; f_2, \; f_1, \; f_2, \; f_3, \tag{8.21}$$

we obtain the representation

$$u = (3\,x - 2\,y)\,f_1 + (2\,x + 1)\,f_2 + f_3 \qquad (8.22)$$

which has remainder 0. □

Example 8.8. Let $f_1 = x\,y - 1$, $f_2 = x^2 + y^2 - 4$, $L = [x, y]$, and

$$u = x^2 + x\,y + y^2 - 5.$$

Using multiple division of u by $[f_1, f_2]$, we obtain the representation

$$u = f_1 + f_2 \qquad (8.23)$$

which has remainder 0. Consider next the division of

$$v = (x\,y^2)\,u = x^3\,y^2 + x^2\,y^3 + x\,y^4 - 5\,x\,y^2 \qquad (8.24)$$

by $[f_1, f_2]$. Since we have simply multiplied u by the monomial $x\,y^2$, we might expect that this division also has remainder 0. Instead, we obtain the representation

$$v = (x^2\,y + x\,y^2 + x + y^3 - 4\,y)\,f_1 + 0 \cdot f_2 + (x + y^3 - 4\,y), \qquad (8.25)$$

where the sum in parentheses on the right is the non-zero remainder. In addition, reversing the order of the divisors does not help because division of v by $[f_2, f_1]$ gives the representation

$$v = (-y)\,f_1 + (y^3 + x\,y^2)\,f_2 + (-y^5 + 4\,y^3 - y) \qquad (8.26)$$

where again the sum in parentheses on the right is the non-zero remainder. In Example 8.16 below, we show how to obtain a remainder 0 for Equation (8.24) using another approach to division. □

In the previous two examples, we start with a division (Equations (8.18) and (8.23)), and then consider a related division where we would like (but do not obtain) a remainder 0 (Equations (8.20) and (8.25)). Since we need a division process that is flexible enough to obtain both of these simplifications, we consider another approach to division.

The new (and more general) division scheme must overcome two problems with multiple division. First, as we saw in Example 8.7, the new process must have some flexibility in the choice of the order of the divisors. Next, as we saw in Example 8.8, simply reordering the divisors does not solve all of our problems. As we shall see in Example 8.16 below, the problem here is that each monomial-based division step actually performs too much simplification. For this reason, we need to change the individual division step as well as the order of the divisions.

The operation in the next definition replaces monomial-based division with a simplified division process.

Definition 8.9. *Let u and $f \neq 0$ be polynomials, and let L be a list of symbols that defines the variable order. For m a monomial of u such that $\mathrm{lm}(f) \mid m$, the* **reduction step** *of u by f using the monomial m is given by*

$$Q = m/\mathrm{lm}(f),$$
$$R = u - Q\,f.$$

The polynomial Q is the **quotient** *of the reduction step, and R is its* **remainder***. We use the notation*

$$u \overset{[f,m]}{\Longrightarrow} R \tag{8.27}$$

to indicate that the reduction step of u by f using the monomial m gives a remainder R.

The reduction step simplifies the process by applying only one step of monomial-based division to a single monomial m of u. Since

$$u = Q\,f + R, \tag{8.28}$$

we view the reduction step as a form of division even though the remainder R may not be reduced with respect to the divisor f. Since a reduction step is used to simplify u with respect to side relation $f = 0$, we have introduced the notation in (8.27) as a way of indicating that the process simplifies u to R in the context of this side relation.

Example 8.10. Let

$$u = x^3\,y^2 + x^2\,y^3 + x\,y^4 - 5\,x\,y^2, \quad f = x^2 + y^2 - 4,$$

$m = x^3\,y^2$, and $L = [x, y]$. Then, $\mathrm{lm}(f) = x^2$ and

$$Q = x^3\,y^2/x^2 = x\,y^2, \quad R = u - Q\,f = x^2\,y^3 - x\,y^2,$$

and so, $u \overset{[f,m]}{\Longrightarrow} x^2\,y^3 - x\,y^2$.   □

The properties of a reduction step are given in the next theorem.

Theorem 8.11. *Let u and $f \neq 0$ be polynomials, and let L be a list of symbols that defines the variable order.*

1. *The remainder R in a reduction step of u by f with the monomial m satisfies $\mathrm{lm}(R) \preceq \mathrm{lm}(u)$.*

2. *If $u \neq 0$ and $m = \mathrm{lm}(u)$, then $\mathrm{lm}(R) \prec \mathrm{lm}(u)$.*

3. $\mathrm{lm}(u) \equiv \max(\{\mathrm{lm}(Q)\,\mathrm{lm}(f), \ \mathrm{lm}(R)\})$

where the maximum is with respect to the monomial order relation.

Proof: To show (1), first, if $u = 0$, the inequality follows immediately from the definition. Next, suppose that $u \neq 0$, and let $f = \mathrm{lm}(f) + f_r$ where f_r represents the lower order monomials. Then

$$
\begin{aligned}
R &= u - \frac{m}{\mathrm{lm}(f)} f \\
&= u - \frac{m}{\mathrm{lm}(f)}\mathrm{lm}(f) - \frac{m}{\mathrm{lm}(f)} f_r \\
&= (u - m) - \left(\frac{m}{\mathrm{lm}(f)} f_r \right).
\end{aligned}
\tag{8.29}
$$

Let's consider the two expressions in parentheses. First,

$$
\mathrm{lm}(u - m) \preceq \mathrm{lm}(u).
\tag{8.30}
$$

In addition,

$$
\mathrm{lm}\left(\frac{m}{\mathrm{lm}(f)} f_r \right) \prec \mathrm{lm}\left(\frac{m}{\mathrm{lm}(f)}\mathrm{lm}(f) \right) = m \preceq \mathrm{lm}(u).
\tag{8.31}
$$

From Equation (8.29) and the relations (8.30) and (8.31), we have

$$
\mathrm{lm}(R) \preceq \max\left(\left\{ u - \mathrm{lm}(u), \mathrm{lm}\left(\frac{m}{\mathrm{lm}(f)} f_r \right) \right\} \right) \preceq \mathrm{lm}(u).
$$

For Part (2), the proof is the same as Part (1) with the relation (8.30) replaced by $\mathrm{lm}(u - \mathrm{lm}(u)) \prec \mathrm{lm}(u)$.

To show (3), first, if $u = 0$, the relationship follows immediately from the definition. Next, if $u \neq 0$ we have, using Equation (8.28),

$$
\mathrm{lm}(u) \preceq \max(\{\mathrm{lm}(Q)\mathrm{lm}(f), \mathrm{lm}(R)\}).
\tag{8.32}
$$

In addition, by Part (1),

$$
\mathrm{lm}(Q\,f) = \mathrm{lm}(u - R) \preceq \max(\{\mathrm{lm}(u), \mathrm{lm}(R)\}) \equiv \mathrm{lm}(u).
\tag{8.33}
$$

From Part (1) and the relation (8.33), we have

$$
\max(\{\mathrm{lm}(Q)\mathrm{lm}(f), \mathrm{lm}(R)\}) \preceq \mathrm{lm}(u).
\tag{8.34}
$$

Therefore, the relations (8.32) and (8.34) imply that

$$\mathrm{lm}(u) \equiv \max(\{\mathrm{lm}(Q)\mathrm{lm}(f), \mathrm{lm}(R)\}). \qquad \square$$

The new division process for a set of divisors is described in the next definition.

Definition 8.12. *Let* $F = \{f_1, \ldots, f_n\}$ *be a set of non-zero polynomials, and let* L *be a list of symbols that defines the variable order. The* **reduction** *of a polynomial* u *by the divisor set* F *is defined by the iteration scheme:*

$$c_1 := c_2 := \cdots := c_n := 0; \quad r := u;$$

> **while** r is not reduced with respect to F **do**
> Select a monomial m in r and a divisor f_i in F
> such that $\mathrm{lm}(f_i, L) \mid m$;
> $Q := m/\mathrm{lm}(f_i, L)$;
> $c_i := c_i + Q$;
> $r := r - Q\, f_i$;

The process terminates when r *is reduced with respect to* F. *When this occurs, the polynomials* $c_1, \ldots, c_n$ *are the* **reduction quotients** *with respect to* F, *and* r *is the* **reduction remainder**. *We use the notation*

$$u \xrightarrow{F} r \qquad (8.35)$$

to indicate that the reduction of u *by* F *gives a remainder* r.

Observe that each pass through the **while** loop involves a single reduction step, and, for each step,

$$r_{old} = Q_{new}\, f_i + r_{new},$$

where r_{old} is the old remainder value, and r_{new} and Q_{new} are the new values. Therefore, since $r = u$ at the start, at each iteration step we obtain the representation

$$u = c_1\, f_1 + \cdots + c_n\, f_n + r, \qquad (8.36)$$

where the polynomials c_i and r are the most recently computed values. At termination, the remainder r is reduced with respect to F.

Since the division process introduced here is used to simplify an expression with respect to side relations $f_1 = 0, \ldots, f_n = 0$, we have introduced the notation in (8.35) as a way of indicating that the process simplifies u to r in the context of these side relations using some reduction step sequence

$$u = r_0 \overset{[f_{i_1}, m_1]}{\Longrightarrow} r_1 \overset{[f_{i_2}, m_2]}{\Longrightarrow} \cdots \overset{[f_{i_q}, m_q]}{\Longrightarrow} = r_q = r. \qquad (8.37)$$

Keep in mind, however, if we chose another reduction step sequence, we may obtain

$$u \overset{F}{\longrightarrow} s$$

where $s \neq r$.

Of course, the process in Definition 8.12 is not really an algorithm until we give a selection scheme for choosing the monomial m and the divisor f_i. One possible scheme is multiple division (see Example 8.13 below), and another one is given in Figure 8.1 below. For now, we assume that some scheme is given (which may vary from example to example) and the process terminates.

A reduction can be obtained with many different reduction step sequences. This point is illustrated in the next example.

Example 8.13. Let

$$u = x^3 + 2\,x^2\,y - 5\,x + y^3 - 3\,y, \quad f_1 = x\,y - 1, \quad f_2 = x^2 + y^2 - 4,$$

where $L = [x, y]$. In Example 8.5 we considered the multiple division of u by $[f_1, f_2]$ which obtains the remainder $x + y^3 - 4\,y$ using the divisor sequence f_1, f_2, f_1 (see Equation (8.13)). From the reduction perspective, this division obtains

$$u \overset{\{f_1, f_2\}}{\Longrightarrow} x + y^3 - 4\,y \qquad (8.38)$$

using the reduction step sequence

$$u \overset{[f_1, 2x^2 y]}{\Longrightarrow} \quad x^3 - 3\,x + y^3 - 3\,y$$
$$\overset{[f_2, x^3]}{\Longrightarrow} \quad -x\,y^2 + x + y^3 - 3\,y \qquad (8.39)$$
$$\overset{[f_1, -x\,y^2]}{\Longrightarrow} \quad x + y^3 - 4\,y.$$

We also obtain the reduction (8.38) using two other reduction step sequences:

$$u \overset{[f_2, x^3]}{\Longrightarrow} \quad 2\,x^2\,y - x\,y^2 - x + y^3 - 3\,y$$
$$\overset{[f_1, 2x^2 y]}{\Longrightarrow} \quad -x\,y^2 + x + y^3 - 3y \qquad (8.40)$$
$$\overset{[f_1, -x\,y^2]}{\Longrightarrow} \quad x + y^3 - 4\,y,$$

$$u \stackrel{[f_2, x^3]}{\Longrightarrow} 2\,x^2\,y - x\,y^2 - x + y^3 - 3\,y$$

$$\stackrel{[f_1, -x\,y^2]}{\Longrightarrow} 2\,x^2 y - x + y^3 - 4\,y \tag{8.41}$$

$$\stackrel{[f_1, 2\,x^2\,y]}{\Longrightarrow} x + y^3 - 4\,y,$$

although neither one of them is obtained by multiple division.

In Example 8.5 we also considered the multiple division of u by $[f_2, f_1]$, which obtains the remainder $-x - y^3 + 4y$ using the divisor sequence f_2, f_1 (see Equation (8.14)). This division corresponds to the reduction

$$u \stackrel{\{f_1, f_2\}}{\longrightarrow} -x - y^3 + 4\,y \tag{8.42}$$

using the reduction step sequence

$$u \stackrel{[f_2, 2x^2 y]}{\Longrightarrow} x^3 - 5\,x - y^3 + 5\,y$$

$$\stackrel{[f_2, x^3]}{\Longrightarrow} -x\,y^2 - x - y^3 + 5\,y \tag{8.43}$$

$$\stackrel{[f_1, -x\,y^2]}{\Longrightarrow} -x - y^3 + 4\,y.$$

In this case, the first two reduction steps correspond to monomial-based division by f_2, and the last step to monomial-based division by f_1.

We can also obtain the reduction (8.42) using two other reduction step sequences (Exercise 4). In fact, these six reduction step sequences are the only ones possible with u, f_1, and f_2.

In Example 8.37 below, we show that u can be simplified to 0 with respect to the side relations $f_1 = 0$, $f_2 = 0$, although this is not obtained with reduction by $\{f_1, f_2\}$. □

Example 8.14. Let $f_1 = x\,y - 1$, $f_2 = x^2 + y^2 - 4$, $L = [x, y]$, and

$$u = 2x^3 + 3x^2 y + x^2 + 2y^3 + 4y + 1.$$

In Example 8.7 we considered the multiple division of u by $F_1 = [f_1, f_2]$ which obtained the remainder $11x + 2y^3 - y^2 + 2y + 5$ (see Equation (8.18)). This corresponds to the reduction

$$u \stackrel{\{f_1, f_2\}}{\longrightarrow} 11x + 2y^3 - y^2 + 2y + 5 \tag{8.44}$$

with reduction step sequence

$$u \stackrel{[f_1, 3x^2 y]}{\Longrightarrow} 2x^3 + x^2 + 3x + 2y^3 + 4y + 1$$

$$\stackrel{[f_2, 2x^3]}{\Longrightarrow} x^2 - 2x\,y^2 + 11x + 2y^3 + 4y + 1 \tag{8.45}$$

$$\overset{[f_2,x^2]}{\Longrightarrow} \quad 2\,y^3 + 4\,y + 5 + 11\,x - 2\,xy^2 - y^2$$

$$\overset{[f_1,-2\,x\,y^2]}{\Longrightarrow} \quad 11x + 2y^3 - y^2 + 2y + 5.$$

We can also obtain the reduction (8.44) with 11 other reduction step sequences, although none of them are obtained with multiple division. In addition, there are 12 reduction step sequences that reduce u to another polynomial, one of which corresponds to multiple division by $[f_2, f_1]$ (Exercise 5).

Let $f_3 = 11x + 2y^3 - y^2 + 2y + 5$ which is the polynomial obtained with the reduction (8.44). In Example 8.7 we observed that u could not be reduced to 0 using multiple division by the polynomials $\{f_1, f_2, f_3\}$ using any divisor order. However, $u \overset{\{f_1, f_2, f_3\}}{\longrightarrow} 0$ using the reduction step sequence (8.45) followed by the reduction step

$$11x + 2y^3 - y^2 + 2y + 5 \overset{[f_3, 11x]}{\Longrightarrow} 0. \qquad \square$$

The next theorem, which plays a role in the implementation of our algorithm, describes the last reduction in the previous example.

Theorem 8.15. *Let $S = \{f_1, \ldots, f_n\}$ be a set of non-zero polynomials, and let L be a list of symbols that defines the variable order. Suppose that there is a reduction step sequence which gives*

$$u \overset{S}{\longrightarrow} r, \tag{8.46}$$

and let $f_{n+1} = r$ and $T = \{f_1, \ldots, f_n, f_{n+1}\}$. Then there is a reduction step sequence which gives

$$u \overset{T}{\longrightarrow} 0. \tag{8.47}$$

Proof: The reduction step sequence for (8.47) is simply the one used to obtain (8.46) followed by the reduction step

$$r \overset{[f_{n+1}, \mathrm{lm}(r)]}{\Longrightarrow} 0. \qquad \square$$

Example 8.16. Consider the reduction of $u = x^2 + x\,y + y^2 - 5$ with respect to $F = \{f_1, f_2\} = \{x\,y - 1, x^2 + y^2 - 4\}$ with $L = [x, y]$. In Example 8.8 we showed that multiple division of u by $[f_1, f_2]$ gives a representation with remainder 0 (see Equation (8.23)). This division is obtained with the reduction step sequence

$$u \overset{[f_1, x\,y]}{\Longrightarrow} (x^2 + y^2 - 4) \overset{[f_2, x^2]}{\Longrightarrow} 0. \tag{8.48}$$

In Example 8.8 we also showed that multiple division of $v = (x\,y^2)u = x^3 y^2 + x^2 y^3 + x\,y^4 - 5x\,y^2$ by $[f_1, f_2]$ obtains remainder $x + y^3 - 4y$ (see Equation (8.25)). This multiple division is obtained with the reduction step sequence

$$v \overset{[f_1, x^3 y^2]}{\Longrightarrow} \left(x^2 y^3 + x^2 y + x\,y^4 - 5x\,y^2\right)$$

$$\overset{[f_1, x^2 y^3]}{\Longrightarrow} \left(x^2 y + x\,y^4 - 4x\,y^2\right)$$

$$\overset{[f_1, x^2 y]}{\Longrightarrow} \left(x\,y^4 - 4x\,y^2 + x\right)$$

$$\overset{[f_1, x y^4]}{\Longrightarrow} \left(-4x\,y^2 + x + y^3\right)$$

$$\overset{[f_1, x]}{\Longrightarrow} \left(x + y^3 - 4y\right).$$

Observe that multiple division has forced all reduction steps to be done with f_1. In other words, so much simplification is done by f_1 there is nothing left to be done by f_2. However, using a reduction step sequence similar to (8.48) we obtain

$$v \overset{[f_1, x^2 y^3]}{\Longrightarrow} \left(x^3 y^2 + x\,y^4 - x\,y^2\right) \overset{[f_2, x^3 y^2]}{\Longrightarrow} 0. \tag{8.49}$$

In each case, the monomial used for reduction is the product of $x\,y^2$ and the corresponding monomial in (8.48).   □

The next theorem, which plays an important role in the theoretical justification of our algorithm, describes the relationship between the sequences (8.48) and (8.49).

Theorem 8.17. Let $F = \{f_1, \ldots, f_n\}$ be a set of non-zero polynomials, and let L be a list of symbols that defines the variable order. Suppose that for a polynomial u, $u \overset{F}{\longrightarrow} g$. Then, for a monomial m, $m\,u \overset{F}{\longrightarrow} m\,g$.

Proof: Since $u \overset{F}{\longrightarrow} g$, there is a sequence of reduction steps

$$u = r_0 \overset{[f_{i_1}, m_1]}{\Longrightarrow} r_1 \overset{[f_{i_2}, m_2]}{\Longrightarrow} \cdots \overset{[f_{i_q}, m_q]}{\Longrightarrow} = r_q = g.$$

where

$$r_{j+1} = r_j - \frac{m_j}{\mathrm{lm}(f_{i_j})} f_{i_j}, \quad j = 1, \ldots, q.$$

Therefore,

$$m\,r_{j+1} = m\,r_j - \frac{m\,m_j}{\mathrm{lm}(f_{i_j})} f_{i_j},$$

and

$$m\,u = m\,r_0 \overset{[f_{i_1},m\,m_1]}{\Longrightarrow} m\,r_1 \overset{[f_{i_2},m\,m_2]}{\Longrightarrow} \cdots \overset{[f_{i_q},m\,m_q]}{\Longrightarrow} m\,r_q = m\,g. \qquad \square$$

The properties of the reduction process are given in the next theorem.

Theorem 8.18. *Let u be a polynomial, $F = [f_1, \ldots, f_n]$ a list of non-zero polynomials, and L a list of symbols that defines the variable order. Then, when reduction terminates*

$$u = c_1 f_1 + \cdots + c_n f_n + r, \tag{8.50}$$

$$\mathrm{lm}(r) \preceq \mathrm{lm}(u), \tag{8.51}$$

$$\mathrm{lm}(u) \equiv \max\left(\{\mathrm{lm}(c_1)\mathrm{lm}(f_1), \ldots, \mathrm{lm}(c_n)\mathrm{lm}(f_n), \mathrm{lm}(r)\}\right) \tag{8.52}$$

where the maximum is with respect to the monomial order relation.

Proof: The representation (8.50) is given in Equation (8.36), and the relation (8.51) follows by applying Theorem 8.11(1) at each reduction step in the process.

We show that Equation (8.52) holds at each reduction step using mathematical induction. When the process begins,

$$c_1 = 0, \ldots, c_n = 0, \quad r = u.$$

At the first reduction step, for some f_i in F, $\mathrm{lm}(f_i) \mid \mathrm{lm}(u)$ and

$$u = Q\,f_i + r, \quad c_i = Q.$$

By Theorem 8.11(3),

$$\mathrm{lm}(u) \equiv \max(\{\mathrm{lm}(Q\,f_i), \mathrm{lm}(r)\}) = \max(\{\mathrm{lm}(c_i)\mathrm{lm}(f_i), \mathrm{lm}(r)\}) \tag{8.53}$$

which is Equation (8.52) at this point.

For the induction step, we assume that Equation (8.52) holds at some point in the process and show that it also holds at the next reduction step. Suppose that at this step

$$r_{new} = r - Q\,f_i \tag{8.54}$$

where r_{new} and Q are the most recently computed values. Since,

$$\begin{aligned} u &= c_1 f_1 + \cdots + c_n f_n + Q\,f_i + r_{new} \\ &= c_1 f_1 + \cdots + c_{i-1} f_{i-1} + (c_i + Q) f_i + c_{i+1} f_{i+1} + \cdots + c_n f_n + r_{new}, \end{aligned}$$

we have

$$\mathrm{lm}(u) \preceq \max(\{\mathrm{lm}(c_1)\mathrm{lm}(f_1), \ldots, \mathrm{lm}(c_{i-1})\mathrm{lm}(f_{i-1}), \quad (8.55)$$
$$\mathrm{lm}(c_i + Q)\mathrm{lm}(f_i), \mathrm{lm}(c_{i+1})\mathrm{lm}(f_{i+1}),$$
$$\ldots, \mathrm{lm}(c_n)\mathrm{lm}(f_n), \mathrm{lm}(r_{new})\}).$$

In addition, the induction hypothesis in Equation (8.54) and Theorem 8.11(3) imply that

$$\mathrm{lm}(u) \equiv \max(\{\mathrm{lm}(c_1)\mathrm{lm}(f_1), \ldots, \mathrm{lm}(c_n)\mathrm{lm}(f_n), \mathrm{lm}(r)\})$$
$$\equiv \max(\{\mathrm{lm}(c_1)\mathrm{lm}(f_1), \ldots, \mathrm{lm}(c_n)\mathrm{lm}(f_n),$$
$$\mathrm{lm}(Q\, f_i), \mathrm{lm}(r_{new})\})$$
$$\succeq \max(\{\mathrm{lm}(c_1)\mathrm{lm}(f_1), \ldots, \mathrm{lm}(c_{i-1})\mathrm{lm}(f_{i-1}), \quad (8.56)$$
$$\mathrm{lm}(c_i + Q)\mathrm{lm}(f_i), \mathrm{lm}(c_{i+1})\mathrm{lm}(f_{i+1}),$$
$$\ldots, \mathrm{lm}(c_n)\mathrm{lm}(f_n), \mathrm{lm}(r_{new})\}).$$

Therefore, the relations (8.55) and (8.56) imply that

$$\mathrm{lm}(u) \equiv \max(\{\mathrm{lm}(c_1)\mathrm{lm}(f_1), \ldots, \mathrm{lm}(c_{i-1})\mathrm{lm}(f_{i-1}),$$
$$\mathrm{lm}(c_i + Q)\mathrm{lm}(f_i), \mathrm{lm}(c_{i+1})\mathrm{lm}(f_{i+1}),$$
$$\ldots, \mathrm{lm}(c_n)\mathrm{lm}(f_n), \mathrm{lm}(r_{new})\}),$$

which is Equation (8.52) for the next reduction step. $\square$

In the next example we show that there are simplifications that cannot be obtained with reduction.

Example 8.19. Let $f_1 = x\,y - 1$, $f_2 = x^2 + y^2 - 4$, and

$$u = (2x + y)\, f_1 + (x - 2y)\, f_2 \qquad (8.57)$$
$$= 2x\,y^2 + 7y - 6x + x^3 - 2y^3, \qquad (8.58)$$

with $L = [x, y]$. Consider the simplification of u with respect to the side relations $f_1 = 0, f_2 = 0$. It is evident from the unexpanded form (8.57) that u simplifies to 0 but not apparent from the expanded form (8.58). However, this simplification cannot be obtained with reduction by $\{f_1, f_2\}$, since there are only two possible reduction step sequences which obtain the same non-zero remainder:

$$u \xRightarrow{[f_1, 2x\,y^2]} 9y - 6x + x^3 - 2y^3$$
$$\xRightarrow{[f_2, x^3]} 9y - 2x - 2y^3 - x\,y^2$$
$$\xRightarrow{[f_1, -x\,y^2]} 8y - 2x - 2y^3,$$

$$u \quad \overset{[f_2, x^3]}{\Longrightarrow} \quad x\,y^2 + 7y - 2x - 2y^3$$

$$\overset{[f_1, x\,y^2]}{\Longrightarrow} \quad 8y - 2x - 2y^3.$$

The problem here is not the reduction process, but the form of the side relations. In Example 8.37 (page 327) we show how this simplification is obtained using reduction with an equivalent set of side relations called the Gröbner basis. □

Reduction Algorithm

In Figure 8.1 we give a procedure *Reduction* that reduces u with respect to F using a selection scheme which obtains the reduction in Theorem 8.15. In this approach the current remainder is a sum $r + R$, where r contains monomials that cannot be reduced by F, and R contains monomials that may be reducible. By doing this we can check for termination by simply comparing R to 0 instead of checking whether the full remainder is reduced with respect to F, as is done in Definition 8.12. These variables are initialized in lines 1–2.

At each pass through the **while** loop, we try to reduce R using a reduction step with monomial $m = \mathrm{lm}(R)$ and a polynomial f from F. The divisibility test is given in line 11. Since a successful division at line 10 results in a monomial which may have a coefficient that is an integer or fraction, we check for this by checking if the denominator of Q is an integer. If the test is successful, we perform the reduction (line 12), update the quotient (line 13), and then continue the process with the first polynomial in F (line 14). By always returning to the first divisor after a successful reduction, we insure that the selection scheme obtains the reduction in Theorem 8.15.

If the test at line 11 fails, we try again with the next divisor in the list (line 16). However, if i is larger than the number of polynomials in F, the current monomial m is not divisible by the leading monomial of any polynomial in F, and so it is removed from R (line 18), added to r (line 19), and the process repeats with the first polynomial in F (line 20). Lines 22-23 are included so that the quotients are returned in the list q.

Theorem 8.20. *Let u be a polynomial, $F = [f_1, \ldots f_n]$ be a list of non-zero polynomials, and L be a list of symbols that defines the variable order. Then, the reduction algorithm in Figure 8.1 terminates.*

Proof: By Theorem 8.11(2), at each execution of line 12, $\mathrm{lm}(R)$ decreases with respect to the $\prec$ relation. On the other hand, if $\mathrm{lm}(f) \nmid \mathrm{lm}(R)$ for all polynomials f in F, then line 18 is executed and again $\mathrm{lm}(R)$ de-

Procedure $Reduction(u, F, L)$;
Input
    u : a polynomial in symbols in L;
    F : a list of non-zero polynomials in symbols in L;
    L : a list of symbols;
Output
    the list $[q, r]$ where q is a list of reduction quotients and r is a
    reduction remainder;
Local Variables
    $R, r, i, c, m, f, lmf, Q, q$;
Begin

```
1     R := u;
2     r := 0;
3     for  i := 1 to  Number_of_operands(F) do
4         c[i] := 0;
5     i := 1;
6     while  R ≠ 0 do
7         m := Leading_monomial(R, L);
8         f := Operand(F, i);
9         lmf := Leading_monomial(f, L);
10        Q := m/lmf;
11        if  Kind(Denominator(Q)) = integer then
12            R := Algebraic_expand(R − Q * f);
13            c[i] := c[i] + Q;
14            i := 1
15        else
16            i := i + 1;
17        if  i = Number_of_operands(F) + 1 then
18            R := R − m;
19            r := r + m;
20            i := 1;
21    q := [ ];
22    for  i from  1 to  Number_of_operands(F) do
23        q := Join(q, [c[i]]);
24    Return([q, r])
```

End

Figure 8.1. An MPL procedure for reduction. (Implementation: Maple (txt), Mathematica (txt), MuPAD (txt).)

creases. Therefore, eventually $R = 0$, and the condition at line 6 terminates the algorithm. At termination, r is reduced with respect to F since each monomial included in r at line 19 is reduced with respect to F.   □

The motivation for introducing the reduction process is to have a division process that is flexible enough to obtain the reductions mentioned in Theorems 8.15 and 8.17. Although the *Reduction* procedure obtains the reduction in Theorem 8.15, it may not obtain the reduction in Theorem 8.17. While Theorem 8.17 is of theoretical importance for the verification of our simplification algorithm, it does not appear as a computational step in the algorithm. For this reason, it is enough to know that the general reduction process in Definition 8.12 can obtain this reduction, even though it is not obtained by the selection scheme used for the procedure *Reduction*.

Exercises

1. Let $u = x^3 + 2x^2y - 5x + y^3 - 3y$, $f_1 = xy - 1$ and $f_2 = x^2 + y^2 - 4$, where x is the main variable. Show that there are infinitely many representations of u of the form (8.12). *Hint:* Find a general representation that contains Equations (8.13) and (8.14).

2. Let $u = xyz + xz^2 - x + y^3 + y^2 - yz^2 + y - 1$, and let

$$f_1 = z^2 - 1, \quad f_2 = y^2 + y - z^2, \quad f_3 = xz - y.$$

Show that u simplifies to 0 using the side relations $f_1 = 0$, $f_2 = 0$, $f_3 = 0$. *Note:* This simplification requires three passes through the **while** loop in the multiple division process in Definition 8.4.

3. Let u be a polynomial, F a list of non-zero polynomials, and let L be a list of variables that defines the variable order. Give a procedure

$$Multiple\_division(u, F, L)$$

that returns a two element list that contains the list of multiple quotients and the multiple remainder.

4. Let

$$u = x^3 + 2x^2y - 5x + y^3 - 3y, \quad f_1 = xy - 1, \quad f_2 = x^2 + y^2 - 4,$$

where $L = [x, y]$.

(a) In Example 8.13 we gave a reduction step sequence for

$$u \xrightarrow{\{f_1, f_2\}} -x - y^3 + 4y.$$

Find two more reduction step sequences that obtain this reduction.

(b) Show that there are six reduction step sequences for u with $\{f_1, f_2\}$.

5. Let $f_1 = xy - 1$, $f_2 = x^2 + y^2 - 4$, $L = [x, y]$, and

$$u = 2x^3 + 3x^2y + x^2 + 2y^3 + 4y + 1.$$

(a) In Example 8.14 we gave a reduction sequence that obtains

$$u \xrightarrow{\{f_1, f_2\}} 11x + 2y^3 - y^2 + 2y + 5.$$

Find two more reduction step sequences that obtain this reduction.

(b) Find a reduction step sequence that reduces u to another polynomial using $\{f_1, f_2\}$.

6. Let $F = \{f_1, f_2\} = \{xy - 1, x^2 + y^2 - 4\}$ and $L = [x, y]$. For

$$u = x^3 + 3x^2 y - 3y^3,$$

find all possible reduction step sequences and indicate which ones correspond to a multiple division and which ones correspond to the algorithm in Figure 8.1.

8.2   Equivalence, Simplification, and Ideals

Let's return now to our main goal, the development of a simplification algorithm that transforms a polynomial u to an equivalent polynomial v that is simplified with respect to the side relations

$$f_1 = 0, \ldots, f_n = 0. \tag{8.59}$$

Polynomial Equivalence with Side Relations

Since simplification involves finding an equivalent simplified expression, we begin by addressing the meaning of equivalence.

Definition 8.21. *Let u and v be polynomials, and let $F = \{f_1, \ldots, f_n\}$ be a set of non-zero polynomials in $\mathbf{Q}[x_1, \ldots, x_p]$. The polynomial u is* **equivalent** *to v with respect to the side relations (8.59) if*

$$u(a_1, \ldots, a_p) = v(a_1, \ldots, a_p)$$

at each point $(x_1, \ldots, x_p) = (a_1, \ldots, a_p)$ that is a solution of the system (8.59). (The values a_i can be real or complex.) We use the notation

$$u \overset{F}{=} v$$

to indicate that u is equivalent to v.

In other words, since the system of equations (8.59) defines a set of points, two polynomials u and v are equivalent if they have the same value as functions at each point in the solution set.

Example 8.22. In Example 8.19 on page 314 we considered the simplification of

$$u \; = \; 2x\,y^2 + 7y - 6x + x^3 - 2y^3$$

with respect to the side relations

$$f_1 = x\,y - 1 = 0, \qquad f_2 = x^2 + y^2 - 4 = 0.$$

These equations have the four solutions

$$x = -\left(1/2\sqrt{6} + 1/2\sqrt{2}\right)^3 + 2\sqrt{6} + 2\sqrt{2}, \quad y = 1/2\sqrt{6} + 1/2\sqrt{2},$$

$$x = -\left(1/2\sqrt{6} - 1/2\sqrt{2}\right)^3 - 2\sqrt{6} + 2\sqrt{2}, \quad y = 1/2\sqrt{6} - 1/2\sqrt{2},$$

$$x = -\left(-1/2\sqrt{6} + 1/2\sqrt{2}\right)^3 + 2\sqrt{6} - 2\sqrt{2}, \quad y = -1/2\sqrt{6} + 1/2\sqrt{2},$$

$$x = -\left(-1/2\sqrt{6} - 1/2\sqrt{2}\right)^3 - 2\sqrt{6} - 2\sqrt{2}, \quad y = -1/2\sqrt{6} - 1/2\sqrt{2}.$$

Since u evaluates to 0 at each of these points, we have $u \stackrel{F}{=} 0$.   □

Consistent Side Relations

Polynomial equivalence is only meaningful if the system (8.59) has at least one solution. This point is illustrated in the next example.

Example 8.23. Consider the side relations

$$f_1 = x\,y + x + 2y + 3 = 0, \tag{8.60}$$

$$f_2 = x\,y + 2y + 1 = 0. \tag{8.61}$$

with $F = \{f_1, f_2\}$. This system of equations does not have a solution for x and y in the complex numbers. To see why, first eliminate y by subtracting Equation (8.61) from Equation (8.60) which gives $x = -2$. Substituting this value into Equation (8.60) gives $1 = 0$, and so the system does not have a solution.

In this situation equivalence is not meaningful because, in a trivial sense, any two polynomials are equivalent. In fact, $0 \stackrel{F}{=} 1$ because these two constant polynomials agree as functions at each point of the empty solution set.   □

In order that our simplification problem be meaningful, we assume that our side relation equations are consistent in the following sense.

Definition 8.24. *A system of equations $f_1 = 0, \ldots, f_n = 0$ that has at least one solution is called a* **consistent** *system of equations. If the system does not have a solution it is called an* **inconsistent** *system.*

In Section 8.3 we show how to test for the consistency of the side relations.

Equivalence and Reduction

Although Definition 8.21 is useful in a theoretical sense, it is not useful in an algorithmic sense. First, consistent systems of equations may have either finitely many or infinitely many solutions. In addition, in most cases it is impossible to find a formula representations for solutions, and even in cases where formulas are available, the verification of equivalence can involve a difficult radical simplification.

A better approach is to avoid the solution of the side relation equations altogether. The next theorem suggests a more fruitful approach that involves reduction.

Theorem 8.25. *Let u and r be polynomials, $F = \{f_1, \ldots, f_n\}$ a set of non-zero polynomials, and L a list that determines variable order. Then, $u \xrightarrow{F} r$ implies that $u \overset{F}{=} r$.*

Proof: Since $u \xrightarrow{F} r$, the polynomial u has a representation

$$u = c_1 f_1 + \cdots c_n f_n + r. \tag{8.62}$$

For each point $(x_1, \ldots, x_p) = (a_1, \ldots, a_p)$ in the solution set of

$$f_1 = 0, \ldots, f_n = 0,$$

we have

$$
\begin{aligned}
u(a_1, \ldots, a_p) &= c_1(a_1, \ldots, a_p) f_1(a_1, \ldots, a_p) + \cdots \\
&\quad + c_n(a_1, \ldots, a_p) f_n(a_1, \ldots, a_p) + r(a_1, \ldots, a_p) \\
&= r(a_1, \ldots, a_p). \qquad \square
\end{aligned}
$$

This theorem suggests that instead of testing equivalence directly, we might use reduction to test for equivalence. Unfortunately, something is lost in the translation, since, as we saw in Example 8.19 (page 314), $u \overset{F}{=} r$ does not imply that $u \xrightarrow{F} r$.

The Zero Equivalence Problem

One aspect of simplification has to do with the *zero equivalence problem* which involves checking whether or not

$$u \overset{F}{=} 0. \tag{8.63}$$

An important class of polynomials that satisfy Equation (8.63) consists of those polynomials that have a representation of the form

$$u = c_1 f_1 + \cdots + c_n f_n \tag{8.64}$$

where $c_1, \ldots, c_n$ are polynomials in $\mathbf{Q}[x_1, \ldots, x_p]$. For this class of polynomials, reduction can determine that $u \overset{F}{=} 0$ although this involves first finding an equivalent set of side relations (the Gröbner basis), and then applying reduction with this new basis. (See Theorem 8.33(b) on page 325.) Unfortunately, there are polynomials that are equivalent to 0 that do not have the form (8.64) for which reduction with a Gröbner basis does not obtain 0. (See Theorem 8.58 on page 345 and Example 8.57 on page 345.)

The Goal of the Simplification Algorithm

In light of the previous discussion, we give two goals for our simplification algorithm.

1. If u has a representation of the form (8.64), the algorithm determines (using reduction) that $u \overset{F}{=} 0$. In addition, we require that this property does not depend on either the order in the symbol list L or the selection scheme used for reduction.

2. For polynomials u in $\mathbf{Q}[x_1, \ldots, x_p]$ that do not have a representation of the form (8.64), the algorithm obtains (using reduction) a polynomial r that is reduced with respect of F such that $u \overset{F}{=} r$.

   In order to describe the algorithm, we introduce some new terminology and theoretical concepts.

Ideals

The collection of polynomials of the form (8.64) is referred to by the terminology in the next definition.

Definition 8.26. *Let*
$$F = \{f_1, \ldots, f_n\}$$

*be a set of non-zero polynomials. The **ideal** I generated by F is the set of polynomials of the form*

$$u = c_1 f_1 + \cdots + c_n f_n, \tag{8.65}$$

*where $c_1, \ldots, c_n$ are polynomials in $\mathbf{Q}[x_1, \ldots, x_p]$. The set F is called a **basis** for I, and the notation*

$$I = \,< f_1, \ldots, f_n >$$

indicates that I is generated by the basis.

In other words, one goal of the simplification algorithm is to obtain the simplified form 0 for all polynomials in the ideal $I = \,< f_1, \ldots, f_n >$.

The next theorem describes the basic algebraic properties of an ideal.

Theorem 8.27. *Let I be an ideal with basis $F = \{f_1, \ldots, f_n\}$.*

1. *If u and v are in I, then $u + v$ is in I.*

2. *If u is in I and c is in $\mathbf{Q}[x_1, \ldots, x_p]$, then cu is in I.*

3. *Suppose that u and $v \neq 0$ are in I, and suppose that m is a monomial of u with $\mathrm{lm}(v) \mid m$. Then, $u \overset{[v,m]}{\Longrightarrow} r$ implies that r is in I.*

4. *If u is in I and $u \overset{F}{\longrightarrow} r$, then r is in I.*

Proof: The proof of Parts (1) and (2) follow directly from the definition of an ideal. The proof of Part (3) follows from the definition of a reduction step together with Parts (1) and (2), and Part (4) follows from (3). The details of the proofs are left to the reader (Exercise 2). □

Example 8.28. Consider the ideal $I = \,< x + y, \ x - y >$ which consists of all polynomials of the form

$$u(x, y) = c_1(x + y) + c_2(x - y) \tag{8.66}$$

where c_1 and c_2 are polynomials. For example, the polynomial $x^2 + y^2$ is in I with $c_1 = x$ and $c_2 = -y$. The representation (8.66) is not unique since $c_1 = y$ and $c_2 = x$ gives another representation for $x^2 + y^2$.

An ideal can have many different bases. For example, $\{x, \ y\}$ is also a basis for I. To show this, we show that the set of polynomials of the form

$$u(x, y) = d_1 x + d_2 y \tag{8.67}$$

is the same as the set of polynomials of the form (8.66). First, since

$$c_1(x + y) + c_2(x - y) = (c_1 + c_2)\, x + (c_1 - c_2)\, y,$$

any polynomial of the form (8.66) has the form (8.67). Conversely, since

$$d_1 x + d_2 y = \frac{d_1 + d_2}{2}(x + y) + \frac{d_1 - d_2}{2}(x - y),$$

any polynomial of the form (8.67) has the form (8.66).

Using the basis $\{x,\ y\}$, we can obtain another simple description of this ideal. From Equation (8.67), $u(0,0) = 0$, and so, each polynomial in I has a zero constant term. Conversely, any polynomial with zero constant term is in I. To show this, suppose that u has a zero constant term, and consider the representation $u = q\,x + r$ where $q = \mathrm{mbquot}(u, x)$ and $r = \mathrm{mbrem}(u, x)$. Notice that $q\,x$ is the sum of all monomials in u that have x as a factor, and r can depend on y but is free of x. Then, since

$$u(0,0) = 0 = q(0,0) \cdot 0 + r(0),$$

we have $r(0) = 0$, which means that r has a zero constant term. Therefore, $r = y\,s(y)$, and u has the required representation $q\,x + s\,y$. $\qquad \square$

In the next example, we show that for single variable polynomials, ideals take a particularly simple form.

Example 8.29. Consider an ideal $I = \,<f_1, \ldots, f_n>$ in $\mathbf{Q}[x]$. Let $f \neq 0$ be a polynomial in the ideal of minimal degree. We show that $\{f\}$ is another basis for the ideal. For any polynomial u in I, polynomial division of u by f gives $u = q\,f + r$ where $\deg(r) < \deg(f)$. However, since $r = u - q\,f$ is in the ideal, the minimal degree property of f implies that $r = 0$. Therefore, $u = q\,f$ which means f is a basis for the ideal. In fact, it can be shown that $I = \,<\gcd(f_1, \ldots, f_n)>$ (Exercise 3(b)). $\qquad \square$

An ideal with a single polynomial for a basis is called a *principal ideal*. The previous example shows that for single variable polynomials, all ideals are principal ideals. For polynomials with more than one variable, however, there are ideals that are not principal (Exercise 4).

Gröbner Bases for Ideals

Reduction with respect to an ideal basis F may not obtain the remainder 0 for polynomials in the ideal. However, if the basis is a special type known as a Gröbner basis,[3] reduction does obtain the remainder 0.

---

[3]Gröbner bases for ideals were invented by Bruno Buchberger who called them Gröbner bases in honor of his Ph.D. dissertation advisor Wolfgang Gröbner. These bases are also known as *standard bases*.

Definition 8.30. *Let* $F = \{f_1, \ldots, f_n\}$ *be a set of non-zero polynomials, and suppose that* $I = <f_1, \ldots, f_n>$. *The set* F *is a* **Gröbner** *basis for* I *if for each* u *in* I, *there is an* f_i *such that* $\mathrm{lm}(f_i) \mid \mathrm{lm}(u)$.

Example 8.31. Consider the ideal $I = <x+y, \ x-y>$ which was considered in Example 8.28, and assume that $L = [x, y]$.

We show first that $F = \{f_1, f_2\} = \{x + y, x - y\}$ is not a Gröbner basis for I. To see why, consider the polynomial

$$u = 2y = f_1 - f_2$$

which is in I. Since both $\mathrm{lm}(f_1, L) = x$ and $\mathrm{lm}(f_2, L) = x$ do not divide $\mathrm{lm}(y, L) = y$, F is not a Gröbner basis.

However, $G = \{g_1, g_2\} = \{x, \ y\}$ is a Gröbner basis for the ideal. To show that G satisfies the Gröbner basis property, from Example 8.28, I consists of all polynomials in x and y that do not have a constant term. Therefore, either $\mathrm{lm}(g_1, L) = x$ or $\mathrm{lm}(g_2, L) = y$ divides the leading monomial of each polynomial in I.

A similar argument shows that $\{x - y, \ 2y\}$ is also a Gröbner basis for I, and so, the Gröbner basis is not unique. $\square$

Example 8.32. Consider the ideal

$$I = <f_1, f_2> = <xy - 1, \ x^2 + y^2 - 4>$$

with $L = [x, y]$. We show that $\mathbf{F} = \{f_1, \ f_2\}$ is not a Gröbner basis for I. Let

$$u = x + y^3 - 4y = -x f_1 + y f_2.$$

Observe that both $\mathrm{lm}(f_1) = xy$ and $\mathrm{lm}(f_2) = x^2$ do not divide $\mathrm{lm}(u) = x$, and, therefore, F is not a Gröbner basis.

A Gröbner basis for this ideal is

$$\mathbf{G} = \{g_1, g_2\} = \{-y^3 + 4y - x, \ y^4 - 4y^2 + 1\}.$$

First, since

$$f_1 = -y\, g_1 - g_2, \qquad f_2 = (-x + y^3 - 4y)g_1 + (y^2 - 4)g_2,$$

and

$$g_1 = x f_1 - y f_2, \qquad g_2 = (-xy + 1)f_1 + y^2 f_2,$$

$\mathbf{G}$ is also a basis for the ideal.

Next, let

$$u = c_1\, g_1 + c_2\, g_2 = c_1(-y^3 + 4y - x) + c_2(-y^4 + 4y^2 - 1)$$

be an arbitrary member of the ideal. If $c_1 \neq 0$, then there is a term in u that contains a power of x, and, since x is the main variable, $\text{lm}(g_1) = -x$ and $\text{lm}(g_1) \mid \text{lm}(u)$. On the other hand, if $c_1 = 0$, then u has a term that is a multiple of y^4, and since $\text{lm}(g_2) = -y^4$, we have $\text{lm}(g_2) \mid \text{lm}(u)$. Therefore, G is a Gröbner basis. $\square$

In the following theorem we give two other conditions that describe a Gröbner basis.

Theorem 8.33. *Let $F = \{f_1, \ldots, f_n\}$, and let $I = \, < f_1, \ldots, f_n >$. The following three conditions are equivalent:*

1. *F is a Gröbner basis for I.*

2. *For each u in I, $u \xrightarrow{F} 0$ for all reduction step sequences.*

3. *For each u in I, there is a representation*

$$u = \sum_{i=1}^{n} c_i f_i \qquad (8.68)$$

where

$$\text{lm}(u) \equiv \max(\{\text{lm}(c_1)\text{lm}(f_1), \ldots, \text{lm}(c_n)\text{lm}(f_n)\}). \qquad (8.69)$$

Proof: To show that (1) implies (2), suppose that F is a Gröbner basis for I. For u in I, suppose $u \xrightarrow{F} r$ where r is reduced with respect to F. By Theorem 8.27 (4), r is in I, and the definition of a reduced polynomial together with the definition of a Gröbner basis implies that $r = 0$.

To show that (2) implies (3), if $u \xrightarrow{F} 0$, then the representation in (3) is given in Theorem 8.18 (page 313) with $r = 0$.

To show that (3) implies (1), observe that Equation (8.69) implies that

$$\text{lm}(u) \equiv \text{lm}(c_j)\text{lm}(f_j)$$

for some j. Therefore, $\text{lm}(f_j) \mid \text{lm}(u)$, which implies that F is a Gröbner basis. $\square$

Although F is a Gröbner basis, Equation (8.69) does not hold for all representations of the form (8.68). This point is illustrated in the next example.

Example 8.34. Let $L = [x, y]$, and consider the ideal with Gröbner basis

$$I = \, <g_1, g_2> \, = \, <x - y, \ 2y>,$$

which was considered in Example 8.31. Let

$$u = -y\, g_1 + (x/2)\, g_2 = y^2.$$

Observe that

$$\mathrm{lm}(u) = y^2 \prec \max(\{\mathrm{lm}(-y)\mathrm{lm}(x - y), \ \mathrm{lm}(x/2)\mathrm{lm}(2y)\}) = x\,y.$$

However, the proof of Theorem 8.33 shows that Equation (8.69) is valid whenever the polynomials $c_1, \ldots, c_n$ are quotients obtained from a reduction. For example, from multiple division of u by $[g_1, g_2]$, we obtain the representation $u = 0 \cdot g_1 + (1/2)y \cdot g_2$ for which Equation (8.69) is valid. $\square$

Another condition that describes a Gröbner basis is given in the following theorem.

Theorem 8.35. $F = \{f_1, \ldots, f_n\}$ *is a Gröbner basis for an ideal I if and only if for each u in $\mathbf{Q}[x_1, \ldots, x_p]$, the polynomial r in the representation*

$$u = c_1 f_1 + \cdots + c_n f_n + r \tag{8.70}$$

is unique.

In this theorem, the uniqueness property for r applies to any representation as in Equation (8.70), not just those that are obtained by reduction. The proof of the theorem is left to the reader (Exercise 9). $\square$

Although the remainders in the previous theorem are unique, the quotients are not. This point is illustrated in the next example.

Example 8.36. Let $L = [x, y]$, and consider the ideal with Gröbner basis

$$I = \, <g_1, g_2> \, = \, <x - y, \ 2y>$$

from Example 8.34. Let

$$u = y^2 x - x\, y^3.$$

Then, multiple division of u by $[g_1, g_2]$ gives the representation

$$u = \left(y^2 - y^3\right) g_1 + \left((1/2)y^2 - (1/2)y^3\right) g_2,$$

while multiple division by $[g_2, g_1]$ gives the second representation

$$u = 0 \cdot g_1 + \left((1/2)\, x\, y - (1/2)\, x\, y^2\right) g_2.$$

In both divisions, we obtain the (unique) remainder 0. $\square$

Simplification with Gröbner Bases

Theorems 8.33(2) and 8.35 have important consequences for the simplification problem. When the polynomials in the side relations

$$g_1 = 0, \cdots, g_n = 0$$

are a Gröbner basis for an ideal I, Theorem 8.33(2) implies that simplification of u in I using reduction obtains 0 for all reduction step sequences. In addition, Theorem 8.35 implies that once a symbol order L is chosen and a Gröbner basis is found, the simplification of polynomials in $\mathbf{Q}[x_1, \ldots, x_p]$ with reduction obtains the same remainder r for all reduction step sequences.

Example 8.37. Let

$$I = <f_1, f_2> = <x\,y - 1, \ x^2 + y^2 - 4>$$

with $L = [x, y]$. In Examples 8.19 and 8.22, we observed that

$$u = 2x\,y^2 + 7y - 6x + x^3 - 2y^3$$

is equivalent to 0 with respect to the side relations $f_1 = 0$, $f_2 = 0$, but this cannot be obtained with reduction by $\{f_1, f_2\}$. However, using the Gröbner basis

$$I = <g_1, g_2> = <-y^3 + 4y - x, \ y^4 - 4y^2 + 1>$$

given in Example 8.32 and the equivalent side relations $g_1 = 0$, $g_2 = 0$, we have $u \overset{\{g_1, g_2\}}{\longrightarrow} 0$ using the ten step reduction sequence

$$u \overset{[g_1, x^3]}{\Longrightarrow} 2xy^2 + 7y - 6x - 2y^3 - x^2y^3 + 4yx^2$$

$$\overset{[g_1, -x^2y^3]}{\Longrightarrow} 2xy^2 + 7y - 6x - 2y^3 + 4yx^2 + xy^6 - 4xy^4$$

$$\overset{[g_1, 4y\,x^2]]}{\Longrightarrow} 18xy^2 + 7y - 6x - 2y^3 + xy^6 - 8xy^4$$

$$\overset{[g_1, x\,y^6]}{\Longrightarrow} 18xy^2 + 7y - 6x - 2y^3 - 8xy^4 - y^9 + 4y^7$$

$$\overset{[g_1, -8x\,y^4]}{\Longrightarrow} 18xy^2 + 7y - 6x - 2y^3 - y^9 + 12y^7 - 32y^5$$

$$\overset{[g_1, 18x\,y^2]}{\Longrightarrow} 7y - 6x + 70y^3 - y^9 + 12y^7 - 50y^5$$

$$\overset{[g_1, -6x]}{\Longrightarrow} -17y + 76y^3 - y^9 + 12y^7 - 50y^5$$

$$\overset{[g_2, -y^9]}{\Longrightarrow} -17y + 76y^3 + 8y^7 - 49y^5$$

$$\overset{[g_2, 8y^7]}{\Longrightarrow} -17y + 68y^3 - 17y^5$$

$$\overset{[g_2, -17y^5]}{\Longrightarrow} 0.$$

Since a Gröbner basis introduces many more opportunities for reduction steps, it usually produces longer reduction sequences and increases dramatically the number of possible sequences.   □

A Computational Test for a Gröbner Basis

In the remainder of this section we develop a computational test for a Gröbner basis. We begin with some new terminology.

Definition 8.38. *Let u and v be non-zero monomials, and let L be a list that defines the variable order. The* **least common multiple** *of u and v is a monomial d that satisfies the following three properties.*

1. $u \mid d$ and $v \mid d$.

2. If $u \mid e$ and $v \mid e$, then $d \mid e$.

3. The coefficient part of d is 1.

The operator $\mathrm{lcm}(u, v, L)$ *represents the least common multiple, and when L is understood from context, we use the simpler notation* $\mathrm{lcm}(u, v)$.

The variable part of the monomial

$$\frac{u\,v}{\gcd(u, v)}$$

satisfies the three properties of the definition (Exercise 10(a)). An operator that obtains the least common multiple is described in Exercise 10(b).

Definition 8.39. *Let u and v be non-zero polynomials, and let L be a list of symbols that defines the variable order. The* **S-polynomial** *of u and v is defined by*

$$S(u, v, L) = \frac{d}{\mathrm{lm}(u, L)}\, u - \frac{d}{\mathrm{lm}(v, L)}\, v, \tag{8.71}$$

where

$$d = \mathrm{lcm}(\mathrm{lm}(u, L), \mathrm{lm}(v, L)).$$

When L is evident from context, we use the simpler notation $S(u, v)$.

In this definition, $d/\mathrm{lm}(u, L)$ and $d/\mathrm{lm}(v, L)$ are monomials, and by Theorem 6.25(1),

$$\mathrm{lm}\left(\frac{d}{\mathrm{lm}(u, L)}u\right) = d = \mathrm{lm}\left(\frac{d}{\mathrm{lm}(v, L)}v\right).$$

Therefore, in Equation (8.71), the leading monomials of $\dfrac{d\,u}{\operatorname{lm}(u,L)}$ and $\dfrac{d\,v}{\operatorname{lm}(v,L)}$ cancel.

Example 8.40. Let $u = x\,y - 1$ and $v = x^2 + y^2 - 4$, and let $L = [x, y]$ define the variable order. Then, $d = \operatorname{lcm}(x^2, x\,y) = x^2 y$, and

$$S(u, v, L) = \frac{x^2\,y}{x\,y}\,u - \frac{x^2\,y}{x^2}\,v = x\,u - y\,v = -x - y^3 + 4\,y.$$

This polynomial is the one used in Example 8.32 to show that a basis was not a Gröbner basis. □

A procedure that obtains an S-polynomial is described in Exercise 10.

Theorem 8.41. *The S-polynomials satisfy the following properties.*

1. $S(u, u) = 0$.

2. $S(u, v) = -S(v, u)$.

3. *If u and v are in an ideal I, then $S(u, v)$ is also in I.*

4. *If m and n are monomials, then,*

$$S(m\,u, n\,v) = \frac{A}{B}S(u, v),$$

   where

$$A = \operatorname{lcm}(\operatorname{lm}(m\,u),\ \operatorname{lm}(n\,v)), \qquad B = \operatorname{lcm}(\operatorname{lm}(u), \operatorname{lm}(v)).$$

5. *Suppose that u and v are non-zero polynomials, and*

$$\operatorname{lm}(u) = \operatorname{lm}(v) = M.$$

   Then:

   (a) $\operatorname{lm}(S(u, v)) \prec M$.
   (b) If
$$f = c\,u + d\,v$$
   where c and d are rational numbers and $\operatorname{lm}(f) \prec M$, then
$$f = k\,S(u, v)$$
   where k is a rational number.

Proof: The proofs of Parts (1), (2), (3), and (4), which follow from the definition, are left to the reader (Exercise 5).

To show (5a), observe in this case $S(u, v) = u - v$. Since M is eliminated from $S(u, v)$, we have $\mathrm{lm}(S(u, v)) \prec M$.

To show (5b), let V be the variable part of M. We have

$$u = aV + r, \qquad v = bV + s,$$

where r and s represent lower order monomials, and a and b are non-zero rational numbers. Since $f = (ca + db)V + cr + ds$ and $\mathrm{lm}(f) \prec M$, we have $ca + db = 0$, and

$$
\begin{aligned}
f &= ca\,(u/a) + db\,(v/b) \\
&= ca\left(\frac{V}{aV}u - \frac{V}{bV}v\right) + \frac{v}{b}(ca + db) \\
&= ca\,S(u, v).
\end{aligned}
$$

$\square$

The next theorem uses S-polynomials to give a computational test that determines if a basis is a Gröbner basis.

Theorem 8.42. *Let $I = \langle g_1, \ldots, g_n \rangle$ where $n \geq 2$, and let L be a list of symbols that defines the variable order. The following conditions are equivalent.*

1. $G = \{g_1, \ldots, g_n\}$ is a Gröbner basis for I.

2. For all $i < j$, $S(g_i, g_j) \xrightarrow{G} 0$ for all reduction step sequences.

3. For all $i < j$, $S(g_i, g_j) \xrightarrow{G} 0$ for at least one reduction step sequence.

Notice that condition (2) refers to *all* reduction step sequences, while condition (3) refers to *at least one* reduction step sequence. By stating the theorem in this way, we can use condition (2) to show that G is not a Gröbner basis by finding just one reduction step sequence which does not obtain 0 and can use the weaker condition (3) to show that G is a Gröbner basis.

Proof of Theorem 8.42: To show that (1) implies (2), if G is a Gröbner basis, then, since $S(g_i, g_j)$ is in I, Theorem 8.33(2) implies that

$$S(g_i, g_j) \xrightarrow{G} 0.$$

It is obvious that (2) implies (3).

We show that (3) implies (1) using the condition in Theorem 8.33(3). Since the proof for this case is quite involved, we illustrate the basic idea for a basis with two polynomials $\{g_1, g_2\}$. We will show that the condition in (3) implies that each f in $I = \langle g_1, g_2 \rangle$ has a representation

$$f = c_1 g_1 + c_2 g_2 \tag{8.72}$$

such that

$$\mathrm{lm}(f) \equiv \max(\{\mathrm{lm}(c_1 g_1), \mathrm{lm}(c_2 g_2)\}). \tag{8.73}$$

Suppose that Equation (8.73) does not hold for a representation where

$$\mathrm{lm}(f) \prec M = \max(\{\mathrm{lm}(c_1 g_1), \mathrm{lm}(c_2 g_2)\}). \tag{8.74}$$

In Example 8.34, we showed that, in this situation, we can find another representation for f where Equation (8.73) holds using multiple division by a Gröbner basis. We take a similar approach here although we can't simply divide f by $[g_1, g_2]$, because the point of the proof is to show that $\{g_1, g_2\}$ is a Gröbner basis. Instead, we apply the reduction process to a related polynomial to find a new representation $f = d_1 g_1 + d_2 g_2$ where

$$M_2 = \max(\{\mathrm{lm}(d_1 g_1), \mathrm{lm}(d_2 g_2)\}) \prec M.$$

If $\mathrm{lm}(f) \equiv M_2$, the proof is complete. If $\mathrm{lm}(f) \prec M_2$, we repeat the process (a finite number of times) until we find a representation that satisfies Equation (8.73).

The inequality (8.74) implies that in Equation (8.72) the leading monomials of $c_1 g_1$ and $c_2 g_2$ must cancel, and, therefore,

$$M = \mathrm{lm}(c_1 g_1) = \mathrm{lm}(c_2 g_2). \tag{8.75}$$

Let V_i be the variable part of $\mathrm{lm}(c_i)$, $i = 1, 2$. We have

$$c_1 = k_1 V_1 + l_1, \quad c_2 = k_2 V_2 + l_2 \tag{8.76}$$

where k_1 and k_2 are rational numbers and l_1 and l_2 represent the lower order monomials with respect to $\prec$. Define

$$g = k_1 V_1 g_1 + k_2 V_2 g_2. \tag{8.77}$$

Observe that

$$f = g + l_1 g_1 + l_2 g_2, \tag{8.78}$$

and since

$$\mathrm{lm}(V_1 g_1) \equiv \mathrm{lm}(c_1 g_1) = M = \mathrm{lm}(c_2 g_2) \equiv \mathrm{lm}(V_2 g_2), \tag{8.79}$$

$$\text{lm}(l_1 g_1) \prec M, \quad \text{lm}(l_2 g_2) \prec M, \tag{8.80}$$

it follows that g contains the part of f that creates the inequality (8.74). In addition, Theorem 8.41(5b) (applied to $V_1 g_1$, $V_2 g_2$, and g), Theorem 8.41(4), and Equation (8.79) imply that

$$g = k\, S(V_1 g_1, V_2 g_2) = k\frac{V_M}{d} S(g_1, g_2), \tag{8.81}$$

where k is a rational number,

$$V_M = \text{lcm}(\text{lm}(V_1 g_1), \text{lm}(V_2, g_2))$$

is the variable part of M, and

$$d = \text{lcm}(\text{lm}(g_1), \text{lm}(g_2)).$$

Equation (8.81), condition (3), and Theorem 8.17 imply that

$$g \xrightarrow{\ G\ } 0$$

for some reduction step sequence. Therefore, Theorem 8.18 implies that there is a representation

$$g = e_1 g_1 + e_2 g_2 \tag{8.82}$$

where

$$\text{lm}(g) \equiv \max(\{\text{lm}(e_1)\text{lm}(g_1), \text{lm}(e_2)\text{lm}(g_2)\}). \tag{8.83}$$

In addition,

$$S(V_1 g_1, V_2 g_2) \quad = \quad \frac{V_M}{\text{lm}(V_1 g_1)} V_1 g_1 - \frac{V_M}{\text{lm}(V_2 g_2)} V_2 g_2. \tag{8.84}$$

Equation (8.79) implies that $V_M/\text{lm}(V_1 g_1)$ and $V_M/\text{lm}(V_2 g_2)$ are rational numbers, and since the leading monomials in

$$\frac{V_M}{\text{lm}(V_{1g1})} V_{1g1}$$

and

$$\frac{V_M}{\text{lm}(V_{2g2})} V_{2g2}$$

cancel, it follows from Equations (8.79) and (8.84) that

$$\text{lm}(g) \equiv \text{lm}(S(V_1 g_1, V_2 g_2)) \prec \max(\{\text{lm}(V_1 g_1), \ \text{lm}(V_2 g_2)\}) \equiv M. \tag{8.85}$$

Finally, from Equations (8.78) and (8.82), we have the new representation

$$f = (e_1 + l_1)g_1 + (e_2 + l_2)g_2,\qquad(8.86)$$

where from the relations (8.80), (8.83), and (8.85),

$$
\begin{aligned}
M_2 &= \max(\{\operatorname{lm}((e_1 + l_1)g_1), \operatorname{lm}((e_2 + l_2)g_2)\}) \\
&\preceq \max(\{\operatorname{lm}(e_1 g_1), \operatorname{lm}(l_1 g_1), \operatorname{lm}(e_2 g_2), \operatorname{lm}(l_2 g_2)\}) \\
&\equiv \max(\{\operatorname{lm}(g), \operatorname{lm}(l_1 g_1), \operatorname{lm}(l_2 g_2), \}) \\
&\prec M.
\end{aligned}
$$

To summarize, we have constructed a new representation for f in Equation (8.86) where $\operatorname{lm}(f) \preceq M_2 \prec M$. If $\operatorname{lm}(f) \equiv M_2$, Theorem 8.33(3) implies that $\{g_1, g_2\}$ is a Gröbner basis. If $\operatorname{lm}(f) \prec M_2$, we reapply the construction a finite number of times until a representation is found such that

$$f = p_1 f_1 + p_2 f_2, \quad \operatorname{lm}(f) \equiv \max(\{\operatorname{lm}(p_1 f_1), \operatorname{lm}(p_2 f_2)\}).$$

Therefore, Theorem 8.33(3) implies that $\{g_1, g_2\}$ is a Gröbner basis. □

Exercises

1. Use a CAS to find a Gröbner basis for some of the ideals in this section. *Note:* In Maple use **gbasis**, in Mathematica use **GroebnerBasis**, and in MuPAD use **groebner::gbasis**. Since computer algebra systems use a number of ordering schemes for monomials, make sure the system uses the lexicographical order. (Implementation: Maple (mws), Mathematica (nb), MuPAD (mnb).)

2. Prove Theorem 8.27.

3. Let I be an ideal in $\mathbf{Q}[x]$.

   (a) If $I = < f_1, f_2 >$, show that $I = < \gcd(f_1, f_2) >$. *Hint:* Use the extended Euclidean algorithm.

   (b) If $I = < f_1, \ldots, f_n >$, show that $I = < \gcd(f_1, \ldots, f_n) >$.

4. Let $I = < x, y >$.

   (a) Show that $< 3x + 2y, 2x + 3y >$ is also a basis for I.

   (b) Show that I is not a principal ideal. *Hint:* Suppose that

   $$I = < f(x, y) >,$$

   and use a degree argument to show that if x and y are multiples of $f(x, y)$, then $f(x, y)$ is a rational number.

5. Suppose that $I = < f >$. Show that $\{f\}$ is a Gröbner basis for the ideal.

6. Suppose that $I = < f_1, \ldots, f_n >$ where

$$Variables(f_i) \cap Variables(f_j) = \emptyset$$

   for $i \neq j$. Show that $\{f_1, \ldots, f_n\}$ is a Gröbner basis for I.

7. Suppose that $I = < f_1, \ldots, f_n >$ where each f_i is a monomial. Show that $\{f_1, \ldots, f_n\}$ is a Gröbner basis for I.

8. Consider the ideal $I = < f_1, f_2 >= < y\,x^2 + 1, \; y^3 + 2y - 1 >$.

   (a) Show that $\{f_1, f_2\}$ is not a Gröbner basis for I.

   (b) Show that $\{g_1, g_2\} = \{x^2 + y^2 + 2, y^3 + 2y - 1\}$ is a Gröbner basis for I using Definition 8.30.

9. Prove Theorem 8.35.

10. (a) Let u and v be non-zero monomials. Show that the polynomial

$$\frac{u\,v}{\gcd(u, v)}$$

   satisfies properties (1) and (2) of Definition 8.38. *Hint:* See Exercise 18, page 36.

   (b) Let L be a list of symbols that defines the variable order, and let u and v be non-zero monomials in the symbols in L with rational number coefficients. Give a procedure

$$Monomial\_lcm(u, v, L)$$

   that obtains the $lcm(u, v, L)$. If $u = 0$ or $v = 0$, return the global symbol **Undefined**.

   (c) Let u and v be polynomials, and let L be a list that defines the variable order. Give a procedure $S\_poly(u, v, L)$ that returns $S(u, v, L)$ in expanded form.

11. Prove Theorem 8.41, Parts (1), (2), (3), and (4).

8.3   A Simplification Algorithm

Let u be a polynomial in $\mathbf{Q}[x_1, \ldots, x_p]$, and consider the simplification of u with respect to the side relations $f_1 = 0, \ldots, f_n = 0$. Our discussion in the previous two sections suggests that our simplification algorithm has two steps.

1. Let $I = < f_1, \ldots, f_n >$, and let L be a list that determines the variable order. Find a Gröbner basis $G = \{g_1, \ldots, g_m\}$ so that $I = <g_1, \ldots, g_m>$.

2. Find the simplified form of u as the (reduced) remainder r of the reduction $u \xrightarrow{G} r$.

The next theorem shows that this algorithm obtains the two goals of our simplification algorithm (page 321).

Theorem 8.43. *Suppose that* $F = \{f_1, \ldots, f_n\}$, *and let* I *be the ideal that is generated by* F. *In addition, suppose that* $G = \{g_1, \ldots, g_m\}$ *is a Gröbner basis for* I.

1. *For* u *in* I, $u \xrightarrow{G} 0$.

2. *If* $u \xrightarrow{G} r$, *then* $u \overset{F}{=} r$.

3. *If* $u \xrightarrow{G} r$, *then* r *is reduced with respect to* F.

Proof: Statement (1) is a restatement of Theorem 8.33(2) (page 325).

To show (2), Theorem 8.25 states that $u \xrightarrow{G} r$ implies $u \overset{G}{=} r$. Since

$$I \ = \ <f_1, \ldots, f_n> \ = \ <g_1, \ldots, g_m>,$$

the two systems $f_1 = 0, \ldots, f_n = 0$ and $g_1 = 0, \ldots, g_m = 0$ have the same solutions, and therefore $u \overset{F}{=} r$.

The proof of (3), which follows from the definition, is left to the reader (Exercise 1). $\square$

Gröbner Basis Algorithm

The first step in a simplification algorithm involves finding a Gröbner basis for an ideal. An algorithm that does this is based on the reduction property for S-polynomials given in Theorem 8.42 on page 330. We illustrate the general approach in the next example.

Example 8.44. Consider the ideal

$$I \ = \ <f_1, f_2> \ = \ <xy + x + 1, \ xy^2 + y + x^2 + 1>,$$

and let $L = [x, y]$. Our strategy is to add the non-zero reductions of S-polynomials to the basis until the third condition in Theorem 8.42 is satisfied. In this example, all reductions are done with the selection scheme given in Figure 8.1 on page 316.

First, $S(f_1, f_2) = x^2 + x - xy^3 - y^2 - y$, and since

$$S(f_1, f_2) \xrightarrow{\{f_1, f_2\}} r = x - 2y - 1,$$

the condition in Theorem 8.42(2) is not satisfied for this S-polynomial. When this happens, however, since r is in I, we can include it in the basis. Therefore, if $f_3 = r$, we have $I = <f_1, f_2, f_3>$, and by Theorem 8.15,

$$S(f_1, f_2) \xrightarrow{\{f_1, f_2, f_3\}} 0. \tag{8.87}$$

To determine if $\{f_1, f_2, f_3\}$ is a Gröbner basis, we must also check that the reduction condition for S-polynomials is satisfied for $S(f_1, f_3)$ and $S(f_2, f_3)$. However,

$$S(f_1, f_3) = x + 2y^2 + y + 1 \xrightarrow{\{f_1, f_2, f_3\}} r = 2y^2 + 3y + 2, \tag{8.88}$$

and since $r \neq 0$, we append $f_4 = r$ to the basis giving $I = <f_1, f_2, f_3, f_4>$. As before, by Theorem 8.15,

$$S(f_1, f_3) \xrightarrow{\{f_1, f_2, f_3, f_4\}} = 0.$$

In addition, a calculation shows that

$$S(f_2, f_3) = xy^2 + 2xy + x + y + 1 \xrightarrow{\{f_1, f_2, f_3, f_4\}} = 0.$$

At this point, since we are using the selection scheme in Figure 8.1, we have the reduction conditions

$$S(f_i, f_j) \xrightarrow{\{f_1, f_2, f_3, f_4\}} = 0, \quad i < j = 1, 2, 3.$$

In fact, some calculation also shows that

$$S(f_i, f_4) \xrightarrow{\{f_1, f_2, f_3, f_4\}} = 0, \quad i = 1, 2, 3,$$

and therefore

$$\{f_1, f_2, f_3, f_4\} = \{xy + x + 1, \ xy^2 + y + x^2 + 1, \ x - 2y - 1, \ 2y^2 + 3y + 2\}$$

is a Gröbner basis for I. $\square$

The algorithm we have followed in the last example is known as Buchberger's algorithm. We show in Theorem 8.52 below that the algorithm terminates and obtains a Gröbner basis for an ideal.

A procedure for Buchberger's algorithm is given in Figure 8.2. The input list F of polynomials generates an ideal, and the input list L of symbols determines the variable order. At line 1, the variable G, which eventually contains the Gröbner basis, is initialized to F. The variable P,

Procedure $G\_basis(F, L)$;
Input
    F : a list of non-zero polynomials in L;
    L : a list of symbols;
Output
    a list with a Gröbner basis;
Local Variables
    G, P, i, j, t, s, r;
Begin

```
1    G := F;
2    P := [ ];
3    for  i := 1 to  Number_of_operands(G) do
4        for  j := i + 1 to  Number_of_operands(G) do
5            P := Join(P, [[Operand(G, i), Operand(G, j)]]);
6    while  P ≠ [ ] do
7        t := Operand(P, 1);
8        P := Rest(P);
9        s := S_poly(Operand(t, 1), Operand(t, 2), L);
10       r := Operand(Reduction(s, G, L), 2);
11       if  r ≠ 0 then
12           for  i from  1 to  Number_of_operands(G) do
13               P := Join(P, [[Operand(G, i), r]]);
14           G := Join(G, [r]);
15   Return(G)
```
End

Figure 8.2. An MPL procedure for Buchberger's algorithm. (Implementation: Maple (txt), Mathematica (txt), MuPAD (txt).)

which is a list of two element lists of polynomials from G, is constructed in lines 2–5. This list, which is used to construct the S-polynomials, changes as the algorithm progresses (lines 8, 12–13). For each pass through the **while** loop, we first choose (and remove) a pair t from P, construct an S-polynomial from t (lines 7–9), and then perform the reduction $s \xrightarrow{G} r$ (line 10). (The S_poly procedure is described in Exercise 10, page 334.) We then check the condition in Theorem 8.42 (line 11). If the reduced polynomial $r \neq 0$, new pairs of polynomials, which are constructed from members of the current basis G with r, are added to P, and r is added to G (lines 12–14). The algorithm terminates when $P = [\ ]$, which means there are no more S-polynomials to test.

In its current form, the *G_basis* procedure returns a basis with some redundant polynomials. The next theorem provides a way to eliminate this redundancy

Theorem 8.45. *Let* $I = < g_1, \ldots, g_n >$ *where* $G = \{g_1, \ldots, g_n\}$ *is a Gröbner basis for* I*, and suppose that* L *is a list of symbols that defines the variable order. If*

$$g_i \xrightarrow{\{g_1, \ldots, g_{i-1}, g_{i+1}, \ldots, g_n\}} 0,$$

then $\{g_1, \ldots, g_{i-1}, g_{i+1}, \ldots, g_n\}$ *is also a Gröbner basis for* I*.*

The proof, which follows from the definition of a Gröbner basis, is left to the reader (Exercise 4).   □

Example 8.46. In Example 8.44 we considered the ideal

$$I = <f_1, f_2> = <x\,y + x + 1, \ x\,y^2 + y + x^2 + 1>$$

where $L = [x, y]$ and showed that

$$\{f_1, f_2, f_3, f_4\} = \{x\,y + x + 1, \ x\,y^2 + y + x^2 + 1, \ x - 2y - 1, \ 2y^2 + 3y + 2\}$$

is a Gröbner basis for the ideal. However, using the selection scheme in Figure 8.1, we have $f_1 \xrightarrow{\{f_2, f_3, f_4\}} 0$ and $f_2 \xrightarrow{\{f_3, f_4\}} 0$. Therefore,

$$\{f_3, f_4\} = \{x - 2y - 1, \ 2y^2 + 3y + 2\}$$

is also a Gröbner basis for I.   □

In Exercise 5 we describe a procedure, based on Theorem 8.45, that eliminates redundant polynomials from a Gröbner basis.

In most cases, finding a Gröbner basis involves significant computation and coefficient explosion, particularly if the coefficients or degrees are large, or there are many variables, or the algorithm encounters many S-polynomials that do not reduce to 0. This point is illustrated in the next example.

Example 8.47. Consider the ideal

$$I = <f_1, f_2> = <2x^3y + 3x^2 + 6, \ 5x^2 + x + 3y^2 + y + 1>$$

where $L = [x, y]$. The *G_basis* procedure forms a basis of eight polynomials:

$$f_1 = 2\,x^3y + 3\,x^2 + 6,$$

$$f_2 = 5\,x^2 + x + 3\,y^2 + y + 1,$$

$$f_3 = -\frac{3}{5}\,xy^3 - \frac{1}{5}\,xy^2 - \frac{4}{25}\,xy - \frac{3}{10}\,x + \frac{3}{25}\,y^3 - \frac{43}{50}\,y^2 - \frac{13}{50}\,y + \frac{27}{10},$$

$$f_4 = \frac{3}{10}\,xy^2 + \frac{73}{750}\,xy - \frac{22}{25}\,x - \frac{3}{25}\,y^5 - \frac{2}{25}\,y^4 - \frac{38}{375}\,y^3 + \frac{23}{750}\,y^2$$
$$+ \frac{1}{250}\,y - \frac{19}{50},$$

$$f_5 = -\frac{3551}{11250}\,xy - \frac{4297}{750}\,x - \frac{6}{25}\,y^7 - \frac{6}{25}\,y^6 - \frac{128}{125}\,y^5$$
$$-\frac{448}{1125}\,y^4 - \frac{16213}{11250}\,y^3 - \frac{6188}{5625}\,y^2 + \frac{13432}{5625}\,y - \frac{389}{250},$$

$$f_6 = -\frac{55134547}{958770}\,x - \frac{647416}{266325}\,y^7 - \frac{540886}{266325}\,y^6 - \frac{8071024}{798975}\,y^5$$
$$-\frac{26558308}{7190775}\,y^4 - \frac{35705726}{2396925}\,y^3 - \frac{1864529}{191754}\,y^2 + \frac{372290599}{14381550}\,y$$
$$\frac{97071797}{4793850},$$

$$f_7 = \frac{1620}{3551}\,y^9 + \frac{2160}{3551}\,y^8 + \frac{2700}{3551}\,y^7 + \frac{1050}{3551}\,y^6 + \frac{95}{53}\,y^5 + \frac{13785}{3551}\,y^4$$
$$-\frac{29295}{3551}\,y^3 - \frac{20785}{3551}\,y^2 + \frac{52785}{3551}\,y + \frac{18675}{3551},$$

$$f_8 = \frac{900}{3551}\,y^8 + \frac{900}{3551}\,y^7 + \frac{1200}{3551}\,y^6 + \frac{550}{10653}\,y^5 + \frac{3475}{3551}\,y^4 + \frac{6500}{3551}\,y^3$$
$$-\frac{55325}{10653}\,y^2 - \frac{5400}{3551}\,y + \frac{31125}{3551}.$$

After removing redundant polynomials from the basis, we obtain a Gröbner basis $\{f_6,\ f_8\}$. $\qquad\qquad\square$

Both the Gröbner basis and the path taken by the algorithm are dependent on the choice of reduction selection scheme, the order of the side relations, and the symbol order. This point is illustrated in the next example.

Example 8.48. Let

$$F = \{f_1, f_2, f_3\} = \{xy + x + z - 3,\ xz + y + x^2 + 1,\ y + z^2 + x + 2\},$$

$L = [x, y, z]$, and let I be the ideal generated by F. For the basis order $[f_1, f_2, f_3]$, the *G_basis* procedure obtains a Gröbner basis with seven members, and, once redundant polynomials are eliminated, we have the Gröbner basis

$$\{y + z^2 + x + 2,$$
$$- 2\,y + 5\,z + z^4 - 2\,z^2 - 10,$$
$$- 5\,z^3 + 13\,z^2 - z^6 - 2\,z^4 + z^5 - 18\,z + 20\}\,.$$

On the other hand, for the basis order $[f_2, f_3, f_1]$, the *G_basis* procedure obtains a Gröbner basis with nine members, and once redundant polynomials are eliminated, we have the Gröbner basis

$$\{y + z^2 + x + 2,$$
$$3\,z + 6\,y - 7\,z^2 - z^4 + 5\,z^3 + z^6 - z^5 + 10,$$
$$5/3\,z^3 - 13/3\,z^2 + 1/3\,z^6 + 2/3\,z^4 - 1/3\,z^5 + 6\,z - 20/3\}\,. \qquad \square$$

Termination of Buchberger's Algorithm

In order to show that Buchberger's algorithm terminates, we must look at some more advanced concepts from the theory of ideals. First, in most textbooks on this subject, an ideal is defined in a more general way than we have given in Definition 8.26 (page 321).

Definition 8.49. *A set $I \subset \mathbf{Q}[x_1, \ldots, x_p]$ is an **ideal** if it satisfies the following two conditions.*

1. *If u and v are in I, then $u + v$ is also in I.*

2. *If u is in I and c is in $\mathbf{Q}[x_1, \ldots, x_p]$, then $c\,u$ is also in I.*

In this definition an ideal is defined as any set of polynomials that satisfies properties (1) and (2) of Theorem 8.27 on page 322. Although this definition appears more general than the one in Definition 8.26, the two definitions are actually equivalent as long as we restrict our attention to the domain $\mathbf{Q}[x_1, \ldots, x_p]$. This fact is the content of the following theorem, known as the Hilbert Basis Theorem.

Theorem 8.50. *Let I be an ideal in the sense of Definition 8.49 which is contained in $\mathbf{Q}[x_1, \ldots, x_p]$. Then, I is generated by a finite basis of polynomials.*

The proof of the theorem is given in the references cited at the end of the chapter. $\qquad \square$

The ideal described in the next definition is used in the proof of the termination of Buchberger's algorithm.

Definition 8.51. *Let $F = \{f_1, \ldots, f_n\}$ be a set of non-zero polynomials that generate an ideal I, and let L be a list of symbols. The **leading monomial ideal** of F is the ideal*

$$\text{LMI}(F) = <\text{lm}(f_1), \ldots, \text{lm}(f_n)> .$$

Theorem 8.52. *Let $F = [f_1, \ldots, f_n]$ be a list of non-zero polynomials that generate an ideal I, and let L be a list of symbols that determines the variable order. Then, the G_basis procedure terminates and returns a Gröbner basis for I.*

Proof: The proof of termination, which is based on the Hilbert Basis Theorem, is by contradiction. If the algorithm does not terminate, there is an infinite sequence of sets

$$\begin{aligned} G_1 &= \{f_1, \ldots, f_n\}, \\ G_2 &= G_1 \cup \{r_1\}, \\ &\;\vdots \\ G_{k+1} &= G_k \cup \{r_k\}, \\ &\;\vdots \end{aligned}$$

where $r_k \neq 0$ is the reduction with respect to G_k of the S-polynomial of two polynomials in G_k. Although

$$G_1 \subset G_2 \subset \cdots G_k \subset G_{k+1} \subset \cdots, \tag{8.89}$$

we still have

$$G_k \subset I, \quad k = 1, 2, \ldots. \tag{8.90}$$

In order to apply the Hilbert Basis Theorem, we need to introduce another ideal into the discussion. From (8.89), we have

$$\text{LMI}(G_1) \subset \text{LMI}(G_2) \subset \cdots \subset \text{LMI}(G_k) \subset \text{LMI}(G_{k+1}) \subset \cdots. \tag{8.91}$$

In addition, since r_k is reduced with respect to G_k, $\text{lm}(r_k)$ is not divisible by the leading monomial of any polynomial in G_k, and therefore

$$\text{LMI}(G_k) \neq \text{LMI}(G_{k+1}). \tag{8.92}$$

Define

$$J = \cup_{k=1}^{\infty} \text{LMI}(G_k),$$

where the notation indicates that J is the union of all the ideals $\mathrm{LMI}(G_k)$, $k = 1, 2, \ldots$. Observe that J is also an ideal (in the sense of Definition 8.49), which is included in $\mathbf{Q}[x_1, \ldots, x_p]$. Therefore, the Hilbert Basis Theorem implies that it has a finite basis

$$J = <p_1, \ldots, p_m>, \tag{8.93}$$

where each polynomial p_j is in some $\mathrm{LMI}(G_{k_j})$ (which depends on j). Since the basis in Equation (8.93) is finite, there is a positive integer k' where

$$\{p_1, \ldots, p_m\} \subset \mathrm{LMI}(G_{k'}),$$

and therefore

$$\mathrm{LMI}(G_k) = \mathrm{LMI}(G_{k'}), \quad \text{for} \quad k \geq k'. \tag{8.94}$$

However, since (8.92) and (8.94) cannot both be true, we reach a contradiction which implies that the G_basis algorithm terminates.

To complete the proof, we must show that the set $G_{k'}$ is a Gröbner basis for I. First, since $F \subset G_{k'}$, Equation (8.90) implies that $G_{k'}$ is a basis for I. In addition, by Equation (8.94), for f, g in $G_{k'}$, $S(f, g) \xrightarrow{G_{k'}} 0$, and therefore Theorem 8.42 implies that $G_{k'}$ is a Gröbner basis for I. $\square$

Simplification

Before we present our simplification procedure, we need to address the problem of side relation consistency. The following theorem shows that this question is easily addressed as part of the simplification process.

Theorem 8.53. Let $F = \{f_1, \ldots, f_n\}$ be a set of non-zero polynomials, and let I be the ideal generated by F. Then, the system $f_1 = 0, \ldots, f_n = 0$ is inconsistent if and only if each Gröbner basis for I contains a non-zero constant polynomial.

Proof: The original system of side relations and the system of side relations determined by a Gröbner basis have the same solutions. Therefore, if the Gröbner basis contains a non-zero constant c, we have a side relation $c = 0$ which implies that the system is inconsistent.

The proof of the implication in the reverse direction, which is more involved, is given in the references cited at the end of the chapter. $\square$

Example 8.54. In Example 8.23 we showed that the system of equations

$$f_1 = x\,y + x + 2y + 3 = 0, \quad f_2 = x\,y + 2y + 1 = 0$$

Procedure *Simplify_side_relations*(u, F, L);
Input
    u : a polynomial in the symbols in L;
    F : a list of non-zero polynomials in the symbols in L;
    L : a list of symbols that determines the variable order;
Output
    the simplified form of u with respect to F or the global
    symbol **Inconsistent**;
Local Variables
    G, H;
Begin
1    $G := G\_basis(F, L)$;
2    $H := Elim\_poly(G, L)$;
3   **if** *Number_of_operands*$(H) = 1$
       and *Kind*$(Operand(H, 1)) \in$ {**integer, fraction**} **then**
4   *Return*(**Inconsistent**)
5   **else**
6   *Return*$(Operand(Reduction(u, H, L), 2))$;
End

Figure 8.3. An MPL procedure for simplifications with side relations. (Implementation: Maple (txt), Mathematica (txt), MuPAD (txt).)

is inconsistent. Applying the *G_basis* procedure to $[f_1. f_2]$, we obtain the Gröbner basis $G = [x\,y + x + 2y + 3, \; x\,y + 2y + 1, \; x + 2, \; 1]$. Since the constant polynomial 1 is in this basis, the side relations are inconsistent. In addition, by applying the process suggested in Theorem 8.45, we obtain a basis with only the constant polynomial 1.   □

A procedure for simplification with side relations is given in Figure 8.3. In line 1 we obtain the Gröbner basis G, and in line 2 we remove redundant polynomials from the basis. (The *Elim_poly* procedure is described in Exercise 5.) At line 3 we check for side relation consistency. This test is based on the observation that if a non-zero constant is in the Gröbner basis, then H only contains a constant. If the side relations are consistent, we return the reduction of u with respect to H (line 6).

Example 8.55. In this example, we consider the simplification of the algebraic number expression described in Example 4.63. Consider the dependent algebraic numbers $\alpha_1 = \sqrt{2}$ and $\alpha_2 = \sqrt{8}$ with minimal polynomials

$p_1(x_1) = x_1^2 - 2$,　$p_2(x_2) = x_2^2 - 8$, and dependence relation

$$h(x_1,\ x_2) = 2\,x_1 - x_2 = 0, \tag{8.95}$$

and consider the expression

$$u = \alpha_1^2 + 2\,\alpha_2^2 + \alpha_1^2\,\alpha_2^2 - 8\,\alpha_1\,\alpha_2 - 2$$

which simplifies to 0. Simplification of u with division by the minimal polynomials $p_1(\alpha_1)$ and $p_2(\alpha_2)$ obtains

$$u = -8\,\alpha_1\,\alpha_2 + 32$$

rather than the simplified form $u = 0$. However,

$$Simplify\_side\_relations(u,\ [p_1(\alpha_1), h(\alpha_1, \alpha_2)],\ [\alpha_1, \alpha_2]) \to 0$$

and

$$Simplify\_side\_relations(u,\ [p_2(\alpha_1), h(\alpha_1, \alpha_2)],\ [\alpha_1, \alpha_2]) \to 0.$$

Observe that, in both cases, we obtain the simplified form using only one of the minimal polynomials together with the dependence relation.　□

Appraisal of the Simplification Algorithm

Practical Limitations. Although the simplification algorithm achieves the goals listed on page 321, it is most useful as a test for zero equivalence at least for expressions that have a representation in the form (8.64). However, when an input expression does not simplify to 0, the algorithm can return a surprisingly complex expression. This point is illustrated in the next example.

Example 8.56. Let

$$I\ =\ <f_1, f_2>\ =\ <2x^3y + 3x^2 + 6,\ 5x^2 + x + 3y^2 + y + 1>$$

with $L = [x, y]$ which was considered in Example 8.47 on page 338, and consider the simplification of

$$\begin{aligned} u\ &=\ f_1 + f_2 + x \tag{8.96}\\ &=\ 2x^3y + 8x^2 + 2x + 3y^2 + y + 7 \end{aligned}$$

with respect to the side relations $f_1 = 0$ and $f_2 = 0$. It is apparent from Equation (8.96) that u simplifies to x with respect to these side relations.

In fact, using the *Reduction* procedure, we obtain $u \overset{\{f_1, f_2\}}{\longrightarrow} x$. However, simplifying with reduction by the Gröbner basis obtained in Example 8.47, we obtain a much more involved expression

$$
\begin{aligned}
r = \ & \frac{372290599}{827018205} y - \frac{9322645}{55134547} y^2 - \frac{71411452}{275672735} y^3 - \frac{3725088}{21205595} y^5 \\
& - \frac{53116616}{827018205} y^4 - \frac{11653488}{275672735} y^7 - \frac{9735948}{275672735} y^6 - \frac{97071797}{275672735}.
\end{aligned}
$$

The problem here is the simple expression x in Equation (8.96) is not reduced with respect to the Gröbner basis. However, reduction with the Gröbner basis can determine that $r - x$ simplifies to 0. □

Theoretical Limitations. In the next example we show that reduction with Gröbner bases cannot always solve the zero equivalence problem.

Example 8.57. Let

$$
F = \{f_1, f_2\} = \{x^2 - 2x + y^2 - 4y + 1, \ x^2 + 4x + y^2 - 4y + 7\}
$$

with $L = [x, y]$, and consider the simplification of $u = 2x + 3y - 4$ with respect to the side relations $f_1 = 0, f_2 = 0$. In this case, the side relations represent two circles that intersect at the single point $(x, y) = (-1, 2)$, and since u evaluates to 0 at this point, $u \overset{F}{=} 0$.

However, in this case we cannot reduce u to 0 even with a Gröbner basis. Indeed, a Gröbner basis for the ideal $I = <f_1, f_2>$ is

$$
G = \{g_1, g_2\} = \{-6x - 6, \ y^2 - 4y + 4\},
$$

and $u \overset{G}{\longrightarrow} 3y - 6$. □

This example shows that the relationship between equivalence and reduction is not as straightforward as we would like. However, a theoretical relationship between the two concepts is given in the following theorem.

Theorem 8.58. *Let $F = \{f_1, \ldots, f_n\}$ be a set of non-zero polynomials in $\mathbf{Q}[x_1, \ldots, x_p]$, and consider the side relations*

$$
f_1 = 0, \ldots, f_n = 0. \tag{8.97}
$$

In addition, let G be a Gröbner basis for $I = <f_1, \ldots, f_n>$. Then, $u \overset{F}{=} 0$ if and only if $u^m \overset{G}{\longrightarrow} 0$ for some positive integer m.

In other words, u simplifies to 0 within the context of the side relations in (8.97) if and only if u^m can be simplified to 0 using reduction with a Gröbner basis. The proof of the theorem, which depends on some involved concepts from the theory of ideals, is given in the references cited at the end of this chapter. □

Example 8.59. Consider the side relations and polynomial u from Example 8.57 above. In this case, $u^2 \overset{G}{\Longrightarrow} 0$ using the reduction step sequence

$$
\begin{array}{rcl}
u^2 & = & 4x^2 + 12x\,y - 16x + 9y^2 - 24y + 16 \\[4pt]
& \overset{[4x^2,\,g_1]}{\Longrightarrow} & 12x\,y - 20x + 9y^2 - 24y + 16 \\[4pt]
& \overset{[12x\,y,\,g_1]}{\Longrightarrow} & -20x + 9y^2 - 36y + 16 \\[4pt]
& \overset{[-20x,\,g_1]}{\Longrightarrow} & 9y^2 - 36y + 16 \\[4pt]
& \overset{[9y^2,\,g_2]}{\Longrightarrow} & 0.
\end{array}
$$

□

Exercises

1. Prove Theorem 8.43(3).

2. Use Buchberger's algorithm to find a Gröbner basis for each of the following ideals. In addition, remove any redundant polynomials from the Gröbner basis. In each case, the variable order is given by $L = [x, y]$.

   (a) $I = \,<x\,y - 1,\; x^2 + y^2 - 4>$.

   (b) $I = \,<x^2\,y - 1,\; x + y^2 - 4>$.

   (c) $I = \,<x^2 + x\,y + y^2,\; x\,y>$.

   (d) $I = \,<x^3\,y + x^2 - 1,\; x + y^2 + 1>$.

   (e) $I = \,<x^3\,y + x^2 - 1,\; x\,y^2 + y + 1>$.

   (f) $I = \,<x^2 - 2,\; x^2 + 2x\,y + 2,\; y^2 - 2>$.

3. Let $F = \{f_1, f_2, f_3\} = \{x\,y - 1,\; x\,z - 1,\; y\,z - 1\}$, and let I be the ideal generated by F.

   (a) Find a Gröbner basis for I.

   (b) Using the simplification algorithm in this section show that

   $$
   3\,x\,y\,z \overset{F}{=} x + y + z.
   $$

4. Prove Theorem 8.45.

5. Let G be a set of polynomials which is a Gröbner basis for an ideal, and let L be a list of symbols that defines the variable order. Give a procedure $Elim\_poly(G, L)$ that eliminates the redundant polynomials from G using the condition described in Theorem 8.45.

6. Suppose that $f_1, \ldots, f_n$ are non-zero polynomials in $\mathbf{Q}[x]$. Suppose that we use the *G_basis* procedure followed by *Elim_poly* procedure (Exercise 5) to find a Gröbner basis for $I = <f_1, \ldots, f_n>$. Explain why this process produces a basis with one polynomial which is a (non-monic) greatest common divisor of $f_1, \ldots, f_n$. *Hint:* See Exercise 3, page 333.

7. Consider the polynomial $u = x + 3y - 7$ and the two side relations

$$f_1 = x\,y - 2x - y + 2 = 0, \quad f_2 = x^4 - 4x^3 + 6x^2 - 4x + y^3 - 6y^2 + 12y - 7 = 0.$$

   (a) Show that $u \overset{\{f_1,f_2\}}{=} 0$ with respect to the side relations, but that this cannot be obtained by reduction with the Gröbner basis G for the ideal $I = <f_1, f_2>$.

   (b) Find a positive integer m such that $u^m \overset{G}{\longrightarrow} 0$.

8. Let $F = [f_1, \ldots, f_n]$ be a list of non-zero polynomials that is a basis for an ideal I, and let L be a list of variables that defines the variable order. Let u be a polynomial in I. Give a procedure

$$Represent\_basis(u, F, L)$$

that finds a representation for u in terms of the basis F. Notice that this can only be done directly with reduction when F is a Gröbner basis. Therefore, if F is not a Gröbner basis, you will need to modify the *G_basis* procedure so that it keeps track of a representation of each member of the Gröbner basis in terms of F, and then uses this information along with the representation of u in terms of a Gröbner basis to find the representation of u in terms of F.

9. Consider the ideal

$$I = <f_1, f_2> = <x\,y - \sqrt{2},\ x^2 + y^2 - 4\sqrt{2}>$$

   which is included in the domain $\mathbf{Q}\left(\sqrt{2}\right)[x, y]$. Find a Gröbner basis for I. *Hint:* $\alpha = \sqrt{2}$ satisfies $\alpha^2 - 2 = 0$.

10. Gröbner basis computation can be used to compute (non-monic) greatest common divisors of polynomials with one variable (Exercise 3, page 333). How can you use a Gröbner basis computation to compute the gcd of $u = x^2 - 2$ and $v = x^2 - 2\sqrt{2}x + 2$ in $\mathbf{Q}\left(\sqrt{2}\right)[x]$?

11. Consider the algebraic number $u = 1 + 2\sqrt{2} + 3\sqrt{3}$. How can you use the simplification algorithm in this section to find the multiplicative inverse of u in the field $\mathbf{Q}(\sqrt{2}, \sqrt{3})$?

Further Reading

In this chapter we have given an introduction to the concept of a Gröbner basis. There is, however, much more to this subject. A good starting point for this material is Geddes, Czapor, and Labahn [39], Chapter 10. Adams and Loustau-nau [1] is a comprehensive introduction to both the theory and applications of Gröbner bases. Our approach in Theorem 8.42, is similar to the one given in this book which contains the complete proof as well as the proofs for Theorems 8.50, 8.53 and 8.58. Becker, Weispfenning, and Kredel [6] gives a more advanced (and theoretical) discussion of Gröbner bases, including an algorithm to compute the power m in Theorem 8.58. Chou [20] describes how Gröbner bases are used for the mechanical development of proofs in geometry. The article by Buchberger [18], the inventor of Gröbner bases, is of historical interest.

9

Polynomial Factorization

The goal of this chapter is the complete description of a modern algorithm for the factorization of polynomials in $\mathbf{Q}[x]$ in terms of irreducible polynomials.

In Section 9.1 we describe an algorithm that obtains a partial factorization of a polynomial. The algorithm can separate factors of different multiplicities as in

$$x^5 + 7\,x^4 + 17\,x^3 + 17\,x^2 + 7\,x + 1 = (x+1)\left(x^2 + 3\,x + 1\right)^2,$$

but is unable to separate factors of the same multiplicity as in

$$x^3 + 4\,x^2 + 4\,x + 1 = (x+1)\left(x^2 + 3\,x + 1\right).$$

This factorization is important, however, because it reduces the factorization problem to polynomials without multiple factors. In Section 9.2 we describe the classical approach to factorization, which is known as Kronecker's algorithm. This algorithm is primarily of historical interest because it is much too slow to be used in practice. In Section 9.3 we describe an algorithm that factors polynomials in $\mathbf{Z}_p[x]$. Although this algorithm is important in its own right, it is included here because it plays a role in the modern approach for factorization in $\mathbf{Q}[x]$. Finally, in Section 9.4 we describe a modern factorization algorithm, known as the Berlekamp-Hensel algorithm, which uses a related factorization in $\mathbf{Z}_p[x]$ together with a lifting algorithm to obtain the factorization in $\mathbf{Q}[x]$.

The efficient factorization of polynomials is a difficult computational problem. For this reason, both the mathematical and computational techniques in this chapter are more involved than those in previous chapters.

9.1   Square-Free Polynomials and Factorization

A *square-free* polynomial is one without any repeated factors. In this section we examine the square-free concept and describe an algorithm that factors a polynomial that is not square-free in terms of square-free factors. This type of factorization is the easiest one to obtain and is an important first step in the irreducible factorization of polynomials.

Square-Free Polynomials

The formal definition of a square-free polynomial reflects the origin of the term "square-free."

Definition 9.1. *Let* $\mathbf{F}$ *be a field. A polynomial* u *in* $\mathbf{F}[x]$ *is* **square-free** *if there is no polynomial* v *in* $\mathbf{F}[\mathbf{x}]$ *with* $\deg(v, x) > 0$ *such that* $v^2 \mid u$.

Although the definition is expressed in terms of a squared factor, it implies that the polynomial does not have a factor of the form v^n with $n \geq 2$.

Example 9.2. The polynomial $u = x^2 + 3\,x + 2 = (x+1)\,(x+2)$ is square-free, while $u = x^4 + 7\,x^3 + 18\,x^2 + 20\,x + 8 = (x + 1)\,(x + 2)^3$ is not. $\qquad\square$

The square-free property can also be described in terms of the irreducible factorization of a polynomial.

Theorem 9.3. *Suppose that* u *is a polynomial in* $\mathbf{F}[x]$ *with an irreducible factorization*

$$u = c\,p_1^{n_1}\,p_2^{n_2}\cdots p_{s-1}^{n_{s-1}}\,p_s^{n_s}.$$

Then, u *is square-free if and only if* $n_i = 1$ *for* $1 \leq i \leq s$.

The proof, which is straightforward, is left to the reader (Exercise 4). $\square$

Theorem 9.5 (below) describes a simple test that determines if a polynomial is square-free. This test involves the derivative of a polynomial q in $\mathbf{F}[x]$, which is defined using the ordinary differentiation rules given in calculus. For a monomial $q = a\,x^n$, the derivative q' is given by

$$q' = \begin{cases} 0, & \text{if } n = 0, \\ \underbrace{(a + \cdots + a)}_{n}\,x^{n-1}, & \text{if } n > 0, \end{cases} \qquad (9.1)$$

where the notation indicates the sum has n copies of a. The derivative of a polynomial is defined as the sum of the derivatives of its monomials.

In Equation (9.1), we have resisted using the simpler expression $q' = n\,a\,x^n$ because the integer n may not be in the field $\mathbf{F}$. For example, for $q = 2\,x^4 + x + 2$ in $\mathbf{Z}_3[x]$,

$$q' = (2 + 2 + 2 + 2)\,x^3 + 1 = 2\,x^3 + 1.$$

With this derivative definition as a staring point, we can obtain the ordinary sum and product rules for differentiation.

In the proof of Theorem 9.5 (below), we use the fact that a polynomial u with positive degree has a derivative which is not the zero polynomial. Unfortunately, for some coefficient fields $\mathbf{F}$, this property is not true. For example, in $\mathbf{Z}_3[x]$, we have

$$(x^3 + 1)' = (1 + 1 + 1)\,x^2 = 0 \cdot x^2 = 0.$$

However, if $\mathbf{F}$ contains the integers $\mathbf{Z}$, the property does hold.[1]

Theorem 9.4. *Let $\mathbf{F}$ be a field that contains $\mathbf{Z}$, and suppose that u is a polynomial in $\mathbf{F}[x]$ with positive degree. Then, u' is not the zero polynomial.*

Proof: Let $u = a_n\,x^n + \cdots + a_0$, where $n > 0$ and $a_n \neq 0$. For the leading monomial $a_n\,x^n$, the derivative is $n\,a_n x^{n-1}$. Since a field has no zero divisors, the coefficient $n\,a_n \neq 0$, and, therefore, the derivative is not the zero polynomial. $\square$

For the remainder of this section, we assume that all fields contain $\mathbf{Z}$.

Theorem 9.5. *Let $\mathbf{F}$ be a field that contains $\mathbf{Z}$, and let u be a polynomial in $\mathbf{F}[x]$. Then, u is square-free if and only if $\gcd(u, u') = 1$.*

Proof: We prove the theorem by proving its negation, namely, u is not square-free if and only if $\gcd(u, u') \neq 1$.

Let's suppose first that u is not square-free, which means there are polynomials v and w, with $\deg(v) > 0$, such that $u = v^2\,w$. Differentiating this expression, we have

$$u' = 2\,v\,v'\,w + v^2\,w'.$$

---

[1] In abstract algebra textbooks, a field $\mathbf{F}$ that contains the integers $\mathbf{Z}$ is called a field of *characteristic* 0.

Since v divides each term of the sum, it also divides u', and since v also divides u, it divides $\gcd(u, u')$ as well. Therefore, $\deg(\gcd(u, u')) \geq \deg(v) > 0$ which implies that $\gcd(u, u') \neq 1$.

We show that the implication goes in the other direction as well. If $\gcd(u, u') \neq 1$, there is an irreducible polynomial v with positive degree that divides both u and u'. Therefore,

$$u = q_1\, v, \qquad u' = q_2\, v, \tag{9.2}$$

and differentiating the first expression, we have

$$u' = q_1\, v' + q_1'\, v.$$

In this expression, v divides both u' and $q_1'\, v$, which implies that

$$v \mid q_1\, v'.$$

Since v is irreducible, it divides one of the terms in this product (Theorem 4.29(2), page 133). However, by Theorem 9.4, $v' \neq 0$, and since $\deg(v') < \deg(v)$, we have $v \nmid v'$. Therefore, $v \mid q_1$ and so, $q_1 = r\, v$. Substituting this expression into u, we obtain $u = r\, v^2$, which implies that u is not square-free. $\qquad\square$

Example 9.6. Consider the polynomial $u = x^3 + 1$ in $\mathbf{Q}[x]$. Then, $u' = 3\, x^2$ and $\gcd(u, u') = 1$ which means that u is square-free. However, u can be factored as $u = (x + 1)(x^2 - x + 1)$. $\qquad\square$

Since the square-free concept is so closely connected to the gcd concept, it is not surprising that the property does not depend on the size of the coefficient field. This is the content of the next theorem.

Theorem 9.7. *Suppose that $\mathbf{F}_1$ and $\mathbf{F}_2$ are fields with $\mathbf{Z}$ contained in $\mathbf{F}_1$ and $\mathbf{F}_1$ contained in $\mathbf{F}_2$. Then, if u is square-free as a polynomial in $\mathbf{F}_1[x]$, it is also square-free as a polynomial $\mathbf{F}_2[x]$.*

Proof: The proof follows from the observation that for u in $\mathbf{F}_1[x]$, the computation of $\gcd(u, u')$ (using differentiation and Euclid's algorithm) involves operations in $\mathbf{F}_1$, and so its value does not change if u is viewed as a polynomial in $\mathbf{F}_2[x]$. $\qquad\square$

Example 9.8. Since the polynomial $u = x^2 - 2$ is square-free in $\mathbf{Q}[x]$, it is also square-free in $\mathbf{Q}(\sqrt{2})[x]$. On the other hand, $x^2 - 2$ is irreducible in $\mathbf{Q}[x]$, but can be factored as $(x - \sqrt{2})(x + \sqrt{2})$ in $\mathbf{Q}(\sqrt{2})[x]$. $\qquad\square$

Square-Free Factorization

Theorem 9.9. *A polynomial u in $\mathbf{F}[x]$ has a unique factorization*

$$u = c\, s_1\, s_2^2 \cdots s_m^m \tag{9.3}$$

*where c is in $\mathbf{F}$ and each s_i is monic and square-free with $\gcd(s_i, s_j) = 1$ for $i \neq j$. The factorization in Equation (9.3) is called the **square-free factorization** of u.*

Example 9.10. The polynomial

$$u = 2\, x^7 - 2\, x^6 + 24\, x^5 - 24\, x^4 + 96\, x^3 - 96\, x^2 + 128\, x - 128$$

has a square-free factorization $2\,(x - 1)\,(x^2 + 4)^3$ where $c = 2$, $s_1 = x - 1$, $s_2 = 1$, and $s_3 = x^2 + 4$. Notice that a square-free factorization may not contain all the powers in Equation (9.3). $\qquad\square$

Example 9.11. A square-free factorization only involves the square-free factors of a polynomial and leaves the deeper structure that involves the irreducible factors intact. For example, in the square-free factorization

$$x^6 - 9\, x^4 + 24\, x^2 - 16 = (x^2 - 1)\,(x^2 - 4)^2,$$

both factors are reducible. $\qquad\square$

Proof of Theorem 9.9: Suppose that u has the factorization in terms of distinct, monic, irreducible polynomials

$$u = c\, p_1^{n_1} \cdot p_2^{n_2} \cdots p_{r-1}^{n_{r-1}} \cdot p_r^{n_r}. \tag{9.4}$$

Define the sequence of polynomials

$$
\begin{aligned}
s_1 &= \text{product of the } p_j \text{ in Equation(9.4) such that } n_j = 1, \\
s_2 &= \text{product of the } p_j \text{ in Equation(9.4) such that } n_j = 2, \\
s_3 &= \text{product of the } p_j \text{ in Equation(9.4) such that } n_j = 3, \\
&\;\;\vdots
\end{aligned}
$$

Since s_i is the product of distinct, irreducible, monic polynomials, it is square-free and has no factors in common with the other s_j ($i \neq j$). In addition, since each of the irreducible factors in Equation (9.4) is a factor of some s_i, the square-free factorization is given by

$$u = c\, s_1\, s_2^2 \cdots s_m^m, \tag{9.5}$$

where $m = \max(\{n_1, \ldots, n_r\})$. Finally, since the irreducible factorization in Equation (9.4) is unique, the square-free factorization is unique as well. $\qquad\square$

A Square-Free Factorization Algorithm

Let $\mathbf{F}$ be a field that contains the integers, and let u be in $\mathbf{F}[x]$. In order to develop the square-free factorization algorithm, we need to express $\gcd(u, u')$ in terms of the square-free factors of u. Let's see what this relationship looks like in a simple case.

Suppose that u has a square-free factorization

$$u = c\,r\,s^2\,t^3, \tag{9.6}$$

where c is in $\mathbf{F}$ and r, s, and t are monic, square-free, and relatively prime polynomials. Differentiating u we have

$$
\begin{aligned}
u' &= c\,(r'\,s^2\,t^3 + 2\,r\,s\,s'\,t^3 + 3\,r\,s^2\,t^2\,t') \\
   &= c\,s\,t^2\,(r'\,s\,t + 2\,r\,s'\,t + 3\,r\,s\,t').
\end{aligned} \tag{9.7}
$$

From Equations (9.6) and (9.7), we see that $s\,t^2$ divides both u and u', and, therefore,

$$s\,t^2 \mid \gcd(u, u'). \tag{9.8}$$

In fact, we can show that $\gcd(u, u') = s\,t^2$. If $\gcd(u, u') \neq s\,t^2$, then the two polynomials

$$u_1 = \frac{u}{s\,t^2} = c\,r\,s\,t,$$

$$u_2 = \frac{u'}{s\,t^2} = c\,(r'\,s\,t + 2\,r\,s'\,t + 3\,r\,s\,t') \tag{9.9}$$

have a common monic, irreducible factor w with $\deg(w) > 0$. However, if $w \mid u_1$, it follows from Theorem 4.29(2) that w divides exactly one of the (relatively prime, square-free) polynomials r, s, and t. Let's suppose that $w \mid r$. (If w divides s or t, the argument is the same.) Since r is square-free, Theorem 9.5 implies that $w \nmid r'$, and since r, s, and t are relatively prime, w does not divide s or t. Therefore, by Theorem 4.29(2),

$$w \nmid c\,r'\,s\,t. \tag{9.10}$$

On the other hand, from Equation (9.9),

$$u_2 - 2\,c\,r\,s'\,t - 3\,c\,r\,s\,t' = c\,r'\,s\,t,$$

where w divides the left side of the equation. Since this statement and (9.10) cannot both be true, the polynomial w does not exist, and therefore,

$$\gcd(u, u') = s\,t^2. \tag{9.11}$$

The general form of Equation (9.11) is given in the following theorem.

Theorem 9.12. *Suppose that* $\mathbf{F}$ *is a field that contains* $\mathbf{Z}$, *and suppose that* u *in* $\mathbf{F}[x]$ *has the square-free factorization* $u = c\, s_1\, s_2^2 \cdots s_m^m$ *with* $m \geq 2$. *Then,*

$$\gcd(u, u') = \prod_{i=2}^{m} s_i^{i-1}. \tag{9.12}$$

Proof: The general proof, which follows the lines of the above argument, is left to the reader (Exercise 7).   □

To motivate the square-free factorization algorithm, suppose that u is a monic polynomial with a square-free factorization

$$u = q\, r^2\, s^3\, t^4,$$

where q, r, s, and t are monic, square-free, and relatively prime polynomials. We show that it is possible to find the factors q, r, s and t using only the polynomial division and greatest common divisor operations.

First, let's see how to obtain the factor q. Using Theorem 9.12, let

$$R = \gcd(u, u') = r\, s^2\, t^3 \tag{9.13}$$

where the exponent of each factor in R is one less than the corresponding exponent in u. Next, using polynomial division, we obtain a polynomial that is the product of each polynomial in the factorization:

$$F = \mathrm{quot}(u, R) = q\, r\, s\, t. \tag{9.14}$$

Let

$$G = \gcd(R, F) = r\, s\, t, \tag{9.15}$$

where all the factors except q appear to the first power. Finally, we obtain q with the division

$$\frac{F}{G} = \mathrm{quot}(q\, r\, s\, t,\ r\, s\, t) = q. \tag{9.16}$$

To obtain the next factor r, we redefine R by dividing the old value of R (from Equation (9.13)) by the value for G (from Equation (9.15)),

$$R = \mathrm{quot}(r\, s^2\, t^3,\ r\, s\, t) = s\, t^2, \tag{9.17}$$

and redefine F as the current value of G in Equation (9.15):

$$F = r\, s\, t. \tag{9.18}$$

Procedure *Square_free_factor*(u, x);
Input
    u : a polynomial in $\mathbf{F}[x]$ where all field operations
        in $\mathbf{F}$ are obtained with automatic simplification
        and $\mathbf{Z}$ is contained in $\mathbf{F}$;
    x : a symbol;
Output
    The square-free factored form of the polynomial u;
Local Variables
    U, c, P, R, F, j, G, s;
Begin

```
1      if  u = 0 then  Return(u)
2      else
3          c := Leading_Coefficient_gpe(u, x);
4          U := Algebraic_expand(u/c);
5          P := 1;
6          R := Polynomial_gcd(U, Derivative(U, x), x);
7          F := Quotient(U, R, x);
8          j := 1;
9          while R ≠ 1 do
10             G := Polynomial_gcd(R, F, x);
11             s := Quotient(F, G, x);
12             P := P * s^j;
13             R := Quotient(R, G, x);
14             F := G;
15             j := j + 1;
16         P := P * F^j;
17         Return(c * P)
```

End

Figure 9.1. The MPL square-free factorization procedure. (Implementation: Maple (txt), Mathematica (txt), MuPAD (txt).)

Then, repeating the operations in Equations (9.15) and (9.16), we obtain the polynomial r:

$$ G \;=\; \gcd(R, F) = s\,t, \qquad \frac{F}{G} = \text{quot}(r\,s\,t, \; s\,t) = r. $$

Continuing in this fashion we find all the components in the factorization.

A procedure that obtains the square-free factorization using these operations is given in Figure 9.1. In lines 3 and 4 we remove the leading coefficient from u. The variable P, which contains the factored form, is initialized at line 5. Lines 6 and 7, which correspond to the operations in Equations (9.13) and (9.14), assign the initial values for R and F. The variable j in line 8 keeps track of the powers of the factors.

Lines 10 and 11, which correspond to the operations in Equations (9.15) and (9.16), obtain values for G and the factor s. The factor s^j is combined with P in line 12, and R, F, and j are redefined in lines 13–15. Notice that the original R (line 6) is a monic polynomial, and when R is redefined by polynomial division (line 13), it is still monic. Eventually $R = 1$, and the loop terminates (line 9).

To see what happens at this point, suppose that the square-free factorization for u is given by $u = s_1 \, s_2^2 \cdots s_m^m$. At the beginning of the last pass through the loop,

$$R = s_m, \qquad F = s_{m-1} \, s_m, \qquad j = m - 1.$$

Therefore, from lines 10–15,

$$G = s_m, \qquad s = s_{m-1}, \qquad P = s_1 \, s_2^2 \cdots s_{m-1}^{m-1},$$
$$R = 1, \qquad F = s_m, \qquad j = m.$$

At this point the loop terminates, and this last factor s_m^m is combined with P in line 16.

Example 9.13. In this example, we trace the procedure for the polynomial

$$u = x^5 + 6\,x^4 + 10\,x^3 - 4\,x^2 - 24\,x - 16.$$

In this case, $c = 1$, $U = u$, $U' = 5\,x^4 + 24\,x^3 + 30\,x^2 - 8\,x - 24$, and

$$R = \gcd(U, U') = x^2 + 4\,x + 4, \qquad F = \mathrm{quot}(U, R) = x^3 + 2\,x^2 - 2\,x - 4.$$

Since $R \neq 1$, the first pass through the loop gives

$$G = \gcd(R, F) = x + 2, \qquad s = \mathrm{quot}(F, G) = x^2 - 2, \quad P = x^2 - 2,$$
$$R = \mathrm{quot}(R, G) = x + 2, \quad F = x + 2, \quad j = 2.$$

On the second pass through the loop,

$$G = x + 2, \qquad s = 1, \qquad P = x^2 - 2,$$
$$R = 1, \quad F = x + 2, \quad j = 3.$$

Since $R = 1$, the loop terminates, and the procedure returns the square-free factorization $c\,P\,F^3 = (x^2 - 2)(x + 2)^3$. $\square$

The square-free factorization algorithm described above does not work for polynomials in $\mathbf{Z}_p[x]$ because the coefficient field does not contain $\mathbf{Z}$. A square-free factorization algorithm for $\mathbf{Z}_p[x]$ is given in Section 9.3.

Square-Free Factorization in Z[x]

Let's consider briefly the square-free factorization of polynomials in $\mathbf{Z}[x]$ using the gcd operations described in Section 6.3. In this case, both the theorems and algorithm are similar to those given above with modifications to account for the form of the gcd in this setting.

Since a polynomial in $\mathbf{Z}[x]$ with positive degree has a non-zero derivative, a theorem similar to Theorem 9.5 is valid. However, since the gcd in $\mathbf{Z}[x]$ is unit normal rather than monic, the theorem takes the following form.

Theorem 9.14. *Let u be a polynomial in $\mathbf{Z}[x]$. Then, u is square-free if and only if* $\deg(\gcd(u, u')) = 0$.

In this setting, u has a unique square-free factorization of the form

$$u = c\, s_1\, s_2^2 \cdots s_m^m, \tag{9.19}$$

where[2]

$$c = \operatorname{sign}(\operatorname{lc}(u))\operatorname{cont}(u), \tag{9.20}$$

and the s_i are square-free, primitive, unit normal, and relatively prime (Exercise 12). In this setting, the analogue of Equation (9.12) is

$$\gcd(u, u') = |c| \prod_{i=2}^{m} s_i^{i-1},$$

where the absolute value guarantees that the gcd is unit normal.

Example 9.15. Let

$$u = -750\, x^4 - 525\, x^3 + 90\, x^2 + 132\, x + 24.$$

Then, $\operatorname{cont}(u) = 3$, and u has the square-free factorization

$$u = -3\, (2\, x - 1)\, (5\, x + 2)^3.$$

In order that each factor be unit normal, it is necessary to take

$$c = \operatorname{sign}(\operatorname{lc}(u))\operatorname{cont}(u) = \operatorname{sign}(-750) \cdot 3 = -3. \qquad \square$$

A procedure that finds the factorization in Equation (9.19) in $\mathbf{Z}[x]$ is similar to the one in Figure 9.1 with two changes. First, c is initialized using Equation (9.20). In addition, the gcd operators in lines 6 and 10 are replaced by the operator *Mv_poly_gcd* (page 249). The details of the procedure are left to the reader (Exercise 13).

---

[2] In Equation (9.20), the sign operator returns $+1$ for a positive rational number, -1 for a negative rational number, and 0 for the number 0.

Exercises

1. Determine whether or not each of the polynomials is square-free.

   (a) $u = x^3/3 + x^2/2 + 1$ in $\mathbf{Q}[x]$.

   (b) $u = 2x^3 + 9x^2 + 12x + 24$ in $\mathbf{Q}[x]$.

   (c) $u = x^3 - 6\sqrt{2}x^2 + 22x - 12\sqrt{2}$ in $\mathbf{Q}(\sqrt{2})[x]$.

2. For which real numbers t is the polynomial $x^3 + tx^2 + 8x + 4$ square-free?

3. Let n be a positive integer, and let $c \neq 0$ be a rational number. Show that $u = x^n - c^n$ is square-free.

4. Prove Theorem 9.3.

5. Let $\mathbf{F}$ be a field that contains the integers, and let u and $v \neq 0$ be polynomials in $\mathbf{F}[x]$.

   (a) For an integer $n \geq 2$, show that if $v^n \mid u$, then $v^{n-1} \mid \gcd(u, u')$.

   (b) For $n \geq 2$ an integer and v irreducible, show that if $v^{n-1} \mid \gcd(u, u')$, then $v^n \mid u$. Give an example that shows this statement may not hold when v is reducible.

   (c) Show that the polynomial $\mathrm{quot}(u, \gcd(u, u'))$ is square-free.

   (d) Show that any irreducible factor of u is also an irreducible factor of
   $$\mathrm{quot}(u, \gcd(u, u')).$$

6. Let $u(x)$ have the square-free factorization $u = c\, s_1\, s_2^2 \cdots s_m^m$. Show that
   $$\deg(u, x) = \deg(s_1, x) + 2\deg(s_2, x) + \cdots + m\deg(s_m, x).$$

7. Prove Theorem 9.12.

8. Prove that u in $\mathbf{Q}[x]$ is square-free if and only if it has no multiple roots that are complex numbers.

9. Using the algorithm described in this section, find a square-free factorization for each of the following polynomials.

   (a) $u = x^3 + x^2 - 8x - 12$ in $\mathbf{Q}[x]$.

   (b) $u = x^3 - 7\sqrt{2}x^2 + 32x - 24\sqrt{2}$ in $\mathbf{Q}(\sqrt{2})[x]$.

10. Give a procedure $Alg\_square\_free\_factor(u, x, p, \alpha)$ that finds the square-free factorization for polynomials in $\mathbf{Q}(\alpha)[x]$ where α is an algebraic number with minimal polynomial p.

11. Another algorithm that obtains the square-free factorization is suggested by the following manipulations. Suppose that u is a monic polynomial in $\mathbf{F}[x]$ with a square-free factorization $u = q\, r^2\, s^3\, t^4$. Let
    $$R = \gcd(u, u') = r\, s^2\, t^3, \qquad F = \mathrm{quot}(u, R) = q\, r\, s\, t \qquad (9.21)$$

$$D = \text{quot}(u', R) - F' = q\,(r'\,s\,t + 2\,r\,s'\,t + 3\,r\,s\,t').$$

We obtain the first factor with the operation

$$s_1 = \gcd(F, D) = q. \tag{9.22}$$

To obtain the next factor, redefine F by dividing the old value of F (from Equation (9.21)) by the value for s_1 to obtain $F = \text{quot}(q\,r\,s\,t,\ q) = r\,s\,t$, and redefine D with

$$D = \text{quot}(D,\ s_1) - F' = r\,(s'\,t + 2\,r\,s\,t'),$$

where the D on the right side of the equation is the old value of D. Then, performing the operation in Equation (9.22), we get the next factor

$$s_2 = \gcd(F, D) = r.$$

Continuing in this fashion, we find all the components in the factorization. Give a procedure that finds the square-free factorization using these operations.

12. Let u be a polynomial in $\mathbf{Z}[x]$. Show that u has a square-free factorization of the form $u = c\,s_1\,s_2^2\,s_m^m$ where the s_i are square-free, primitive, unit normal, and relatively prime and $c = \text{sign}(\text{lc}(u))\,\text{cont}(u)$.

13. Give a procedure *Square_free_factor_Z*(u, x) that obtains the square-free factorization of u in $\mathbf{Z}[x]$ using the approach described on page 358.

9.2 Irreducible Factorization: The Classical Approach

In this section we describe a simple (but highly inefficient) algorithm that obtains the irreducible factorization for polynomials in $\mathbf{Q}[x]$.

First, we show that it is sufficient to give a factorization algorithm for polynomials in the simpler domain $\mathbf{Z}[x]$. Let

$$u = u_n x^n + \cdots + u_0$$

be a polynomial in $\mathbf{Q}[x]$, and let a_i and b_i be the numerator and denominator of the coefficients u_i. To obtain an equivalent factorization in $\mathbf{Z}[x]$, let M be the least common multiple[3] of the denominators,

$$M = \text{lcm}(b_0, b_1, \ldots, b_n), \tag{9.23}$$

---

[3]A procedure that obtains the least common multiple of two integers is described in Exercise 18 on page 36. The least common multiple of integers $b_0, \ldots, b_n$ is given recursively by

$$\text{lcm}(b_0, \ldots, b_n) = \text{lcm}(\text{lcm}(b_0, \ldots, b_{n-1}), b_n).$$

and define the polynomial in $\mathbf{Z}[x]$:

$$v = M\,u = (M\,a_n/b_n)\,x^n + \cdots + (M\,a_0/b_0). \qquad (9.24)$$

Then,

$$u = \frac{v}{M} = \frac{\mathrm{cont}(v)}{M}\,\mathrm{pp}(v), \qquad (9.25)$$

where $\mathrm{pp}(v)$ is a primitive polynomial in $\mathbf{Z}[x]$. In addition, if $\mathrm{pp}(v)$ has the irreducible factorization (in $\mathbf{Z}[x]$)

$$\mathrm{pp}(v) = p_1(x)\cdots p_r(x),$$

then a factorization of u is given by

$$u = \frac{\mathrm{cont}(v)}{M}\,p_1(x)\cdots p_r(x). \qquad (9.26)$$

Although the polynomials p_i are not necessarily monic, this form is similar to the one obtained by the factor operator in most computer algebra systems.

Example 9.16. If $u = (3/2)\,x^2 - 3/8$, then

$$M = 8, \qquad v = 12\,x^2 - 3, \qquad \mathrm{cont}(v) = 3, \qquad \mathrm{pp}(v) = 4\,x^2 - 1.$$

We have $u = (3/8)\,(4\,x^2 - 1) = (3/8)\,(2\,x - 1)(2\,x + 1)$. $\qquad\qquad \square$

Before describing a factorization algorithm, we need to verify a technical point. Although each polynomial p_i in Equation (9.26) is irreducible in $\mathbf{Z}[x]$, we need to show that it is also irreducible in the larger domain $\mathbf{Q}[x]$. This point is addressed in the following theorem.

Theorem 9.17. *Let u be an irreducible polynomial in $\mathbf{Z}[x]$ with positive degree. Then, u is also irreducible in $\mathbf{Q}[x]$.*

Proof: The proof is by contradiction. Suppose that $u = p\,q$, where p and q are in $\mathbf{Q}[x]$ with positive degree. We use p and q to find a factorization in $\mathbf{Z}[x]$.

Let M be the least common multiple of the denominators of the coefficients of p, and let N be the least common multiple of the denominators of q. Then, the polynomials

$$p_0 = M\,p, \qquad q_0 = N\,q$$

are in $\mathbf{Z}[x]$ with positive degree, and

$$(M \cdot N)\,u = p_0\,q_0.$$

In addition, if the integer $M \cdot N$ has the irreducible factorization

$$M \cdot N = i_1\,i_2 \cdots i_s,$$

then

$$i_1\,i_2 \cdots i_s\,u = p_0\,q_0.$$

Now, by Theorem 6.55 (page 242), i_1 divides either all of the coefficients of p_0 or all of the coefficients of q_0. If i_1 divides all of the coefficients p_0, we form a new polynomial p_1 in $\mathbf{Z}[x]$ by dividing i_1 into each coefficient of p_0. Therefore,

$$i_2\,i_3 \cdots i_s\,u = p_1\,q_0.$$

Continuing in this fashion, we can eliminate each integer factor i_j from the left side of the equation and represent u as a product of polynomials in $\mathbf{Z}[x]$ with positive degree. This contradicts the hypothesis that u is irreducible in $\mathbf{Z}[x]$. $\square$

Kronecker's Factorization Algorithm in Z[x]

Let u be a polynomial in $\mathbf{Q}[x]$ with positive degree. Using the algorithm described in Section 9.1 and Equation (9.25), the factorization problem is reduced to the factorization of primitive, square-free polynomials in $\mathbf{Z}[x]$. The algorithm described in this section also works, however, when u is not primitive or square-free. The approach we describe below is attributed to the German mathematician Leopold Kronecker (1823–1891).

Let $u(x)$ be a polynomial of degree $n \geq 2$, and suppose that $u(x)$ can be factored in $\mathbf{Z}[x]$ as

$$u(x) = q(x)\,v(x) \tag{9.27}$$

where both factors have positive degree. Since $\deg(q(x)) + \deg(v(x)) = n$, one of the factors (say $q(x)$) has degree $\leq n/2$, and since $n/2$ may not be an integer, let $s = \lfloor n/2 \rfloor$. The algorithm either finds a factor $q(x)$ with positive degree $\leq s$ or shows that $q(x)$ cannot be found. Let

$$q(x) = q_s\,x^s + q_{s-1}\,x^{s-1} + \cdots + q_0. \tag{9.28}$$

Since $q(x)$ has $s + 1$ unknown coefficients, we first determine its value at $s + 1$ values of x and then find its coefficients by solving a system of linear equations. We use the original polynomial $u(x)$ to find the x values and the corresponding $q(x)$ values.

Since $u(x)$ has degree n, the equation $u(x) = 0$ has at most n roots, and there are integers $x_1, x_2, \ldots, x_{s+1}$ such that

$$u(x_i) \neq 0, \quad i = 1, 2, \ldots, s + 1.$$

Although we don't know the value of $q(x_i)$, we do know from Equation (9.27) that $q(x_i) \mid u(x_i)$. If we let S_i represent the set of (positive and negative) divisors of $u(x_i)$, then

$$q(x_i) \in S_i, \quad i = 1, 2, \ldots, s + 1. \tag{9.29}$$

Since we don't know the values $q(x_i)$, the best that we can do is try the various possibilities from the sets S_i. To determine a possible candidate for $q(x)$, we choose a collection of values

$$r_i \in S_i, \quad i = 1, 2, \ldots, s + 1,$$

and find the polynomial $q(x)$ that passes through the points

$$(x_1, r_1), (x_2, r_2), \ldots, (x_{s+1}, r_{s+1}).$$

If $q(x)$ is a divisor of $u(x)$ with positive degree, then we have found a factor of $u(x)$. If not, we repeat this process for another set of values $r_1, \ldots, r_{s+1}$. We keep doing this until we either have found a factor or have exhausted all possible choices of the values r_i.

If we don't find a factor by this method, the original polynomial is irreducible. If we do find a factor, then we divide $q(x)$ into $u(x)$ and obtain the factorization $u(x) = q(x) v(x)$. However, since either $q(x)$ or $v(x)$ may have coefficients that are fractions (rather than integers), we first find an equivalent factorization in terms of polynomials with integer coefficients (Exercise 3), and then complete the factorization by applying the process recursively to these new factors.

Example 9.18. Consider the polynomial

$$u(x) = x^4 + 2x^3 + 2x^2 + 2x + 1.$$

Since u has degree 4, we search for a factor with degree $s = \lfloor 4/2 \rfloor = 2$:

$$q(x) = q_2 x^2 + q_1 x + q_0. \tag{9.30}$$

To find the coefficients q_1, q_2, and q_3, we need three values of x where $u(x) \neq 0$. These values are given by $u(0) = 1$, $u(1) = 8$, and $u(2) = 45$, and therefore,

$$\begin{aligned}
S_1 &= \{\pm 1\}, \\
S_2 &= \{\pm 1, \pm 2, \pm 4, \pm 8\}, \\
S_3 &= \{\pm 1, \pm 3, \pm 5, \pm 9 \pm 15, \pm 45\}.
\end{aligned}$$

Next choose three values r_i, one from each S_i, substitute $(0, r_1)$, $(1, r_2)$, and $(2, r_2)$ in Equation (9.30), and solve the linear system for the coefficients q_0, q_1, and q_2. For example, for $r_1 = 1$, $r_2 = -1$, and $r_3 = 3$, we obtain $q(x) = 3x^2 - 5x + 1$ which (by polynomial division) is not a factor of u. However, for $r_1 = 1$, $r_2 = 2$, and $r_3 = 3$, we obtain $q(x) = x + 1$ which is a factor, and, by polynomial division

$$u = (x+1)(x^3 + x^2 + x + 1). \tag{9.31}$$

The cubic polynomial in Equation (9.31) can be factored further with a recursive application of the process as

$$(x^3 + x^2 + x + 1) = (x+1)(x^2 + 1).$$

The quadratic polynomial $x^2 + 1$ cannot be factored further, and so the factorization is given by $u(x) = (x+1)^2 (x^2 + 1)$. $\square$

Procedures for Kronecker's algorithm are given in Figures 9.2 and 9.3. The main procedure is the *Kronecker* procedure shown in Figure 9.2. At

---

> **Procedure** *Kronecker*(u, x);
> **Input**
> u : a polynomial in $\mathbf{Z}[x]$;
> x : a symbol;
> **Output**
>   an irreducible factorization of u in $\mathbf{Z}[x]$;
> **Local Variables**
> n, s, x_u_values, S_sets;
> **Begin**
> 1 $n := Degree\_gpe(u, x)$;
> 2   **if** $n \leq 1$ **then**
> 3 $Return(u)$
> 4   **else**
> 5 $s := Floor(n/2)$;
> 6 $x\_u\_values := Find\_x\_u\_values(u, s, x)$;
> 7 $S\_sets := Find\_S\_sets(x\_u\_values)$;
> 8 $Return(Kronecker\_factors(S\_sets, u, x))$
> **End**

Figure 9.2. The main MPL procedure for Kronecker's polynomial factorization algorithm for polynomials in $\mathbf{Z}[x]$. (Implementation: Maple (txt), Mathematica (txt), MuPAD (txt).)

Procedure $Kronecker\_factors(S\_sets, u, x)$;
Input
    u : a polynomial in $\mathbf{Z}[x]$;
    S_sets : a list of lists of the form $[x_i, S_i]$ where S_i is
           the set of positive and negative divisors of $u(x_i)$;
    x : a symbol;
Output
    an irreducible factorization of u in $\mathbf{Z}[x]$;
Local Variables $c, m, N, i, j, S, points, X, q, v, k, w, g, p$;
Begin

```
1    N := 1;
2    for  i := 1 to  Number_of_operands(S_sets) do
3        c[i] := 1;
4        m[i] := Number_of_operands(Operand(Operand(S_sets, i), 2));
5        N := N * m[i];
6    k := 1;
7    while  k ≤ N do
8        points := [ ];
9        for  j := 1 to  Number_of_operands(S_sets) do
10           w := Operand(S_sets, j);
11           X := Operand(w, 1);
12           S := Operand(w, 2);
13           points := Adjoin([X, Operand(S, c[j])], points);
14       q := Lagrange_polynomial(points, x);
15       if  Degree_gpe(q, x) > 0 then
16           g := Polynomial_division(u, q, x);
17           if  Operand(g, 2) = 0 then
18               v := Operand(g, 1);
19               p := Find_integer_factors(q, v, x);
20               Return(Kronecker(Operand(p, 1), x)
                         * Kronecker(Operand(p, 2), x));
21       j := 1;
22       while  j ≤ Number_of_operands(S_sets) do
23           if  c[j] < m[j] then
24               c[j] := c[j] + 1;
25               j := Number_of_operands(S_sets) + 1;
26           else
27               c[j] := 1;  j := j + 1;
28       k := k + 1;
29    Return(u)
```
End

Figure 9.3. The MPL $Kronecker\_factors$ procedure. (Implementation: Maple (txt), Mathematica (txt), MuPAD (txt).)

line 6, the procedure *Find_x_u_values* obtains a list of lists

$$[[x_1, u(x_1)], \ldots, [x_{s+1}, u(x_{s+1})]], \tag{9.32}$$

where x_i is an integer and $u(x_i) \neq 0$. At line 7, the procedure *Find_S_sets* obtains another list

$$[[x_1, S_1], \ldots, [x_{s+1}, S_{s+1}]] \tag{9.33}$$

where S_i is the set of positive and negative divisors of $u(x_i)$. The details of both of these procedures are left to the reader (Exercises 4 and 5).

The *Kronecker_factors* procedure, which is the heart of the algorithm, is invoked at line 8. This procedure is given in Figure 9.3. The procedure searches for a factor of u by choosing, in a systematic way, values u_i from the sets S_i and creating and testing a trial factor that passes through the points

$$[[x_1, u_1], \ldots, [x_{s+1}, u_{s+1}]].$$

The array $c[i]$, $i = 1, 2, \ldots, s + 1$ keeps track of which divisor is selected from each S_i, while the array $m[i]$ contains the number of divisors in S_i. Both of these arrays are given values in lines 2–5. Since each $c[i]$ is initialized to 1, the first trial factor passes through the point $[x_i, Operand(S_i, 1)]$ for $1 \leq i \leq s + 1$. The variable N, which contains the number of ways a trial factor can be chosen, is also initialized in this loop.

The main loop of the procedure is given in lines 7–28. In lines 8–13, we construct a list called *points* that is used to find a trial divisor q in line 14. (The *Lagrange_polynomial* procedure is described in Exercise 6.) In lines 15–17, we check if q is a divisor with positive degree and, if so, divide q into u at line 16. If the remainder of this division is 0 (line 17), then v is the quotient (line 18). Since q and v may have coefficients that are not integers, at line 19 we first find the equivalent factors with integer coefficients using the *Find_integer_factors* procedure (Exercise 3), and then apply the *Kronecker* procedure recursively to the new polynomials and return the product of the factors (line 20). In lines 21–29, we increment the $c[j]$ values and j in preparation for trying another trial factor q. If we complete the main loop without finding a divisor, then we return the irreducible polynomial u (line 29).

Computational Efficiency of Kronecker's Algorithm

Although Kronecker's algorithm shows that polynomial factorization is an algorithmic process, it is notoriously slow. In fact, the algorithm is only practical for low degree polynomials with small coefficients, since in other cases the number of trial factors is prohibitively large.

Since polynomial factorization is such an important operation for computer algebra, it has attracted the attention of research mathematicians and computer scientists who have developed a number of more efficient (and much more involved) approaches to the problem. We describe a modern factorization algorithm in Section 9.4.

Kronecker's Algorithm for Multivariate Polynomials

Kronecker's algorithm has a natural extension to multivariate polynomials which we illustrate with an example.

Example 9.19. Consider the polynomial

$$u(x, y) = 3\,x\,y - y\,x^3 - x^2 + 2\,x + xy^2 + x^2\,y + y^2\,x^2.$$

Let's first express $u(x, y)$ as a polynomial in y with coefficients that are polynomials in x:

$$u(x, y) = \left(2\,x - x^2\right) + \left(3\,x + x^2 - x^3\right)\,y + \left(x + x^2\right)\,y^2.$$

We can simplify the process with

$$\mathrm{cont}(u, y) = \gcd\left(2\,x - x^2,\ 3\,x + x^2 - x^3,\ x + x^2\right) = x$$

and the preliminary factorization

$$u = \mathrm{cont}(u, y)\,\mathrm{pp}(u, y) = x \cdot \left((2 - x) + \left(3 + x - x^2\right)\,y + (1 + x)y^2\right).$$
$$(9.34)$$

In this case, $\mathrm{cont}(u, y)$ is irreducible. If this were not so, it could be factored with the univariate version of Kronecker's algorithm.

Let

$$w(x, y) = \mathrm{pp}(u, y) = \left((2 - x) + \left(3 + x - x^2\right)\,y + (1 + x)y^2\right).$$

Observe that $w(x, y)$ is a quadratic in y, and so, if it can be factored, both of the factors must be linear in y. Let

$$q(x, y) = m(x)\,y + b(x) \qquad (9.35)$$

be one of the linear factors, and suppose that

$$w(x, y) = q(x, y)\,v(x, y)$$

where $v(x, y)$ is also linear in y. To find the two coefficients $m(x)$ and $b(x)$, we imitate what was done in the univariate case, but now the unknown coefficients are polynomials in x. To do this, we need to find the values of $q(x, y)$ at two values $y = y_1$ and $y = y_2$, and then, by substituting $(y_1, q(x, y_1))$ and $(y_2, q(x, y_2))$ into Equation (9.35), we obtain two linear equations for the two unknown polynomials $m(x)$ and $b(x)$. We use $w(x, y)$ to find the two y values and the corresponding $q(x, y)$ values.

As in the univariate case, we need two y values with $w(x, y) \neq 0$. In this case,

$$w(x, 0) = 2 - x, \qquad w(x, 1) = 6 + x - x^2$$

will do. Since $q(x, y)$ is a divisor of $w(x, y)$, we must find all the divisors of $w(x, 0)$ and $w(x, 1)$. Since these expressions are polynomials in x, we can find the divisors of these polynomials by first factoring them (using Kronecker's algorithm for univariate polynomials) and then finding all divisors of the polynomials (Exercise 12, page 125). Therefore,

$$q(x, 0) \quad \in \quad S_1 = \{\pm 1, \ \pm (2 - x)\},$$
$$q(x, 1) \quad \in \quad S_2 = \{\pm 1, \ \pm (2 + x), \ \pm (3 - x), \ \pm (2 + x)(3 - x)\}.$$

To find a trial factor $q(x, y)$, we choose an expression r_1 from S_1 and an expression r_2 from S_2. Therefore, substituting

$$y = 0, \quad q(x, y) = r_1$$

and

$$y = 1, \quad q(x, y) = r_2$$

into Equation (9.34), we obtain two equations for $m(x)$ and $b(x)$. If the solution $q(x, y)$ is a divisor of $w(x, y)$, then $q(x, y)$ is a factor. If not we repeat the process with other values for u_1 and u_2. For example, if we choose

$$r_1 = 2 - x, \quad r_2 = 3 - x,$$

we obtain the two equations

$$2 - x = m(x) \cdot 0 + b(x), \quad 3 - x = m(x) \cdot 1 + b(x).$$

Solving these equations we obtain $b(x) = 2 - x$ and $m(x) = 1$, and so a trial factor is $q(x, y) = y + (2 - x)$. In this case $q(x, y)$ is a factor, and by dividing $q(x, y)$ into $w(x, y)$ we obtain the factorization

$$w(x, y) = (y + 2 - x)(x y + y + 1).$$

Therefore, using Equation (9.34), the original polynomial $u(x, y)$ can be factored as

$$u(x, y) = x(y + 2 - x)(x y + y + 1). \qquad \square$$

Exercises

1. Factor each of the following using Kronecker's algorithm.

    (a) $x^3 + 2x^2 + 2x + 1$

    (b) $x^3 + x^2 + x + 1$

2. Use Kronecker's algorithm to show that $u = x^2 + 1$ cannot be factored in $\mathbf{Z}[x]$.

3. Let u be a polynomial in $\mathbf{Z}[x]$, and let $u = v\,w$ be a factorization in terms of rational coefficients. Give a procedure

$$Find\_integer\_factors(v, w, x)$$

    that returns a list $[v_1, w_1]$ where $u = v_1 \cdot w_1$ is a factorization in $\mathbf{Z}[x]$.

4. Let u be a polynomial in $\mathbf{Z}[x]$ with positive degree, x a symbol, and $s > 0$ an integer. Give a procedure

$$Find\_x\_u\_values(u, x, s)$$

    that returns the list (9.32), where x_i is an integer and $u(x_i) \neq 0$.

5. Let L be a list of the form (9.32). Give a procedure

$$Find\_S\_sets(L)$$

    that returns the list (9.33). (Exercise 6 on page 34 is useful in this exercise.)

6. Let $P = [[x_1, y_1], \ldots, [x_{r+1}, y_{r+1}]]$ be a list of 2 element lists, where x_i and y_i are rational numbers. The *Lagrange interpolation polynomial* that passes through these points is given by

$$L(x) = \sum_{i=1}^{r+1} y_i L_i(x).$$

    where

$$L_i(x) = \frac{(x - x_1) \cdots (x - x_{i-1})(x - x_{i+1}) \cdots (x - x_{r+1})}{(x_i - x_1) \cdots (x_i - x_{i-1})(x_i - x_{i+1}) \cdots (x_i - x_{r+1}).}$$

    Give a procedure *Lagrange_polynomial*(P, x) that returns the polynomial $L(x)$. For example,

$$Lagrange\_polynomial([[1, 1], [2, -1]], x) \to -2x + 3.$$

7. Give a procedure

$$Find\_Z\_poly(u, x)$$

    that returns the list $[\text{cont}(v)/M, \text{pp}(v)]$ where M is given by Equation (9.23) and v is given by Equation (9.24).

8. Give a procedure

$$Factor\_sv(u, x)$$

that finds the irreducible factorization of a polynomial u in $\mathbf{Q}[x]$. (The suffix "sv" stands for "single variable.") First, find a square-free factorization of u in $\mathbf{Q}[x]$. Then, use Kronecker's algorithm to factor the primitive part of the integer-coefficient version of each square-free factor of u (see Equations (9.23)-(9.26)). (Exercise 7 is useful in this exercise.)

In Exercise 18 on page 429, we suggest a modification of this procedure that finds the irreducible factorization in $\mathbf{Z}[x]$ using the approach described in Section 9.4.

9. Use Kronecker's algorithm to factor $u = y^2 + x\,y + y + x$.

9.3 Factorization in $Z_p[x]$

Let $p > 1$ be a prime number. In this section we describe an algorithm that finds the irreducible factorization of a polynomial u in $\mathbf{Z}_p[x]$. First, we obtain a square-free factorization of u by using a modification of the approach described in Section 9.1, and then complete the process by finding the irreducible factorization of each of the square-free factors. Since the finite field $\mathbf{Z}_p$ has a richer algebraic structure than the infinite integral domain $\mathbf{Z}$, the irreducible factorization problem is simpler in $\mathbf{Z}_p[x]$ than in $\mathbf{Z}[x]$.

Although the factorization problem in $\mathbf{Z}_p[x]$ is important in its own right, we introduce it here because it plays an essential role in the factorization algorithm in $\mathbf{Z}[x]$.

Notation Convention. In this section, to simplify the notation, we use the ordinary algebraic operators ($+$ and $\cdot$) for addition and multiplication of polynomials rather than the more cumbersome notation $\oplus_p$ and $\otimes_p$ introduced for $\mathbf{Z}_p$ in Section 2.3. For example, for $u(x)$ and $v(x)$ in $\mathbf{Z}_p[x]$, $u(x) + v(x)$ refers to the addition of polynomials where the arithmetic of corresponding coefficients of the polynomials is performed in $\mathbf{Z}_p$.

Algebraic Relationships in $Z_p[x]$

We begin by describing a number of algebraic relationships that apply to polynomials in $\mathbf{Z}_p[x]$.

Theorem 9.20. *Let $p > 1$ be a prime number. Then, in $\mathbf{Z}_p[x]$,*

$$x^p - x = x\,(x - 1)\cdots(x - (p - 2))\,(x - (p - 1)). \tag{9.36}$$

Proof: By Fermat's little theorem (Theorem 2.43(3), page 54), each b in $\mathbf{Z}_p$ is a root of $x^p - x = 0$. Since both sides of Equation (9.36) are monic and have the same degree and roots, the relationship follows. □

The relationship in the next theorem is the polynomial version of Theorem 2.43(2) on page 54. The proof is similar to the proof for that theorem.

Theorem 9.21. *Let $p > 1$ be a prime number, and let $u(x)$ and $v(x)$ be polynomials in $\mathbf{Z}_p[x]$. Then, $(u(x) + v(x))^p = u(x)^p + v(x)^p$.* □

The relationship in the next theorem plays a key role in the development of both the square-free and irreducible factorization algorithms in $\mathbf{Z}_p[x]$.

Theorem 9.22. *Let $p > 1$ be a prime number, and let $u(x)$ be in $\mathbf{Z}_p[x]$. Then, $u(x)^p = u(x^p)$.*

Proof: The proof uses mathematical induction on $n = \deg(u)$. For $n = 0$, the polynomial u is a constant, and so the theorem follows from Fermat's little theorem (page 54). Suppose that $n > 0$, and assume the induction hypothesis that the theorem holds for polynomials with degree $\leq n - 1$. We have

$$u(x) = u_n x^n + v(x)$$

where v represents all the terms in $u(x)$ that have degree $\leq n - 1$. Using Theorem 9.21 and the induction hypothesis applied to v, we have

$$
\begin{aligned}
u(x)^p &= (u_n x^n + v(x))^p = (u_n x^n)^p + v(x)^p \\
&= u_n^p \cdot (x^n)^p + v(x^p) = u_n \cdot (x^p)^n + v(x^p) \\
&= u(x^p).
\end{aligned}
$$
□

Square-Free Factorization Algorithm in $Z_p[x]$

In Section 9.1 we described an algorithm that obtains the square-free factorization for polynomials in $\mathbf{F}[x]$ when the field $\mathbf{F}$ contains the integers. In that setting, both the mathematics and the algorithm use the property that a polynomial with positive degree cannot have a zero derivative (Theorem 9.4, page 351). Since this property does not hold for all polynomials in $\mathbf{Z}_p[x]$ (e.g., for $p = 3$, $(x^3 + 1)' = 0$), the mathematical development and algorithm must be modified. Fortunately, there is a simple way to describe polynomials in $\mathbf{Z}_p[x]$ which have a zero derivative.

Theorem 9.23. *Let $p > 1$ be a prime number, and let u be a polynomial in $\mathbf{Z}_p[x]$. Then, $u' = 0$ if and only if $u = f(x^p)$, where $f(x)$ is in $\mathbf{Z}_p[x]$.*

Proof: If $u = f(x^p)$, then p divides the coefficient of each monomial in the derivative, and so $u' = 0$. Conversely, suppose that $u = a_n x^n + \cdots + a_0$ and $u' = 0$. This implies that p divides the coefficient of each monomial $a_i\, i\, x^{i-1}$ of u', and since p is a prime, by Theorem 2.16(2), we have $p \mid a_i$ or $p \mid i$. If $p \mid a_i$, then since $0 \le a_i \le p - 1$, we have $a_i = 0$. On the other hand, if $p \mid i$, the monomial in u has the form $a_i (x^p)^q$ where $q = \mathrm{iquot}(i, p)$. Since each non-zero monomial has this form, u has the form $f(x^p)$. $\square$

Using the last theorem and Theorem 9.22, we obtain the analogue of Theorem 9.5 in $\mathbf{Z}_p[x]$.

Theorem 9.24. *Let $p > 1$ be a prime number, and let u be a polynomial in $\mathbf{Z}_p[x]$. Then, u is square-free if and only if $\gcd(u, u') = 1$.*

Proof: The proof is similar to the one for Theorem 9.5 (page 351), with the following modification to the second half of the proof when u has a factor with positive degree and zero derivative. As in the earlier proof, suppose that $u = q_1 v$ with $\deg(v) > 0$, but now suppose that $v' = 0$. Then, by Theorems 9.22 and 9.23, $v = (f(x))^p$, which implies that u is not square-free. With this observation added to the proof of Theorem 9.5, we obtain a proof of the present theorem. $\square$

Suppose now that u has the square-free factorization

$$u = c \prod_{1 \le i \le m} s_i^i \tag{9.37}$$

where the polynomials s_i are monic, square-free, and relatively prime. An algorithm that obtains this factorization is similar to the one given in Section 9.1 with a modification that accounts for the possibility that some of the square-free factors may have zero derivative. Let

$$v = \prod_{\substack{1 \le i \le m \\ s_i' \ne 0}} s_i^i, \qquad w = \prod_{\substack{1 \le i \le m \\ s_i' = 0}} s_i^i,$$

where v and w are monic and

$$u = c\,v\,w. \tag{9.38}$$

Since $w' = 0$, we have $u' = c\,v'\,w$, and since each square-free factor of v has non-zero derivative, $\gcd(v, v')$ is obtained with manipulations similar to those given in Section 9.1. Therefore (using Exercise 11, page 144),

$$
\begin{aligned}
\gcd(u, u') &= \gcd(c\,v\,w, c\,v'\,w) \\
&= w \gcd(v, v') \\
&= w \prod_{\substack{2 \le i \le m \\ s_i' \ne 0}} s_i^{i-1}.
\end{aligned} \tag{9.39}
$$

A procedure that obtains the square-free factorization in $\mathbf{Z}_p[x]$ is similar to the one in Figure 9.1 (page 356) with three modifications. The following discussion refers to the variables in this procedure.

First, Equation (9.39) shows that the variable R assigned at line 6 of this procedure has w as a factor. In addition, this factor is not removed from R by the manipulations in the loop (lines 10–15). To see why, suppose that at line 4, $U = v\,w$. Then at lines 6–7,

$$
R = w \gcd(v, v'), \qquad F = \frac{U}{R} = \frac{v}{\gcd(v, v')}.
$$

In addition, in the **while** loop that begins at line 9, G at line 10 does not have w as a factor, and so R at line 13 does have w as a factor. Therefore, if line 9 is replaced by

while $Derivative(R, x) \ne 0$ **do**,

then at line 16, P contains the square-free factorization of v and $R = w$.

The second modification of the procedure obtains the square-free factorization of w. By Theorems 9.22 and 9.23, we can represent w as

$$
w(x) = f(x^p) = f(x)^p, \tag{9.40}
$$

where $f(x)$ is obtained by substituting $x^{1/p}$ for x in w. We obtain the square-free factorization of f with a recursive application of the procedure and then obtain the factorization for w by taking the factorization for f to the power p.

Finally, the square-free factorization for u is obtained using the factorizations for v and w and Equation (9.38). The details of the procedure are left to the reader (Exercise 1).

Irreducible Factorization of Square-Free Polynomials in $\mathbf{Z}_p[x]$

To complete the factorization problem, we must obtain the irreducible factorization of each of the monic, square-free polynomials s_i in Equation (9.37).

Let u be a monic square-free polynomial in $\mathbf{Z}_p[x]$ with $n = \deg(u)$. A simple approach to the factorization problem is obtained by mimicking the trial-factors approach used by Kronecker's algorithm. If u is reducible, it must have a monic factor of degree m with $1 \leq m \leq \mathrm{iquot}(n, 2)$, and since $\mathbf{Z}_p$ is finite, there are a finite number of trial polynomials of this type. Using polynomial division, we check each trial polynomial (starting with polynomials of degree 1), and if a factor f is found, we obtain the complete factorization by applying the algorithm recursively to quot(u,f). If none of the trial polynomials is a factor, then u is irreducible.

Although, the approach is not difficult to implement (Exercise 10), like Kronecker's algorithm, it is highly inefficient. Indeed, the number of trial factors is given by

$$\frac{p^{\mathrm{iquot}(n,2)+1} - p}{p - 1}$$

which is large even for small values of p (Exercise 2). For example, for $p = 11$ and $n = 16$, there are $235, 794, 768$ trial factors with degree $1 \leq m \leq 8$. There has got to be a better way!

A better approach is an algorithm, discovered in the late 1960s by E. R. Berlekamp, that obtains the factorization by reducing the problem to the solution of a system of linear equations followed by a number of gcd calculations.

Auxiliary Polynomials

The first step in the development of the algorithm is the formal definition of the auxiliary polynomials that are used in the gcd calculations. The Chinese remainder theorem (Theorem 4.41, page 140) plays a central role here since it guarantees (in a theoretical sense) the existence of these polynomials. However, the Chinese remainder procedure that finds the solution to the remainder equations has no computational role in the algorithm.

The auxiliary polynomials are defined in the following way. Let $u(x)$ be a monic and square-free polynomial in $\mathbf{Z}_p[x]$ with an irreducible factorization

$$u = u_1 \cdots u_r, \tag{9.41}$$

where each u_i is monic and has positive degree. Suppose that $a_1, \ldots, a_r$ are in $\mathbf{Z}_p$, and consider the polynomial $h(x)$ that satisfies the system of remainder equations and degree condition

$$\mathrm{rem}(h(x), u_i) = a_i \text{ for } 1 \leq i \leq r, \quad \deg(h) < \deg(u). \tag{9.42}$$

Since u is square-free, the u_i are distinct, and so the Chinese remainder theorem guarantees that there is a unique $h(x)$ that satisfies these conditions.

Example 9.25. Suppose that $p = 5$ and $u = x^6 + x^5 + x + 4$. Using the algorithm described later in this section, the irreducible factorization for u in $\mathbf{Z}_5[x]$ is given by

$$u = u_1\, u_2\, u_3 = (x^2 + x + 2)\,(x^2 + 2\,x + 3)\,(x^2 + 3\,x + 4).$$

Each choice of a_1, a_2, and a_3 gives a unique polynomial $h(x)$. For example, if $a_1 = 1$, $a_2 = 2$, and $a_3 = 3$, then using the algorithm suggested in the proof of the Chinese remainder theorem, we have

$$h = 4\,x^5 + 4\,x. \tag{9.43}$$

In a similar way, if $a_1 = 1$, $a_2 = 1$, and $a_3 = 2$, then

$$h = 2\,x^5 + x^4 + 4\,x^3 + 4\,x^2 + 2\,x + 1. \tag{9.44}$$

Finally, if $a_1 = 1$, $a_2 = 1$, and $a_3 = 1$, then

$$h = 1. \tag{9.45}$$

$\square$

The next theorem gives two important properties of the auxiliary polynomials $h(x)$.

Theorem 9.26. *Let $p > 1$ be a prime number, and let $u(x)$ be a monic, square-free polynomial in $\mathbf{Z}_p[x]$ with the irreducible factorization*

$$u = u_1 \cdots u_r.$$

1. *There are p^r distinct polynomials $h(x)$ that satisfy the remainder equations (9.42) where each sequence $a_1, \ldots, a_r$ gives a unique $h(x)$.*

2. *For each $h(x)$ that satisfies the remainder equations (9.42),*

$$u(x) = \prod_{j=0}^{p-1} \gcd(u(x), h(x) - j) \tag{9.46}$$

where the factors in the product are relatively prime.

Proof: To show (1), for each sequence $a_1, a_2, \ldots, a_r$, the Chinese remainder theorem shows that there is a unique $h(x)$ that satisfies the remainder equations (9.42). In addition, since $0 \le a_i \le p - 1$, there are p^r such sequences, and since each sequence gives a distinct polynomial, there are p^r such polynomials.

To show (2), first observe that the factors in Equation (9.46) are relatively prime since the polynomials $h(x) - j$, $j = 0, 1, \ldots, p-1$ are relatively prime (Exercise 11, page 125). Since both sides of Equation (9.46) are monic, we need only show that each side divides the other side. Let

$$H(x) = \prod_{j=0}^{p-1} \gcd(u(x), h(x) - j).$$

Since $h(x)$ satisfies the remainder equations (9.42), each irreducible factor u_i of u divides some $h(x) - j$, which implies that each $u_i \mid H(x)$. In addition, since the polynomials $u_1, \ldots, u_r$ are relatively prime, Theorem 4.29(3) (page 133) implies that $u(x) \mid H(x)$. On the other hand, each of the relatively prime factors of $H(x)$ divides $u(x)$, and so $H(x) \mid u(x)$. Therefore, we obtain Equation (9.46).   □

The relationship in Equation (9.46) shows that the auxiliary polynomials provide a way to obtain a factorization using gcd calculations. However, for some j, the expression $\gcd(u(x), h(x) - j)$ may be 1, and for other j, it may be reducible. Both of these points are illustrated in the next example.

Example 9.27. Suppose that $p = 5$, and consider the polynomial

$$u = x^6 + x^5 + x + 4$$

in $\mathbf{Z}_p[x]$ from Example 9.25. Using the h in Equation (9.43) and the procedure *Poly_div_p* (Exercise 9, page 125), the factorization in Equation (9.46) is

$$u = 1 \cdot (x^2 + x + 2)(x^2 + 2x + 3)(x^2 + 3x + 4) \cdot 1.$$

In this case, we obtain all three irreducible factors of u. However, using the h in Equation (9.44), we obtain

$$u = 1 \cdot 1 \cdot (x^4 + 3x^3 + 2x^2 + 2x + 1)(x^2 + 3x + 4) \cdot 1.$$

In this case, two of the irreducible factors of u coalesce into one factor. Finally, using the h given in Equation (9.45),

$$u = 1 \cdot (x^6 + x^5 + x + 4) \cdot 1 \cdot 1 \cdot 1.$$

In this case, Equation (9.46) does not obtain any of the irreducible factors of u.   □

Computation of $h(x)$

You may have noticed that there is something amiss about the polynomials $h(x)$. Although we have said that the polynomials are useful for finding the factors of u, their formal definition is in terms of the irreducible factors of u which are unknown. This predicament is addressed in the next theorem which provides another view of $h(x)$ in terms of the unfactored form of $u(x)$.

Theorem 9.28. Let $p > 1$ be a prime number, and let $u(x)$ be a square-free polynomial in $\mathbf{Z}_p[x]$ with positive degree. In addition, suppose that $h(x)$ is in $\mathbf{Z}_p[x]$ with $\deg(h) < \deg(u)$. Then, $h(x)$ satisfies the remainder equations (9.42) for some sequence $a_1, \dots, a_r$ if and only if

$$u \mid h(x)^p - h(x). \tag{9.47}$$

Proof: First, let's assume that $h(x)$ satisfies the remainder equations (9.42) and show that it satisfies the condition (9.47). We have by (9.42),

$$h(x) = q_i\, u_i + a_i, \quad \text{for } 1 \le i \le r, \tag{9.48}$$

where $q_i = \text{quot}(h, u_i)$ and $a_i = \text{rem}(h, u_i)$. Therefore, by Theorem 9.21 and Theorem 2.43(3) (page 54),

$$
\begin{aligned}
h(x)^p &= (q_i\, u_i + a_i)^p \\
&= q_i^p\, u_i^p + a_i^p \\
&= (q_i^p\, u_i^{p-1})\, u_i + a_i. \tag{9.49}
\end{aligned}
$$

From Equations (9.48) and (9.49), we have

$$h(x)^p - h(x) = (q_i^p\, u_i^{p-1} - q_i)\, u_i$$

which implies that $u_i \mid h(x)^p - h(x)$ for $1 \le i \le r$. Since the polynomials $u_1, \dots, u_r$ are relatively prime, Theorem 4.29(3) (page 133) implies that $u(x) \mid h(x)^p - h(x)$.

To show the converse statement, we assume that the condition (9.47) is true, and show that $h(x)$ satisfies the remainder equations (9.42). Substituting $h(x)$ for x in Equation (9.36) we obtain

$$h(x)^p - h(x) = h(x)\,(h(x) - 1) \cdots (h(x) - (p-1)) \tag{9.50}$$

where the factors in the product are relatively prime. Now Equations (9.47) and (9.50) imply that for each irreducible factor u_i of u,

$$u_i \mid h(x)\,(h(x) - 1) \cdots (h(x) - (p-1)),$$

and since u_i is irreducible, it divides exactly one factor on the right of this expression. Let a_i be the member of $\mathbf{Z}_p$ determined by this factor. Therefore, $u_i \mid h - a_i$ for $i = 1, \ldots, r$, and the uniqueness property for polynomial division implies that h satisfies the remainder equations $\mathrm{rem}(h, u_i) = a_i$. $\square$

A Linear System of Equations for $h(x)$

We show next how the alternate description of h in Theorem 9.28 is used to obtain a system of (dependent) linear equations which has as solutions all the auxiliary polynomials $h(x)$.

Let $n = \deg(u) \geq 1$. Since $\deg(h) < n$, each $h(x)$ has the form

$$h(x) = h_0 + h_1 x + \cdots + h_{n-1} x^{n-1}. \tag{9.51}$$

Let's set up a system of equations for the coefficients of h. First, by Theorem 9.22,

$$h(x)^p = h(x^p) = \sum_{j=0}^{n-1} h_j x^{p\,j},$$

and

$$h(x)^p - h(x) = \sum_{j=0}^{n-1} h_j \left(x^{p\,j} - x^j \right). \tag{9.52}$$

By polynomial division,

$$x^{p\,j} = Q_j(x)\, u(x) + r_j(x) \tag{9.53}$$

where $\deg(r_j(x)) \leq n - 1$. Substituting this expression into Equation (9.52), we have

$$
\begin{aligned}
h(x)^p - h(x) &= \sum_{j=0}^{n-1} h_j \left(Q_j(x)\, u(x) + r_j(x) - x^j \right) \\
&= \left(\sum_{j=0}^{n-1} h_j\, Q_j(x) \right) u(x) + \sum_{j=0}^{n-1} h_j \left(r_j(x) - x^j \right).
\end{aligned}
$$

Observe that the sum on the far right has degree $< n$, and therefore, using the condition (9.47) and the uniqueness property of polynomial division, we have

$$0 = \mathrm{rem}(h(x)^p - h(x), u(x)) = \sum_{j=0}^{n-1} h_j \left(r_j(x) - x^j \right). \tag{9.54}$$

Now, suppose that

$$r_j(x) = \sum_{i=0}^{n-1} r_{ij} x^i, \tag{9.55}$$

and let

$$\delta_{ij} = \begin{cases} 0, & i \neq j, \\ 1, & i = j. \end{cases}$$

Substituting Equation (9.55) into Equation (9.54), we obtain

$$\begin{aligned}
0 &= \sum_{j=0}^{n-1} h_j \left(\sum_{i=0}^{n-1} (r_{ij} x^i) - x^j \right) \\
&= \sum_{j=0}^{n-1} h_j \left(\sum_{i=0}^{n-1} (r_{ij} x^i - \delta_{ij} x^i) \right) \\
&= \sum_{i=0}^{n-1} \left(\sum_{j=0}^{n-1} h_j (r_{ij} - \delta_{ij}) \right) x^i,
\end{aligned}$$

where in the last sum the coefficient of each x^i is 0. This gives the following system of n linear equations for the coefficients of $h(x)$:

$$\sum_{j=0}^{n-1} h_j (r_{ij} - \delta_{ij}) = 0, \quad i = 0, 1, \ldots, n-1.$$

In matrix form the system of equations is

$$R\bar{h} = \bar{0}, \tag{9.56}$$

where

$$R = \begin{vmatrix} r_{0,0} - 1 & r_{0,1} & \cdots & r_{0,(n-1)} \\ r_{1,0} & r_{1,1} - 1 & \cdots & r_{1,(n-1)} \\ \vdots & \vdots & & \vdots \\ r_{(n-1),0} & r_{(n-1),1} & \cdots & r_{(n-1),(n-1)} - 1 \end{vmatrix} \tag{9.57}$$

and

$$\bar{h} = \begin{vmatrix} h_0 \\ h_1 \\ \vdots \\ h_{n-1} \end{vmatrix}, \quad \bar{0} = \begin{vmatrix} 0 \\ 0 \\ \vdots \\ 0 \end{vmatrix}. \tag{9.58}$$

The system of equations (9.56) is not independent. For one thing, since $\deg(u) > 0$, $r_0(x) = \operatorname{rem}(x^{p \cdot 0}, u) = 1$, and so all entries in the first column are 0. This means that h_0 does not appear in the equations, and so h_0 can be any member of $\mathbf{Z}_p$. In fact, there are exactly p^r distinct solutions to the system (9.56). This follows since each solution to this system corresponds to a unique polynomial $h(x)$ that satisfies Equation (9.42) for some sequence $a_1, \ldots, a_r$, and there are p^r ways to choose this sequence.

The next theorem shows that we obtain useful information about the factorization of u by obtaining the general solution to Equation (9.56).

Theorem 9.29. *Let $u(x)$ be a square-free polynomial in $\mathbf{Z}_p[x]$ that has an irreducible factorization with r factors. Then, the system of equations (9.56) has r linearly independent solutions $\bar{h} = \bar{b}_1, \ldots, \bar{h} = \bar{b}_r$, and any solution to Equation (9.56) is a linear combination of these solutions.*

The sequence of solution vectors $\bar{b}_1, \ldots, \bar{b}_r$ described in the theorem is called a *basis* for the solutions.

Proof: Since the right side of Equation (9.56) is $\bar{0}$, and since the system has more than one solution, we know from linear algebra that we can find a basis for the solutions. All we need to show is the basis has r vectors.

Suppose that Equation (9.56) has a basis with m vectors $\bar{b}_1, \ldots, \bar{b}_m$ so that any solution can be expressed as

$$\bar{h} = c_1 \bar{b}_1 + \cdots + c_m \bar{b}_m \tag{9.59}$$

where c_i is in $\mathbf{Z}_p$. Since each c_i can have p different values, there are p^m distinct expressions of the form (9.59), and since there are p^r solutions to Equation (9.56), we have $m = r$. $\qquad\square$

Example 9.30. Let $p = 5$, and let's consider again the polynomial from the previous examples, $u = x^6 + x^5 + x + 4$. In this case, the remainders in Equation (9.53) are

$$
\begin{aligned}
r_0 &= \operatorname{rem}(x^{p\cdot 0}, u) = 1, \\
r_1 &= \operatorname{rem}(x^{p\cdot 1}, u) = x^5, \\
r_2 &= \operatorname{rem}(x^{p\cdot 2}, u) = 3\,x^5 + 2\,x^4 + 3\,x^3 + 2\,x^2 + 3\,x + 1, \\
r_3 &= \operatorname{rem}(x^{p\cdot 3}, u) = 3\,x^5 + 4\,x^4 + x^2 + 3\,x, \\
r_4 &= \operatorname{rem}(x^{p\cdot 4}, u) = x^5 + 2\,x^4 + 2\,x^3 + 2\,x^2 + x + 1, \\
r_5 &= \operatorname{rem}(x^{p\cdot 5}, u) = x,
\end{aligned}
$$

which gives

$$R = \begin{vmatrix} 1-1 & 0 & 1 & 0 & 1 & 0 \\ 0 & 0-1 & 3 & 3 & 1 & 1 \\ 0 & 0 & 2-1 & 1 & 2 & 0 \\ 0 & 0 & 3 & 0-1 & 2 & 0 \\ 0 & 0 & 2 & 4 & 2-1 & 0 \\ 0 & 1 & 3 & 3 & 1 & 0-1 \end{vmatrix}$$

$$= \begin{vmatrix} 0 & 0 & 1 & 0 & 1 & 0 \\ 0 & 4 & 3 & 3 & 1 & 1 \\ 0 & 0 & 1 & 1 & 2 & 0 \\ 0 & 0 & 3 & 4 & 2 & 0 \\ 0 & 0 & 2 & 4 & 1 & 0 \\ 0 & 1 & 3 & 3 & 1 & 4 \end{vmatrix} \qquad (9.60)$$

where all arithmetic is done in $\mathbf{Z}_5[x]$. Using an approach described later in this section, a basis for the solutions is

$$\bar{b}_1 = \begin{vmatrix} 1 \\ 0 \\ 0 \\ 0 \\ 0 \\ 0 \end{vmatrix}, \quad \bar{b}_2 = \begin{vmatrix} 0 \\ 0 \\ 4 \\ 4 \\ 1 \\ 0 \end{vmatrix}, \quad \bar{b}_3 = \begin{vmatrix} 0 \\ 1 \\ 0 \\ 0 \\ 0 \\ 1 \end{vmatrix}. \qquad (9.61)$$

Since the basis has three vectors, we know that u has three irreducible factors. These solutions to Equation (9.56) correspond to the three linearly independent solutions to Equation (9.42):

$$b_1(x) = 1, \qquad b_2(x) = x^4 + 4x^3 + 4x^2, \qquad b_3(x) = x^5 + x,$$

and any auxiliary polynomial $h(x)$ is a linear combination of these polynomials. For example, $h(x)$ in Equation (9.44) is given by

$$h = 2x^5 + x^4 + 4x^3 + 4x^2 + 2x + 1 = 1 \cdot b_1 + 1 \cdot b_2 + 2 \cdot b_3. \qquad \square$$

Separating the Factors of $u(x)$ with Basis Polynomials

The final step in the factorization process involves finding the irreducible factors of u using the auxiliary polynomials $h(x)$. The basic tool for this factorization is the relation in Equation (9.46) on page 375. However, as we saw in Example 9.27, for some $h(x)$, Equation (9.46) may not obtain the complete irreducible factorization. In fact, if u has more than

p factors, Equation (9.46) cannot obtain the complete factorization with one $h(x)$, and so more than one auxiliary polynomial is needed. The next theorem and subsequent discussion show how the basis polynomials $b_1(x), b_2(x), \ldots, b_r(x)$ can be used to separate the factors.

Theorem 9.31. *Suppose that u is a square-free polynomial in $\mathbf{Z}_p[x]$ with $r \geq 2$ factors, and let u_i and u_j be two distinct irreducible factors of u.*

1. *There is a basis polynomial $b_k(x)$ such that $\mathrm{rem}(b_k, u_i) \neq \mathrm{rem}(b_k, u_j)$.*

2. *There is an element d in $\mathbf{Z}_p$ such that*

$$u_i \mid \gcd(u, b_k - d), \quad u_j \nmid \gcd(u, b_k - d)$$

where b_k is obtained in Part (1) of the theorem.

Proof: To show (1), from the definition of the auxiliary polynomials, there is an h that satisfies Equation (9.42) such that

$$\mathrm{rem}(h, u_i) = a_i \neq a_j = \mathrm{rem}(h, u_j). \tag{9.62}$$

In addition, we can represent h as

$$h = c_1 \, b_1 + \cdots + c_r \, b_r$$

where c_i is in $\mathbf{Z}_p$. We claim that it is not possible to have

$$\mathrm{rem}(b_k, u_i) = \mathrm{rem}(b_k, u_j), \quad 1 \leq k \leq r.$$

If this were so,

$$
\begin{aligned}
\mathrm{rem}(h, u_i) &= c_1 \, \mathrm{rem}(b_1, u_i) + \cdots + c_r \, \mathrm{rem}(b_r, u_i) \\
&= c_1 \, \mathrm{rem}(b_1, u_j) + \cdots + c_r \, \mathrm{rem}(b_r, u_j) \\
&= \mathrm{rem}(h, u_j)
\end{aligned}
$$

which contradicts the conditions in (9.62).

To show (2), let b_k be the polynomial obtained in Part (1), and let $d = \mathrm{rem}(b_k, u_i)$. Then, $u_i \mid b_k - d$, which implies that $u_i \mid \gcd(u, b_k - d)$. Next, if $u_j \mid \gcd(u, b_k - d)$, then $u_j \mid b_k - d$, and the uniqueness property of polynomial division implies that $\mathrm{rem}(b_k, u_j) = d$ which contradicts Part (1) of the theorem. $\qquad\square$

Theorem 9.31(2) gives a way to determine each factor u_i of u. The theorem implies that for each of the other factors

$$u_j, \quad j = 1, \ldots, i-1, i+1, \ldots, r,$$

there is a basis polynomial $b_{\theta(j)}$ in $\{b_1, \ldots, b_r\}$ and an element d_j in $\mathbf{Z}_p$ such that

$$u_i \mid \gcd(u, b_{\theta(j)} - d_j), \qquad u_j \nmid \gcd(u, b_{\theta(j)} - d_j), \tag{9.63}$$

for $j = 1, \ldots, i - 1, i + 1, \ldots, r$. (The function notation $\theta(j)$ indicates the choice of the basis polynomial depends on the factor u_j.) If we let

$$e_j = \gcd(u, b_{\theta(j)} - d_j), \tag{9.64}$$

then

$$u_i = \gcd(e_1, \ldots, e_{i-1}, e_{i+1}, \ldots, e_r). \tag{9.65}$$

Indeed, by (9.63), u_i divides all the e_j, and so it divides the gcd on the right. However, since

$$\gcd(e_1, \ldots, e_{i-1}, e_{i+1}, \ldots, e_r) \mid e_j \mid u,$$

and since no other factor of u divides all the e_j, we obtain Equation (9.65).

The relationship in Equation (9.65) is important in a theoretical sense because it shows that by performing a sufficient number gcd calculations of the form

$$\gcd(u, b_k - j), \qquad k = 1, \ldots, r, \qquad j = 0, \ldots, p - 1,$$

we have enough information to obtain each factor of u. However, the formula is not useful in a computational sense because it overstates the work required. Later in this section, we give an algorithm that obtains the factorization from the basis polynomials.

Procedures for Berlekamp's Algorithm

Let u be a square-free and monic polynomial in $\mathbf{Z}_p[x]$. Berlekamp's algorithm obtains the irreducible factorization of u using the following three steps.

1. Find the matrix R.

2. Using the matrix R, find a list $S = [b_1, \ldots, b_r]$ that contains a basis of auxiliary polynomials. When there are r polynomials in S, the polynomial u has r irreducible factors.

3. Use the basis polynomials in S to find the irreducible factors of u.

Procedure *Berlekamp_factor*(u, x, p);
Input
    u : a monic, square-free polynomial in $\mathbf{Z}_p[x]$;
    x : a symbol;
    p : a prime > 1;
Output
    The set of irreducible factors of u;
Global
    R : an n by n matrix with entries in $\mathbf{Z}_p$;
Local Variables
    n, S;
Begin
1      $n := Degree\_gpe(u, x)$;
2   **if** $n = 0$ **or** $n = 1$ **then** *Return*($\{u\}$)
3   **else**
4         $R\_matrix(u, x, n, p)$;
5         $S := Auxiliary\_basis(x, n, p)$;
6   **if** *Number_of_operands*(S) $= 1$ **then**
7   *Return*($\{u\}$)
8   **else**
9   *Return*(*Find_factors*(u, S, x, p))
End

Procedure *R_matrix*(u, x, n, p);
Input
    u : a monic, square-free polynomial in $\mathbf{Z}_p[x]$;
    x : a symbol;
    n : a positive integer representing deg(u);
    p : a prime > 1;
Output
    Creates the global matrix R
Global
    R : an n by n matrix with entries in $\mathbf{Z}_p$;

The body of the procedure is left to the reader (Exercise 5).

Figure 9.4. The MPL procedure *Berlekamp_factor* and the heading of the procedure *R_matrix* in Berlekamp's algorithm. The *Berlekamp_factor* procedure returns the set of irreducible factors of u. (Implementation: Maple (txt), Mathematica (txt), MuPAD (txt).)

The Berlekamp_factor Procedure. Figure 9.4 gives the *Berlekamp_factor* procedure that obtains this factorization. The three procedure calls that correspond to these steps are in lines 4, 5, and 9. Observe that the matrix R, which is created by the call on *R_matrix* at line 4 and then used by *Auxiliary_basis* at line 5, is a global variable.

The R_matrix Procedure. The heading of the *R_matrix* procedure is given in Figure 9.4. There are a number of ways to find the coefficients of R. One possibility is to use the polynomial divisions that define the remainder coefficients r_{ij} using Equations (9.53) and (9.55). Another approach is based on a recurrence relation that is described in Exercise 5. We leave the details of this procedure to the reader.

The Auxiliary_basis Procedure. The second step in Berlekamp's algorithm involves finding a basis of solutions to a linear system of equations. In the next example we illustrate an approach to this problem. The general algorithm and procedure are described after the example.

Example 9.32. Let $p = 5$, and consider the linear system of equations

$$R\,\bar{y} = \bar{0}, \tag{9.66}$$

where

$$R = \begin{vmatrix} 0 & 3 & 1 & 2 \\ 2 & 0 & 2 & 1 \\ 2 & 1 & 4 & 0 \\ 4 & 1 & 1 & 1 \end{vmatrix}, \quad \bar{y} = \begin{vmatrix} y_1 \\ y_2 \\ y_3 \\ y_4 \end{vmatrix}, \quad \bar{0} = \begin{vmatrix} 0 \\ 0 \\ 0 \\ 0 \end{vmatrix}. \tag{9.67}$$

To find a basis of solutions, we obtain an equivalent system with the same solutions from which we can determine a basis. Our approach is similar to the solution algorithms encountered in linear algebra, where linear operations are applied to the rows of the matrix to eliminate variables from equations. Since the right side of Equation (9.66) is $\bar{0}$, our manipulations do not modify this side of the equation, and so we only show how the matrix R changes over the course of the computations. At any point in the computation, R_i represents the ith row of the current matrix R, and C_j represents the jth column.

To begin, let's eliminate y_1 from all but one of the equations. To do this, we use the first non-zero entry $r_{21} = 2$ in C_1 to eliminate the other entries in this column. This entry is called the *pivot* entry for column C_1. First, we multiply row R_2 of R by the multiplicative inverse of the pivot

$$r_{21}^{-1} = 2^{-1} = 3 \quad (\text{in } \mathbf{Z}_5)$$

to obtain the new second row

$$R_2' = r_{21}^{-1} R_2 = |1^* \; 0 \; 1 \; 3| \tag{9.68}$$

where the * superscript indicates the location of the pivot entry. Next, we modify the other three rows so that the value in the first column of each row is zero:

$$\begin{aligned}
R_1' &= R_1 - r_{11} R_2' = |0 \; 3 \; 1 \; 2|, \\
R_3' &= R_3 - r_{31} R_2' = |0 \; 1 \; 2 \; 4|, \\
R_4' &= R_4 - r_{41} R_2' = |0 \; 1 \; 2 \; 4|.
\end{aligned} \tag{9.69}$$

Although the first entry in R_1 is already 0, we have applied the transformation to R_1 to illustrate the general approach. With these manipulations, we obtain the new the system of equations

$$\begin{vmatrix} 0 & 3 & 1 & 2 \\ 1^* & 0 & 1 & 3 \\ 0 & 1 & 2 & 4 \\ 0 & 1 & 2 & 4 \end{vmatrix} \begin{vmatrix} y_1 \\ y_2 \\ y_3 \\ y_4 \end{vmatrix} = \begin{vmatrix} 0 \\ 0 \\ 0 \\ 0 \end{vmatrix}, \tag{9.70}$$

where now R represents the matrix on the left. Notice that the variable y_1 only appears in the second equation, and this new system has the same solutions as Equation (9.66). The modification of R using Equations (9.68) and (9.69) is called the *elimination step* associated with y_1.

Next, we eliminate the variable y_2 from all but one of the equations in (9.70). To do this, using the new R, we find a non-zero pivot entry in C_2 in a row which has not been used for an earlier pivot. In this case, a pivot is located in R_1. Multiplying R_1 by $r_{12}^{-1} = 3^{-1} = 2$, we obtain a new first row

$$R_1' = |0 \; 1^* \; 2 \; 4|,$$

and then form the new rows

$$\begin{aligned}
R_2' &= R_2 - r_{22} R_1' = |1 \; 0 \; 1 \; 3|, \\
R_3' &= R_3 - r_{32} R_1' = |0 \; 0 \; 0 \; 0|, \\
R_4' &= R_4 - r_{42} R_1' = |0 \; 0 \; 0 \; 0|.
\end{aligned}$$

This gives a new system of equations

$$\begin{vmatrix} 0 & 1^* & 2 & 4 \\ 1^* & 0 & 1 & 3 \\ 0 & 0 & 0 & 0 \\ 0 & 0 & 0 & 0 \end{vmatrix} \begin{vmatrix} y_1 \\ y_2 \\ y_3 \\ y_4 \end{vmatrix} = \begin{vmatrix} 0 \\ 0 \\ 0 \\ 0 \end{vmatrix}, \tag{9.71}$$

where now R represents the matrix on the left. Notice that the variable y_2 only appears in the first equation, and this system has the same solutions as Equation (9.66).

Next, using the new R, we try to eliminate y_3 from all but one equation by finding a pivot in C_3. However, since the only non-zero entries in C_3 are in rows that already contain pivots, we cannot find a pivot. Instead, this situation is an opportunity to construct the first member of the basis. Since there are now only two independent equations, we can assign values to two of the variables (say $y_3 = 1$ and $y_4 = 0$) and find a basis vector of the form

$$\bar{S}_3 = \begin{vmatrix} y_1 \\ y_2 \\ 1 \\ 0 \end{vmatrix}. \tag{9.72}$$

The subscript 3 refers to the current column, and the two unknown values y_1 and y_2 are determined using the two non-zero equations in (9.71). (We will see below why this is a useful form.) Using the first equation in (9.71), we have $y_2 + 2 = 0$ or $y_2 = -2 = 3$ (in $\mathbf{Z}_5$), and using the second equation, we have $y_1 + 1 = 0$ or $y_1 = -1 = 4$. Therefore, the first basis vector is

$$\bar{S}_3 = \begin{vmatrix} 4 \\ 3 \\ 1 \\ 0 \end{vmatrix}. \tag{9.73}$$

Finally, we try to find a pivot for y_4, and again find that the only non-zero entries in C_4 are in rows that already contain pivots. Instead, we find a second basis solution of the form

$$\bar{S}_4 = \begin{vmatrix} y_1 \\ y_2 \\ 0 \\ 1 \end{vmatrix}. \tag{9.74}$$

Using the first two non-zero equations in (9.71), we obtain

$$\bar{S}_4 = \begin{vmatrix} 2 \\ 1 \\ 0 \\ 1 \end{vmatrix}. \tag{9.75}$$

Let's show that the set $S = \{\bar{S}_3, \bar{S}_4\}$ is a basis for the solutions. First, the solutions in S are linearly independent since one solution is not a constant multiple of the other solution. To show that any solution $\bar{y}$ of Equation (9.67) is a linear combination of the basis, we use Equation (9.71) to obtain y_1 and y_2 in terms of y_3 and y_4 (in $\mathbf{Z}_5$) as

$$y_1 = -y_3 - 3\,y_4 = 4\,y_3 + 2\,y_4, \qquad y_2 = -2\,y_3 - 4\,y_4 = 3\,y_3 + y_4.$$

Therefore,

$$\bar{y} = \begin{vmatrix} y_1 \\ y_2 \\ y_3 \\ y_4 \end{vmatrix} = \begin{vmatrix} 4\,y_3 + 2\,y_4 \\ 3\,y_3 + y_4 \\ y_3 \\ y_4 \end{vmatrix} = y_3\bar{S}_3 + y_4\bar{S}_4.$$

Notice that the forms used for the basis vectors lead to this simple relationship. □

The General Approach. Let's consider the general system of equations (9.66), where R is an n by n matrix with entries in $\mathbf{Z}_p$, and $\bar{y}$ and $\bar{0}$ are n-dimensional vectors. An algorithm that obtains a basis of solutions uses an approach similar to the one in the last example. First, we try to find a pivot entry in column C_1 by selecting a row R_i where $r_{i1} \neq 0$. If a pivot is found, we modify R using the elimination step given below in Equations (9.76) and (9.77) with $j = 1$. On the other hand, if all the entries in C_1 are zero, then a pivot cannot be found, and we define the first solution basis vector

$$\bar{S}_1 = \begin{vmatrix} 1 \\ 0 \\ \vdots \\ 0 \end{vmatrix}.$$

In this case, y_1 does not appear in any equation, and so $\bar{S}_1$ is a solution to the system of equations.

In general, let R be the matrix obtained after applying the elimination step (if possible) to C_{j-1}. We search for a pivot in the next column C_j. A pivot occurs in this column at row R_i if the following conditions are satisfied.

1. $r_{ij} \neq 0$.

2. R_i does not contain the pivot entry for some column C_l with $l < j$.

Let's suppose a pivot is found in row R_i. We modify the matrix R using the following elimination step.

1. Replace row R_i by a new row

$$R_i' = r_{ij}^{-1}\,R_i. \tag{9.76}$$

2. For each of the other rows R_k $(k \neq i)$, form a new row

$$R'_k = R_k - r_{kj} R'_i. \tag{9.77}$$

We claim that this elimination step at C_j with Equations (9.76) and (9.77) does not change any element in column C_l with $l < j$. This follows, since in the pivot row R_i, $r_{il} = 0$ for $l < j$. Indeed, if C_l has a pivot, since R_i contains the pivot location for C_j (and not for C_l), the elimination step for C_l implies that $r_{il} = 0$. On the other hand, if C_l does not have a pivot, then we also have $r_{il} = 0$, because otherwise R_i would contain a pivot for C_l.

Once the pivot operation for column C_j is done, in the new matrix R,

$$r_{kj} = \begin{cases} 1, & \text{for } k = i, \\ 0, & \text{for } k \neq i \end{cases}$$

which implies that the variable y_j appears only in the ith equation. In addition, the new system $R\bar{h} = \bar{0}$ has the same solutions as the original system.

To keep track of which rows contain pivots, we define a sequence of non-negative integers P_i, $i = 1, \dots, n$ with the following rules.

1. At the start of the algorithm, $P_i = 0$, $i = 1, \dots, n$.

2. If the pivot in column C_j is in row R_i, then $P_i = j$.

Observe that the condition $P_i = 0$ provides a simple check that row R_i does not contain a pivot.

Next, let's suppose that a pivot is not found in C_j. In this case, we define the next basis vector

$$\bar{S}_j = \begin{vmatrix} s_{1j} \\ \vdots \\ s_{nj} \end{vmatrix} = \begin{vmatrix} s_{1j} \\ \vdots \\ s_{(j-1)j} \\ 1 \\ 0 \\ \vdots \\ 0 \end{vmatrix},$$

where

$$s_{lj} = \begin{cases} -r_{ej}, & \text{if } l < j \text{ and } P_e = l \text{ for some (unique) } 1 \leq e \leq n, \\ 0, & \text{if } l < j \text{ and } P_e \neq l \text{ for all } 1 \leq e \leq n, \\ 1, & \text{if } l = j, \\ 0, & \text{if } j < l. \end{cases} \tag{9.78}$$

In other words, for $l < j$, $s_{lj} = -r_{ej}$ whenever column C_l has a pivot in row R_e, and $s_{lj} = 0$ when C_l does not have a pivot.

Example 9.33. Let's consider again the computation in Example 9.32 where $p = 5$ and

$$R = \begin{vmatrix} 0 & 3 & 1 & 2 \\ 2 & 0 & 2 & 1 \\ 2 & 1 & 4 & 0 \\ 4 & 1 & 1 & 1 \end{vmatrix}.$$

At the beginning of the computation $P_i = 0$ for $i = 1, 2, 3, 4$. After the pivot for C_1 is found in R_2, we set $P_2 = 1$, and after the pivot for C_2 is found in R_1, we set $P_1 = 2$.

When $j = 3$, since

$$R = \begin{vmatrix} 0 & 1^* & 2 & 4 \\ 1^* & 0 & 1 & 3 \\ 0 & 0 & 0 & 0 \\ 0 & 0 & 0 & 0 \end{vmatrix},$$

C_3 does not have a pivot, and Equation (9.78) gives

$$\bar{S}_3 = \begin{vmatrix} -r_{23} \\ -r_{13} \\ 1 \\ 0 \end{vmatrix} = \begin{vmatrix} -1 \\ -2 \\ 1 \\ 0 \end{vmatrix} = \begin{vmatrix} 4 \\ 3 \\ 1 \\ 0 \end{vmatrix}.$$

In a similar way, for $j = 4$, we have

$$\bar{S}_4 = \begin{vmatrix} -r_{24} \\ -r_{14} \\ 0 \\ 1 \end{vmatrix} = \begin{vmatrix} -3 \\ -4 \\ 0 \\ 1 \end{vmatrix} = \begin{vmatrix} 2 \\ 1 \\ 0 \\ 1 \end{vmatrix}. \qquad \square$$

Returning to the general approach, let's show that $\bar{S}_j$ defined by Equation (9.78) is a solution to the system of equations. At this point in the computation,

$$R = \begin{vmatrix} 0 & \cdots & 1^* & \cdots & r_{ek} & \cdots & r_{ej} & r_{e,j+1} & \cdots & r_{en} \\ \vdots & & \vdots & & \vdots & & \vdots & \vdots & & \vdots \\ 0 & \cdots & 0 & \cdots & 0 & \cdots & 0 & r_{f,j+1} & \cdots & r_{fn} \\ \vdots & & \vdots & & \vdots & & \vdots & \vdots & & \vdots \\ & & \uparrow & & \uparrow & & \uparrow & & & \\ & & l & & k & & j & & & \end{vmatrix} \qquad (9.79)$$

where row R_e is a typical row with a pivot and row R_f is a typical row without a pivot. To show that $R_e \cdot \bar{S}_j = 0$, suppose that in Equation (9.79) column l contains the pivot and column k is another typical column to the left of C_j. From Equation (9.78), the vector $\bar{S}_j$ has the form

$$
\bar{S}_j = \left|
\begin{array}{c|l}
\vdots & \\
-r_{ej} & \leftarrow l \\
\vdots & \\
0 & \leftarrow k \\
\vdots & \\
1 & \leftarrow j \\
0 & \\
\vdots & \\
0 &
\end{array}
\right| . \tag{9.80}
$$

Using the corresponding entries in R_e and $\bar{S}_j$, we have

$$
\begin{aligned}
R_e \cdot \bar{S}_j &= 1 \cdot (-r_{ej}) + \cdots + r_{ek} \cdot 0 + r_{ej} \cdot 1 + \cdots \\
&\quad + r_{e,j+1} \cdot 0 + \cdots + r_{en} \cdot 0 \\
&= 0.
\end{aligned}
$$

In row R_f, which does not contain a pivot, all entries in positions $\leq j$ are 0. Indeed, entries in columns with pivots are 0, and entries in other columns are 0 because R_f does not contain a pivot. Therefore,

$$
R_f \cdot \bar{S}_j = 0
$$

and $\bar{S}_j$ is a solution to the system of equations. In fact, $\bar{S}_j$ is a solution to the system of equations at the end of the computation because all future changes in R occur in columns $j+1$ to n, and the entries of $\bar{S}_j$ are 0 for all positions from $j+1$ to n.

Using the above scheme, we obtain a set of solutions to Equation (9.66)

$$
S = \{\bar{S}_j \mid C_j \text{ does not have a pivot}\} \tag{9.81}
$$

where the notation indicates a new $\bar{S}_j$ is formed whenever C_j does not have a pivot.

To show that S is a basis we must show that: (1) the solutions in S are linearly independent, and (2) each solution to Equation (9.66) is a linear combination of the solutions in S. The linear independence follows directly from the definition for $\bar{S}_j$ and is left to the reader (Exercise 6).

To show (2), we show that for any solution $\bar{y}$ to Equation (9.66), we have

$$
\bar{y} = \sum_{\substack{j \\ C_j \text{ does not} \\ \text{have a pivot}}} y_j \bar{S}_j .
$$

We verify this by showing that for each component y_l of $\bar{y}$,

$$y_l = \sum_{\substack{j \\ C_j \text{ does not} \\ \text{have a pivot}}} y_j\, s_{lj}. \tag{9.82}$$

To show this, first assume that C_l (the column associated with the variable y_l) does not have a pivot. With this assumption, Equation (9.78) implies that

$$s_{lj} = \begin{cases} 0, & \text{if } l \neq j, \\ 1, & \text{if } l = j, \end{cases}$$

and so Equation (9.82) follows.

On the other hand, suppose that C_l has a pivot with $P_e = l$, and consider the entries in row R_e at the end of the computation. First, the elimination steps give $r_{el} = 1$, and $r_{ej} = 0$ for all other columns C_j that have pivots. We also have $r_{ej} = 0$ when $j < l$, where C_j does not have a pivot because, if this were not so, R_e would contain a pivot for one of these columns C_j. Therefore, substituting $\bar{y}$ into the eth equation, we obtain

$$y_l + \sum_{\substack{j > l \\ C_j \text{ does not} \\ \text{have a pivot}}} y_j\, r_{ej} = 0. \tag{9.83}$$

However, since C_l has a pivot in R_e, Equation (9.78) implies that

$$s_{lj} = \begin{cases} -r_{ej}, & \text{if } l < j, \\ 1, & \text{if } l = j, \\ 0, & \text{if } j < l, \end{cases} \tag{9.84}$$

and Equations (9.83) and (9.84) imply

$$y_l = \sum_{\substack{j > l \\ C_j \text{ does not} \\ \text{have a pivot}}} y_j\, (-r_{ej})$$

$$= \sum_{\substack{j \\ C_j \text{ does not} \\ \text{have a pivot}}} y_j\, s_{lj}.$$

The procedure *Auxiliary_basis* that obtains a basis is given in Figures 9.5 and 9.6. Since the procedure is used by the *Berlekamp_factor* procedure, the basis is returned as a list S of polynomials rather than as a list of vectors.

Procedure *Auxiliary_basis*(x, n, p);
Input
    x : a symbol;
    n : a positive integer;
    p : a prime > 1;
Output
    a list containing a basis of one or more auxiliary polynomials;
Global
    R : an n by n matrix with entries in $\mathbf{Z}_p$;
Local Variables
    $i, P, S, j, pivot\_found, a, k, f, l, s, e, c$;
Begin

```
1      for  i := 1 to  n do
2          P[i] := 0;
3      S := [ ];
4      for  j := 1 to  n do
5          i := 1;
6          pivot_found := false;
7          while not pivot_found and  i ≤ n do
8              if  R[i, j] ≠ 0 and  P[i] = 0 then
9                  pivot_found := true
10             else  i := i + 1;
11         if  pivot_found then
12             P[i] := j;
13             a := Multiplicative_inverse_p(R[i, j], p);
14             for  l := 1 to  n do
15                 R[i, l] := Irem(a * R[i, l], p);
16             for  k := 1 to  n do
17                 if  k ≠ i then
18                     f := R[k, j];
19                     for  l := 1 to  n do
20                         R[k, l] := Irem(R[k, l] − f * R[i, l], p)
```

Continued in Figure 9.6.

Figure 9.5. An MPL procedure that finds a basis of auxiliary polynomials. (Implementation: Maple (txt), Mathematica (txt), MuPAD (txt).)

In lines 1–2, the variables $P[i]$ (which correspond to P_i in the above discussion) are initialized to 0. In the j loop that begins at line 4, we examine each column of the current matrix R to determine if it contains a pivot (lines 5–10). If a pivot is found, we perform the elimination step in Equations (9.76) and (9.77) (lines 12–20). The procedure *Multiplicative_inverse_p* in line 13 is described in Exercise 11 on page 59.

```
21          elseif not pivot_found then
22              s := x^{j-1};
23              for l := 1 to j − 1 do
24                  e := 0;
25                  i := 1;
26                  while e = 0 and i ≤ n do
27                      if l = P[i] then e := i
28                      else  i := i + 1;
29                  if e > 0 then
30                      c := Irem(−R[e, j], p);
31                      s := s + c * x^{l-1};
32              S := Join(S, [s]);
33      Return(S)
    End
```

Figure 9.6.   Continuation of Figure 9.5.

If a pivot is not found, we construct a basis solution (lines 22–31). Although the basis polynomial is defined by the vector in Equation (9.78), the component s_{lj} corresponds to the coefficient of x^{l-1} in a basis polynomial (see Equations (9.51) and (9.58)). Based on the property that $s_{lj} = 1$ when $l = j$, at line 22 we initialize the polynomial to x^{j-1}. The loop starting at line 23 determines the coefficients of smaller powers of x. The statements in lines 24–28 determine if a column C_l has a pivot row, and, if one is found, the row of the pivot is represented by the integer e. If C_l has a pivot, the coefficient of the polynomial is determined (in $\mathbf{Z}_p$) at line 31, while if a pivot is not found, the coefficient is 0, and so no action is taken.

The matrices that arise in Berlekamp's algorithm always have zeroes in column C_1. Therefore, when *Auxiliary_basis* is applied to these matrices, the first basis polynomial is always the polynomial 1. The procedure *Find_factors* described below makes use of this observation.

The *Find_factors* Procedure.   The procedure *Find_factors*, that finds the irreducible factors using the list S of basis polynomials, is shown in Figure 9.7.   The procedure is based on the relationships in the following theorem.

Theorem 9.34. *Let u be a square-free, monic polynomial in $\mathbf{Z}_p[x]$, and let h be an auxiliary polynomial that satisfies the remainder equations (9.42) on page 374.*

1. *If w is monic and $w \mid u$, then*

$$w = \prod_{j=0}^{p-1} \gcd(w, h - j). \tag{9.85}$$

2. *Suppose that w is monic with $w \mid u$, and let $g = \gcd(w, h - j)$ with $\deg(g) > 0$. Then,*

$$\gcd(g, h - i) = \begin{cases} g, & i = j, \\ 1, & i \neq j, \end{cases} \tag{9.86}$$

and for $q = \text{quot}(w, g)$,

$$q = \prod_{\substack{i=0 \\ i \neq j}}^{p-1} \gcd(q, h - i). \tag{9.87}$$

The proofs of these properties are left to the reader (Exercise 8). □

The relationship in Equation (9.85) shows that the basic factor relationship in Equation (9.46) also holds for divisors w of u even though h is defined in terms of u. This means that we can obtain the irreducible factors of u using operations of the form

$$\gcd(w, b_k(x) - j), \tag{9.88}$$

where w is either u or some (reducible or irreducible) divisor of u obtained earlier in the computation. The relationships in Equations (9.86) and (9.87) are used below to justify the logic of the algorithm. We illustrate the approach in the next example.

Example 9.35. Let $p = 5$, and consider the polynomial $u = x^6 + x^5 + x + 4$. Three basis polynomials were obtained in Example 9.30 (page 380), and so u has three factors. Since $b_1 = 1$, we do not obtain new factors with this basis polynomial. Using the second basis polynomial $b_2 = x^4 + 4x^3 + 4x^2$, we have

$$\gcd(u, b_2 - j) = 1, \quad j = 0, 1. \tag{9.89}$$

However,

$$g = \gcd(u, b_2 - 2) = x^4 + 4x^3 + 4x^2 + 3$$

Procedure *Find_factors*(u, S, x, p);
Input
    u : a monic square-free polynomial in $\mathbf{Z}_p[x]$;
    S : a list containing a basis of one or more auxiliary
         polynomials where the first polynomial in S is $b = 1$;
    x : a symbol;
    p : a prime ≥ 2;
Output
    The set of irreducible factors of u;
Local Variables
    $r, factors, k, b, old\_factors, i, w, j, g, q$;
Begin
1     $r := Number\_of\_operands(S)$;
2     $factors := \{u\}$;
3   **for** $k := 2$ **to** r **do**
4        $b := Operand(S, k)$;
5        $old\_factors := factors$;
6   **for** $i := 1$ **to** $Number\_of\_operands(old\_factors)$ **do**
7            $w := Operand(old\_factors, i)$;
8            $j := 0$;
9   **while** $j \leq p - 1$ **do**
10               $g := Poly\_gcd\_p(b - j, w, x, p)$;
11   **if** $g = 1$ **then** $j := j + 1$
12   **elseif** $g = w$ **then** $j := p$
13   **else**
14                  $factors := factors \sim \{w\}$;
15                  $q := Operand(Poly\_div\_p(w, g, x, p), 1)$;
16                  $factors := factors \cup \{g, q\}$;
17   **if** $Number\_of\_operands(factors) = r$ **then** $Return(factors)$
18   **else**
19                      $j := j + 1$
20                      $w := q$;
End

Figure 9.7. The MPL procedure *Find_factors* for Berlekamp's algorithm. (Implementation: Maple (txt), Mathematica (txt), MuPAD (txt).)

is a factor of u, and we obtain another factor with

$$q = \text{quot}(u, g) = x^2 + 2x + 3.$$

Although $u = q\,g$, this is not the irreducible factorization since there are only two factors. Instead of continuing the gcd calculations with u, we apply the process to the factors obtained so far. However, by Equation (9.86), we do not obtain any new factors using the auxiliary polynomial b_2 with the factor g, and we continue the process with b_2 and q. The relationship in Equation (9.87) suggests we search for factors of q with calculations of the form $\gcd(q, b_2 - i)$ with $i \neq 2$. However, by Equation (9.89), when $i = 0, 1$ we do not obtain new factors of q and so continue the process with $i = 3, 4$. We obtain

$$\gcd(q, b_2 - 3) = 1, \qquad \gcd(q, b_2 - 4) = q,$$

and so no new factors are obtained with b_2.

At this point we use the next basis polynomial $b_3 = x^5 + x$, and apply the process to both of the current factors in

$$u = (x^4 + 4x^3 + 4x^2 + 3)(x^2 + 2x + 3). \tag{9.90}$$

For $w = x^4 + 4x^3 + 4x^2 + 3$, we obtain $\gcd(w, b_3 - j) = 1$ for $j = 0, 1$, while

$$g = \gcd(w, b_3 - 2) = x^2 + 3x + 4.$$

This implies that w is reducible and can be replaced in the factorization (9.90) by the product of g and

$$q = \operatorname{quot}(w, g) = x^2 + x + 2.$$

Therefore,

$$u = (x^2 + 3x + 4)(x^2 + x + 2)(x^2 + 2x + 3),$$

and since there are three distinct factors, this is the irreducible factorization and the process terminates. $\qquad\square$

The algorithm used by the *Find_factors* procedure (Figure 9.7) is similar to the one illustrated in the last example. To begin, at line 1 we determine the number of irreducible factors r as the size of the basis list S. The variable *factors*, which first appears at line 2, contains the set of factors at each point in the computation. This variable is initialized to $\{u\}$, and the procedure terminates when *factors* contains r polynomials (line 17).

For each traversal of the k loop beginning at line 3, a basis polynomial b in S is selected at line 4 for use in gcd calculations. Notice that this loop starts with the second basis vector because the first basis vector is the polynomial 1, which does not give new factors. At line 5, another variable *old_factors* is given the value of *factors*. When $k = 2$, *old_factors* has the value $\{u\}$, and on subsequent passes through the k loop it contains the divisors obtained using the previous basis polynomials. We introduce this new variable because we reference each of the old factors at line 7, and the value of *factors* is changed whenever a new factor is found.

In the i loop beginning at line 6, we select an old factor w at line 7 and try to find a new factor by applying the gcd operation in line 10 for various values of j. Notice that when $g = 1$ (line 11), we have not found a new factor, and so we increment j by 1 which continues the computation. If $g = w$, then Equation (9.85) implies that we will not find any new factors with this w using the current basis function, and so we assign j to p, which terminates the j loop.

If g is any other polynomial, then we have found a factor of w. At this point, we remove w from *factors* (line 14), add g and $q = \text{quot}(w, g)$ to *factors* (line 16), and check for termination at line 17. If this condition is not satisfied, then by Equation (9.86), there is no reason to continue the gcd calculations with g and the current basis polynomial b because larger values of j obtain a gcd of 1. In addition, Equation (9.87) implies that any factors of q obtained with the current b are obtained with j values different from the current j. However, none of the factors of the current q are obtained with earlier j values because they would have been obtained earlier in the computation, and we would have a different q at this point. Therefore, we are justified in using larger values of j, and so we increment j by 1 (line 10), assign the value of q to w (line 20), and continue looking for factors of the new w at line 9.

The process described in the previous paragraph is applied (using the i loop) to each of the old factors of u. If we don't obtain all the factors with the current basis polynomial, we continue the process with the next basis polynomial (lines 3, 4). Continuing in this fashion, we eventually obtain all factors of u.

Exercises

1. Give a procedure $Square\_free\_factor\_p(u, x, p)$ that obtains the square-free factorization of u using the recursive approach that is described beginning on page 373. Explain why the recursion terminates. (Exercise 9, page 125 and Exercise 4, page 142 are useful in this problem.)

2. Let u be a monic polynomial in $\mathbf{Z}_p[x]$ with $\deg(u) = n$. Show that there are $(p^{\text{iquot}(n,2)+1} - p)/(p-1)$ trial factors of degree m with $1 \leq m \leq \text{iquot}(n, 2)$.

3. Let A and U be lists of $r > 0$ polynomials in $\mathbf{Z}_p[x]$ that satisfy the conditions in the Chinese Remainder theorem (Theorem 4.41, page 140). Give a procedure $Polynomial\_Chinese\_rem\_p(A, U, x, p)$.

4. (a) Show that $h(x) = c$ (a constant polynomial in $\mathbf{Z}_p$) if and only if $a_1 = a_2 = \cdots = a_r = c$.

(b) Suppose that u is irreducible. Show that $h(x)$ is constant.

(c) Suppose that $h(x)$ has positive degree. Explain why the factorization in Equation (9.46) has at least two distinct factors of positive degree.

5. In this exercise, we describe a recurrence relation that determines the remainders $\text{rem}(x^{pj}, u)$. Suppose that $u = x^n + v$ where $n = \deg(u)$, and v contains all terms with a degree $\leq n - 1$. Let $y_k = \text{rem}(x^k, u)$. We can represent y_k as $y_k(x) = c_k x^{n-1} + z_k(x)$, where c_k is in $\mathbf{Z}_p$ and $z_k(x)$ contains all terms with degree $\leq n - 2$.

   (a) Derive the recurrence relation

   $$y_{k+1} = -c_k v + x z_k.$$

   (b) Give a procedure $R\_matrix(u, x, n, p)$ that creates the matrix R associated with u. Use the recurrence relation in (a) to obtain the remainders. The matrix R created by this procedure is a global variable. For this reason, do not return an expression from the procedure.

6. Show that the set of solutions S in (9.81) is linearly independent.

7. Use the algorithm that finds an auxiliary basis to show that the matrix in Equation (9.60) has the basis given in Equation (9.61).

8. Prove Theorem 9.34. *Hint:* Exercise 12, page 144 is useful in this problem.

9. Let $p = 5$, and consider the square-free polynomial $x^3 + 2x + 4$. Show that this polynomial is irreducible using Berlekamp's algorithm.

10. Give a procedure $Trial\_factor\_p(u, x, p)$ that finds an irreducible factorization of u in $\mathbf{Z}_p[x]$ using the trial-factors approach described on page 373. First, obtain a square-free factorization of u, and then apply the trial factors approach to each of the monic, square-free factors. Be sure to use only trial factors that are monic, and arrange the computation to minimize the number of trial divisors.

11. Give a procedure $Factor\_p(u, x, p)$ that obtains the irreducible factorization of u in $\mathbf{Z}_p[x]$. The procedure should first obtain a square-free factorization of u and then apply the $Berlekamp\_factor$ procedure to each of the square-free factors.

9.4 Irreducible Factorization: A Modern Approach

Let u be a polynomial in $\mathbf{Q}[x]$. To find the irreducible factorization of u, first, we obtain the square-free factorization using the algorithm described in Section 9.1. Since each of the polynomials in the square-free factorization is monic and square-free, the problem is reduced to polynomials with these properties. Next, using Equation (9.24) on page 361, the problem is simplified again to primitive, square-free polynomials in $\mathbf{Z}[x]$. The last step in the process involves the factorization of these polynomials.

The algorithm described in this section obtains the irreducible factorization of a square-free, monic polynomial u in $\mathbf{Z}[x]$ by factoring a related polynomial in $\mathbf{Z}_p[x]$ (for a suitable prime p) and then using the factors of this new polynomial to obtain the factors of u.

Notation Conventions. In this section, we perform polynomial operations in both $\mathbf{Z}[x]$ and $\mathbf{Z}_m[x]$. To distinguish the operations in the two contexts, we adopt the following notation conventions.

1. For polynomial operations in $\mathbf{Z}[x]$, we use the ordinary infix symbols for addition $(+)$ and multiplication $(\cdot)$.

2. For polynomial operations in $\mathbf{Z}_m[x]$, we use the symbols $\oplus_m$ for addition and $\otimes_m$ for multiplication that were introduced in Section 2.3. In this section, m is either a prime number p (in which case $\mathbf{Z}_p$ is a field) or a positive integer power of a prime.

3. To simplify the notation, polynomials in both $\mathbf{Z}[x]$ and $\mathbf{Z}_m[x]$ are represented using ordinary infix notation.

For example, $u = x^2 + 3x + 2$ and $v = x^3 + 4x + 4$ are polynomials in both $\mathbf{Z}[x]$ and $\mathbf{Z}_5[x]$. We have, in $\mathbf{Z}[x]$,

$$u + v = x^3 + x^2 + 7x + 6,$$

and, in $\mathbf{Z}_5[x]$,

$$u \oplus_5 v = x^3 + x^2 + 2x + 1.$$

Reduction to Monic Polynomials

The algorithm in this section applies to monic, square-free polynomials in $\mathbf{Z}[x]$. Since the transformation in Equation (9.24) obtains a non-monic, primitive polynomial in $\mathbf{Z}[x]$, the first order of business is to restate the factorization problem in terms of monic polynomials. Simply dividing a polynomial by its leading coefficient does not work, however, because some coefficients of the new polynomial may not be integers. Instead, we proceed as follows. Let

$$u = \sum_{i=0}^{n} a_i x^i$$

be a primitive polynomial in $\mathbf{Z}[x]$. Our approach is to define a monic polynomial v associated with u, factor v, and then use the factors of v to obtain the factors of u. Let

$$v = a_n^{n-1} u(y/a_n) \tag{9.91}$$

$$= a_n^{n-1} \sum_{i=0}^{n} a_i (y/a_n)^i \tag{9.92}$$

$$= \sum_{i=0}^{n-1} a_n^{n-1-i} a_i y^i + y^n,$$

where v is a monic polynomial in $\mathbf{Z}[y]$. Suppose that v has the irreducible factorization

$$v = p_1(y) \cdots p_r(y).$$

One way to obtain a factorization for u is to invert the transformation in Equation (9.92):

$$u = \left(1/a_n^{n-1}\right) v(a_n x) = \left(1/a_n^{n-1}\right) p_1(a_n x) \cdots p_r(a_n x) \tag{9.93}$$

although in this factorization, when $a_n \neq \pm 1$, the leading constant is in $\mathbf{Q}$ but not necessarily in $\mathbf{Z}$. However, since

$$v(a_n x) = a_n^{n-1} u$$

and since u is primitive, we have

$$\mathrm{cont}(v(a_n x), x) = a_n^{n-1} \mathrm{cont}(u, x) = a_n^{n-1}.$$

Therefore, using Theorem 6.57, u has the factorization

$$
\begin{aligned}
u &= \left(1/a_n^{n-1}\right) v(a_n x) \\
&= \left(1/a_n^{n-1}\right) \mathrm{cont}(v(a_n x), x) \, \mathrm{pp}(v(a_n x), x) \\
&= \mathrm{pp}(v(a_n x), x) \\
&= \mathrm{pp}(p_1(a_n x), x) \cdots \mathrm{pp}(p_r(a_n x), x).
\end{aligned}
\tag{9.94}
$$

Example 9.36. Consider the primitive polynomial

$$u(x) = 8 x^5 - 48 x^4 + 90 x^3 - 90 x^2 + 117 x - 27.$$

Then,

$$v = 8^4 u(y/8) = y^5 - 48 y^4 + 720 y^3 - 5760 y^2 + 59904 y - 110592$$

which has the factorization

$$v = p_1 p_2 p_3 = (y - 24) \left(y^2 + 96\right) \left(y^2 - 24 y + 48\right).$$

Inverting the transformation using Equation (9.93), we obtain

$$u = (1/4096) (8 x - 24) \left(64 x^2 + 96\right) \left(64 x^2 - 192 x + 48\right).$$

However, using Equation (9.94), we have

$$
\begin{aligned}
u &= \mathrm{pp}(p_1(8x), x) \, \mathrm{pp}(p_2(8x), x) \, \mathrm{pp}(p_3(8x), x) \\
&= (x - 3) \left(2 x^2 + 3\right) \left(4 x^2 - 12 x + 3\right).
\end{aligned}
$$

$\square$

The Symmetric Representation for Z_m

Let $m \geq 2$ be an integer. In Section 2.3 we considered the set

$$\mathbf{Z}_m = \{0, 1, \ldots, m - 1\} \tag{9.95}$$

with the two binary operations

$$a \oplus_m b = \text{irem}(a + b, \ m), \qquad a \otimes_m b = \text{irem}(a \cdot b, \ m).$$

With these operations, $\mathbf{Z}_m$ satisfies the field axioms F-1 through F-11, and satisfies axiom F-12 if and only if m is prime (Theorem 2.41, page 53). The representation (9.95) is called the *non-negative representation* for $\mathbf{Z}_m$ because all elements are non-negative integers.

Our algorithm factors a polynomial u in $\mathbf{Z}[x]$ by factoring a related polynomial in $\mathbf{Z}_m[x]$, and then using the factors in $\mathbf{Z}_m[x]$ to obtain the factors of u. Since u and its factors may have negative coefficients, we need another representation for $\mathbf{Z}_m$ that includes some negative integers. This alternate representation is defined as

$$\mathbf{Z}_m = \{-\text{iquot}(m - 1, 2), \ldots, 0, \ldots, \text{iquot}(m, 2)\}. \tag{9.96}$$

For example,

$$\mathbf{Z}_5 = \{-2, -1, 0, 1, 2\}, \qquad \mathbf{Z}_6 = \{-2, -1, 0, 1, 2, 3\}.$$

For m odd, this representation is symmetric about 0, while for m even, it is nearly symmetric with one more integer to the right of 0 than to the left. For this reason, the representation (9.96) is called the *symmetric* representation for $\mathbf{Z}_m$.

The connection between the two representations is defined as follows; each b in the non-negative representation is associated with the unique integer in the symmetric representation given by

$$S_m(b) = \begin{cases} b, & \text{if } 0 \leq b \leq \text{iquot}(m, 2), \\ b - m, & \text{if } \text{iquot}(m, 2) < b < m. \end{cases} \tag{9.97}$$

For example, for $m = 5$, $S_5(3) = -2$ and $S_5(4) = -1$.

For a and b in $\mathbf{Z}_m$ in symmetric form, define the binary operations

$$a \oplus_m b = S_m(\text{irem}(a + b, \ m)), \qquad a \otimes_m b = S_m(\text{irem}(a \cdot b, \ m)).$$

For $m = 5$, the operations are given by the following tables.

$\oplus_5$	0	1	2	-2	-1
0	0	1	2	-2	-1
1	1	2	-2	-1	0
2	2	-2	-1	0	1
-2	-2	-1	0	1	2
-1	-1	0	1	2	-2

$\otimes_5$	0	1	2	-2	-1
0	0	0	0	0	0
1	0	1	2	-2	-1
2	0	2	-1	1	-2
-2	0	-2	1	-1	2
-1	0	-1	-2	2	1

The Projection from $Z[x]$ to $Z_m[x]$

The projection operator in the next definition defines the relationship between $\mathbf{Z}[x]$ and $\mathbf{Z}_m[x]$.

Definition 9.37. *Let $m \geq 2$ be an integer, and let $u = a_n\, x^n + \cdots + a_0$ be in $\mathbf{Z}[x]$.*

1. *For the non-negative representation of $\mathbf{Z}_m$, define*

$$T_m(u) = \mathrm{irem}(a_n,\, m)\, x^n + \cdots + \mathrm{irem}(a_0,\, m). \qquad (9.98)$$

2. *For the symmetric representation of $\mathbf{Z}_m$, define*

$$T_m(u) = S_m(\mathrm{irem}(a_n,\, m))x^n + \cdots + S_m(\mathrm{irem}(a_0,\, m)). \qquad (9.99)$$

In either case, $T_m(u)$ is called the **projection** *of u in $\mathbf{Z}_m[x]$.*

Example 9.38. For $u = x^2 + 5\,x + 8$, using the non-negative representation of $\mathbf{Z}_5$, we have $T_5(u) = x^2 + 3$, while using the symmetric representation, we have $T_5(u) = x^2 - 2$. Notice that there are infinitely many polynomials with the projection $x^2 - 2$. For example, using the symmetric representation, we also have $T_5(6\,x^2 + 10\,x + 13) = x^2 - 2$. □

To simplify our notation, we use the same operator $T_m(u)$ for both the non-negative and symmetric representations for $\mathbf{Z}_m$. In our mathematical discussions, it will be clear from the context whether (9.98) or (9.99) is appropriate. In our procedures, however, we use the projection operators $Tnn(u, x, m)$ for the non-negative representation (9.98) and $Ts(u, x, m)$ for the symmetric representation (9.99) (Exercise 2).

The projection operator satisfies the following properties.

Theorem 9.39. *Let $m \geq 2$ be an integer, and let v and w be polynomials in $\mathbf{Z}[x]$. Then, the projection operator satisfies the following properties.*

1. *$\deg(T_m(v)) \leq \deg(v)$ where the operator $<$ applies when $m \mid \mathrm{lc}(v)$.*

2. *$T_m(v) = 0$ if and only if m divides each coefficient of v.*

3. *For v in $\mathbf{Z}_m[x]$, $T_m(v) = v$.*

4. *$T_m(T_m(v)) = T_m(v)$.*

5. *$T_m(v + w) = T_m(v) \oplus_m T_m(w)$.*

6. *$T_m(v \cdot w) = T_m(v) \otimes_m T_m(w)$.*

Proof: We prove the theorem using the non-negative, representation of $\mathbf{Z}_m$ and leave the proof of the symmetric case to the reader (Exercise 3).

Parts (1), (2), and (3) follow directly from Definition 9.37, and Part (4) follows from Part (3).

To show (5), first for integers a and b, integer division gives

$$a = q_1\, m + T_m(a), \qquad b = q_2\, m + T_m(b). \qquad (9.100)$$

Therefore,

$$
\begin{aligned}
T_m(a+b) &= \operatorname{irem}(a+b,\, m) \\
&= \operatorname{irem}(q_1 m + T_m(a) + q_2 m + T_m(b),\, m) \\
&= \operatorname{irem}(T_m(a) + T_m(b),\, m) \\
&= T_m(a) \oplus_m T_m(b). \qquad (9.101)
\end{aligned}
$$

Part (5) follows for polynomials by applying this relationship to the corresponding coefficients of v and w.

To show (6), using (9.100), we obtain

$$
\begin{aligned}
T_m(a \cdot b) &= \operatorname{irem}\left(q_1 q_2\, m^2 + T_m(a)q_2 m + T_m(b)q_1 m + T_m(a)T_m(b), m\right) \\
&= \operatorname{irem}(T_m(a) \cdot T_m(b),\, m) \\
&= T_m(a) \otimes_m T_m(b). \qquad (9.102)
\end{aligned}
$$

Part (6) follows for polynomials by applying Equations (9.101) and (9.102) to the coefficients of $v \cdot w$. $\qquad\qquad\square$

As a consequence of Theorem 9.39(5),(6), the polynomial operations $\oplus_m$ and $\otimes_m$ can be obtained by performing the corresponding operations in $\mathbf{Z}[x]$ (with expansion and automatic simplification) and then projecting the results into $\mathbf{Z}_m[x]$. This point is illustrated in the next example.

Example 9.40. In $\mathbf{Z}_7[x]$ (using the symmetric representation), we have

$$
\begin{aligned}
(2\,x - 3) &\otimes_7 (3\,x - 1) \otimes_7 \left(x^2 + 4\,x - 1\right) \\
&= T_7(2\,x - 3) \otimes_7 T_7(3\,x - 1) \otimes_7 T_7\left(x^2 + 4\,x - 1\right) \\
&= T_7\left((2\,x - 3)(3\,x - 1)(x^2 + 4\,x - 1)\right) \\
&= T_7\left(6\,x^4 + 13\,x^3 - 47\,x^2 + 23\,x - 3\right) \\
&= -x^4 - x^3 + 2\,x^2 + 2\,x - 3.
\end{aligned}
$$
$\qquad\qquad\square$

The next theorem gives some simple relationships between polynomials in $\mathbf{Z}[x]$ and $\mathbf{Z}_m[x]$ that are useful in our manipulations.

Theorem 9.41. *Let u and v be polynomials in $\mathbf{Z}_m[x]$. Then, there are polynomials h and k in $\mathbf{Z}[x]$ such that*

$$
\begin{aligned}
u \oplus_m v &= u + v + m\,h, \\
u \otimes_m v &= u \cdot v + m\,k, \\
T_m(u) &= u + m\,h,
\end{aligned}
$$

where the operations on the right sides of the equations are performed in $\mathbf{Z}[x]$.

Proof: The theorem, which holds for either the non-negative or symmetric representations of $\mathbf{Z}_m[x]$, follows directly from the definitions of the arithmetic operations in $\mathbf{Z}_m$ (Exercise 8). $\qquad\square$

Relationship of Factorization of u and $T_m(u)$

Let's return now to the factorization problem. Theorem 9.39(6) shows that a factorization of u in $\mathbf{Z}[x]$ is reflected in a factorization of $T_m(u)$ in $\mathbf{Z}_m[x]$. For example, in $\mathbf{Z}[x]$, we have

$$ u = x^2 + 11x + 28 = (x+7)(x+4), $$

and in $\mathbf{Z}_5[x]$, we have the symmetric representation

$$ T_5(u) = x^2 + x - 2 = (x+2) \otimes_5 (x-1). $$

In some cases, however, the projection has more factors than u does. For example, in $\mathbf{Z}[x]$,

$$ u = x^3 + x^2 + x + 1 = (x+1)(x^2 + 1), $$

while, in $\mathbf{Z}_5[x]$,

$$ T_5(u) = (x+1) \otimes_5 (x+2) \otimes_5 (x-2). $$

However, when $m \nmid \mathrm{lc}(u)$, $T_m(u)$ cannot have fewer factors (of positive degree) than u. Indeed, if

$$ u = u_1 \cdots u_r $$

where each factor has positive degree, then Theorem 9.39(6) implies that

$$ T_m(u) = T_m(u_1) \otimes_m \cdots \otimes_m T_m(u_r). $$

In addition, by Theorem 4.1 (page 112), $m \nmid \mathrm{lc}(u_i)$. and so Theorem 9.39(1) implies that each polynomial $T_m(u_i)$ has positive degree.

On the other hand, if $m \mid \mathrm{lc}(u)$, the projection may have fewer factors. For example, in $\mathbf{Z}[x]$,

$$u = 5x^2 + 11x + 2 = (5x + 1)(x + 2)$$

while $T_5(u) = x + 2$.

We summarize this discussion in the next theorem.

Theorem 9.42. *Let u be a polynomial in $\mathbf{Z}[x]$ such that $m \nmid \mathrm{lc}(u)$.*

1. *Suppose that the irreducible factorization of u has r factors with positive degree and the irreducible factorization of $T_m(u)$ has s factors with positive degree. Then, $s \geq r$.*

2. *If $T_m(u)$ is irreducible in $\mathbf{Z}_m[x]$, then u is irreducible in $\mathbf{Z}[x]$.*

Part (2) of the theorem is an important special case of Part (1).

Example 9.43. The converse of Theorem 9.42(2) is not true. For example, $u = x^2 + x + 4$ is irreducible in $\mathbf{Z}[x]$, while

$$T_5(u) = x^2 + x - 1 = (x - 2) \otimes_5 (x - 2)$$

in $\mathbf{Z}_5[x]$. In fact, $T_5(u)$ is not even square-free.   □

Example 9.44. Let's show that the polynomial

$$u = x^3 + 6x^2 + 16x + 1$$

is irreducible using Theorem 9.42(2). First, we try

$$T_2(u) = x^3 + 1 = (x + 1) \otimes_2 (x^2 + x + 1).$$

Since the projection is reducible, this gives no information about the reducibility of u. In a similar way,

$$T_3(u) = x^3 + x + 1 = (x^2 + x - 1) \otimes_3 (x - 1)$$

and

$$T_5(u) = x^3 + x^2 + x + 1 = (x + 1) \otimes_5 (x + 2) \otimes_5 (x - 2),$$

and so we still cannot arrive at a conclusion. However,

$$T_7(u) = x^3 - x^2 + 2x + 1$$

is irreducible in $\mathbf{Z}_7[x]$, and therefore, u is irreducible in $\mathbf{Z}[x]$.   □

The previous examples show that the irreducible factorizations of u and $T_m(u)$ can look quite different. The question is then: how can the irreducible factorization of $T_m(u)$ help obtain the factors of u?

Here is one approach. Let's suppose that u has the irreducible factorization

$$u = u_1 \cdots u_r \tag{9.103}$$

in $\mathbf{Z}[x]$, and suppose that m is large enough so that the coefficients of u and each of its factors u_i are in the symmetric representation of $\mathbf{Z}_m[x]$. Then, since $T_m(u) = u$ and $T_m(u_i) = u_i$,

$$u = u_1 \otimes_m \cdots \otimes_m u_r$$

is a factorization of u in $\mathbf{Z}_m[x]$. However, if some of the u_i are reducible in $\mathbf{Z}_m[x]$, this is not the irreducible factorization for u in this setting.

On the other hand, if

$$u = v_1 \otimes_m \cdots \otimes_m v_s$$

is the irreducible factorization of u in $\mathbf{Z}_m[x]$, then each u_i is either some v_j or the product (in $\mathbf{Z}_m[x]$) of a number of v_j. This suggests that each v_j or the product of a number of v_j is a possible factor of u in $\mathbf{Z}[x]$ and, by checking all such polynomials, we can recover the factors u_i in Equation (9.103). We illustrate this approach in the next example.

Example 9.45. Consider the polynomial

$$u = x^4 + 5x^3 + 6x^2 + 5x + 1$$

where u and each factor of u is in $\mathbf{Z}_{13}[x]$. Since 13 is prime, we factor the square-free polynomial u using Berlekamp's algorithm which gives (in the symmetric representation for $\mathbf{Z}_{13}$)

$$u = (x + 4) \otimes_{13} (x + 6) \otimes_{13} (x - 2) \otimes_{13} (x - 3).$$

However, none of these linear factors is a factor of u in $\mathbf{Z}[x]$.

Let's consider the trial factors obtained as the product of two of the linear factors. For example, we try

$$v = (x + 4) \otimes_{13} (x + 6) = x^2 - 3x - 2$$

but find, using polynomial division in $\mathbf{Z}[x]$, that v is not a factor. On the other hand,

$$v = (x + 4) \otimes_{13} (x - 3) = x^2 + x + 1$$

is a factor, and

$$\mathrm{quot}(u, v) = x^2 + 4x + 1$$

is also a factor. Therefore, since u has no linear factors, its irreducible factorization in $\mathbf{Z}[x]$ is

$$u = (x^2 + x + 1)(x^2 + 4x + 1). \tag{9.104}$$

$\square$

In order to give an algorithm that uses this approach, we must address two issues.

1. How do we determine m so that u and all of its (unknown) irreducible factors have coefficients in the symmetric representation of $\mathbf{Z}_m$? We cannot simply use for m the maximum of the absolute value of the coefficients of u because there are polynomials that have factors with coefficients larger than any coefficient of the original polynomial. (For example, see Equation (9.105) below.)

2. In Example 9.45, we used a prime $m = 13$ so that Berlekamp's algorithm obtains a factorization in $\mathbf{Z}_m[x]$. For a large prime, however, Berlekamp's algorithm is time consuming, and so we prefer to use only small primes. But if m is too small, the coefficients of u or one of its factors may not be in $\mathbf{Z}_m$.

We consider both of these issues below.

Coefficient Bounds on Factors of a Polynomial

The next definition and theorem provide a way to determine m so that all of the coefficients of u and its factors are in $\mathbf{Z}_m$.

Definition 9.46. *Let* $u = a_n x^n + \cdots + a_0$ *be in* $\mathbf{Z}[x]$. *The* **height** *of the polynomial* u *is defined as*

$$||u|| = \max(\{|a_n|, \ldots, |a_0|\}).$$

In most cases, for $v \mid u$, we have $||v|| \leq ||u||$. This point is illustrated by the factorization in Equation (9.104). There are, however, some polynomials where $||v|| > ||u||$. For example,

$$u = x^4 - 2x^3 - x^2 - 2x + 1 = (x^2 + x + 1)(x^2 - 3x + 1) \tag{9.105}$$

where $3 = ||x^2 - 3x + 1|| > ||u|| = 2$.

The next theorem gives a bound on the coefficients of a factor of u.

Theorem 9.47. *Let u and v be in $\mathbf{Z}[x]$ where $v \mid u$, and let $n = \deg(u)$. Then, $\|v\| \leq B$ where*

$$B = 2^n \sqrt{n+1}\,\|u\|. \tag{9.106}$$

We do not prove this theorem, since the proof depends on some involved mathematics that is well beyond the scope of this book. The general theory that leads to this theorem is cited in the references at the end of this chapter. $\square$

The upper bound B is usually considerably greater than the coefficients of v. For example, for u in Equation (9.105),

$$B = 2^4 \sqrt{5} \cdot 2 \approx 71.55,$$

while the coefficients of both factors of u are actually bounded by 3. However, B provides a way to obtain an m so that u and all of its irreducible factors are in the symmetric representation $\mathbf{Z}_m[x]$. Since each b in $\mathbf{Z}_m[x]$ satisfies $|b| \leq m/2$ (see Equation (9.96)), we choose an integer m with

$$2\,B \leq m. \tag{9.107}$$

For example, for u in Equation (9.105),

$$m = 144 \geq 2\,B \approx 143.11.$$

Example 9.48. Consider the polynomial

$$u = x^5 - 48x^4 + 720x^3 - 5760x^2 + 59904x - 110592.$$

Then, $\|u\| = 110592$ and $B = 2^5 \sqrt{6}\,110592$. Therefore, for

$$m \geq 2^6 \sqrt{6} \cdot 110592,$$

all coefficients of u and its irreducible factors are in $\mathbf{Z}_m[x]$. The actual value of m used in the factorization of u is determined in Example 9.57 below. $\square$

Hensel's Lifting Theorem

We have described an approach that factors u in $\mathbf{Z}[x]$ by obtaining a factorization in $\mathbf{Z}_m[x]$ for a suitably large positive integer m. Although Berlekamp's algorithm obtains the factorization when m is prime, it is not efficient when m is large. Hensel's lifting theorem (Theorem 9.52 below) provides a more efficient strategy. In this approach, we choose a small prime p, factor $T_p(u)$ in $\mathbf{Z}_p[x]$, and then use this factorization and the lifting process to obtain a factorization in $\mathbf{Z}_{p^k}[x]$ for $k > 1$. From the inequality (9.107), when

$$2\,B \le p^k,$$

the symmetric representation of $\mathbf{Z}_{p^k}$ contains the coefficients of both u and all of its factors.

The Hensel lifting process is based on the following generalization of the extended Euclidean algorithm.

Theorem 9.49. *Let $\mathbf{F}$ be a field, and let $v_1, \ldots, v_s$ be two or more polynomials in $\mathbf{F}[x]$ with positive degree such that $\gcd(v_i, v_j) = 1$ for $i \ne j$. Let F be a polynomial in $\mathbf{F}[x]$ with*

$$\deg(F) < \sum_{i=1}^{s} \deg(v_i), \tag{9.108}$$

and let

$$v = v_1 \cdots v_s. \tag{9.109}$$

Then, there are polynomials $r_1, \ldots, r_s$ with $\deg(r_i) < \deg(v_i)$ such that

$$F = \sum_{i=1}^{s} r_i g_i, \tag{9.110}$$

where

$$g_i = v/v_i. \tag{9.111}$$

Proof: We first show that, for the special case $F = 1$, there are polynomials $\sigma_1, \ldots, \sigma_s$ such that

$$\sum_{i=1}^{s} \sigma_i g_i = 1. \tag{9.112}$$

The proof uses mathematical induction. First, for the base case $s = 2$, let

$$v = v_1 v_2, \quad g_1 = v/v_1 = v_2, \quad g_2 = v/v_2 = v_1.$$

In this case, Theorem 4.30 (page 133) implies that there are polynomials A and B such that

$$A\,g_1 + B\,g_2 = A\,v_2 + B\,v_1 = 1, \tag{9.113}$$

and so $\sigma_1 = A$, $\sigma_2 = B$.

For the induction step, assume that Equation (9.112) is true for the $s - 1$ polynomials $v_1, \ldots, v_{s-1}$. Define v and g_i by Equations (9.109) and (9.111), and let

$$h_i = \frac{v}{v_s v_i}, \quad i = 1, \ldots, s - 1. \tag{9.114}$$

By the induction hypothesis, there are polynomials $\tau_1, \ldots, \tau_{s-1}$ such that

$$\sum_{i=1}^{s-1} \tau_i h_i = 1. \tag{9.115}$$

The polynomials $g_s = v/v_s$ and v_s are relatively prime, and so, by Theorem 4.30, there are polynomials A and B such that

$$A v_s + B g_s = 1. \tag{9.116}$$

In addition, from Equations (9.111), (9.114), and (9.115), we have

$$v_s = v_s \sum_{i=1}^{s-1} \tau_i h_i = \sum_{i=1}^{s-1} \left(\frac{\tau_i v_s g_s}{v_i} \right) = \sum_{i=1}^{s-1} \tau_i g_i. \tag{9.117}$$

Therefore, using Equations (9.116) and (9.117), we obtain

$$1 = A \sum_{i=1}^{s-1} \tau_i g_i + B g_s = \sum_{i=1}^{s} \sigma_i g_i \tag{9.118}$$

where

$$\sigma_i = \begin{cases} A \tau_i, & i = 1, \ldots, s - 1, \\ B, & i = s. \end{cases} \tag{9.119}$$

We now use Equation (9.112) to derive the representation (9.110) for a general polynomial F that satisfies the condition (9.108). From Equation (9.112), we have

$$\sum_{i=1}^{s} F \sigma_i g_i = F, \tag{9.120}$$

and by polynomial division

$$F \sigma_i = q_i v_i + r_i \tag{9.121}$$

where

$$q_i = \operatorname{quot}(F \sigma_i, \, v_i), \quad r_i = \operatorname{rem}(F \sigma_i, \, v_i), \tag{9.122}$$

and $\deg(r_i) < \deg(v_i)$. Using Equations (9.111), (9.120), and (9.121), we have

$$\sum_{i=1}^{s} r_i g_i = \sum_{i=1}^{s} F \sigma_i g_i - \sum_{i=1}^{s} q_i v_i g_i = F - v \sum_{i=1}^{s} q_i. \tag{9.123}$$

To show Equation (9.110), we must show that $\sum_{i=1}^{s} q_i = 0$. First, using Equation (9.111),

$$\deg\left(\sum_{i=1}^{s} r_i g_i\right) \quad \leq \quad \max_{1 \leq i \leq s}\left(\{\deg(r_i) + \deg(g_i)\}\right)$$

$$< \quad \max_{1 \leq i \leq s}\left(\{\deg(v_i) + \deg(g_i)\}\right) \qquad (9.124)$$

$$= \quad \deg(v).$$

Now, if $\sum_{i=1}^{s} q_i \neq 0$, Theorem 4.1 (page 112) and the inequality (9.108) imply that

$$\deg\left(F - u \sum_{i=1}^{s} q_i\right) \quad = \quad \max\left(\left\{\deg(F), \ \deg(v) + \deg\left(\sum_{i=1}^{s} q_i\right)\right\}\right)$$

$$= \quad \deg(v) + \deg\left(\sum_{i=1}^{s} q_i\right)$$

$$\geq \quad \deg(v). \qquad (9.125)$$

Since the inequalities (9.124) and (9.125) cannot both be true, we have

$$\sum_{i=1}^{s} q_i = 0.$$

Therefore, Equation (9.123) implies Equation (9.110). □

In Exercise 11, we describe procedures that obtain the polynomial coefficients σ_i and r_i when $\mathbf{F} = \mathbf{Z}_p$.

Example 9.50. Let $\mathbf{F}[x] = \mathbf{Z}_3[x]$ (with the symmetric representation for $\mathbf{Z}_3$), and consider the polynomials

$$v_1 = x, \quad v_2 = x - 1, \quad F = x + 1.$$

We have

$$v = x \otimes_3 (x - 1), \quad g_1 = v/v_1 = x - 1, \quad g_2 = v/v_2 = x.$$

Applying the extended Euclidean algorithm to $g_1 = v_2$ and $g_2 = v_1$, there are polynomials $\sigma_1 = -1$ and $\sigma_2 = 1$ such that $(\sigma_1 \otimes_3 g_1) \oplus_3 (\sigma_2 \otimes_3 g_2) = 1$. Now with

$$r_1 = \mathrm{rem}(F \otimes_3 \sigma_1, \ v_1) = -1, \quad r_2 = \mathrm{rem}(F \otimes_3 \sigma_2, \ v_2) = -1,$$

we have

$$(r_1 \otimes_3 g_1) \oplus_3 (r_2 \otimes_3 g_2) = ((-1) \otimes_3 (x-1)) \oplus_3 ((-1) \otimes_3 x) = x+1. \quad \square$$

Example 9.51. Let $F[x] = \mathbf{Z}_{11}[x]$ (with the symmetric representation for $\mathbf{Z}_{11}$), and let

$$v_1 = x^2 - 2x + 4, \quad v_2 = x+5, \quad v_3 = x-5, \quad v_4 = x-2,$$

$$F = -4x^4 + x^3 - 4x^2 - 3x + 4.$$

Observe that F satisfies the degree condition (9.108). By repeatedly applying the extended Euclidean algorithm (as in the proof of Theorem 9.49), we have

$$\begin{array}{rcl}
\sigma_1 & = & -2x^6 - 2x^5 - 2x^4 - 4x^3 - 5x^2 + 3x + 2, \\
\sigma_2 & = & 2x^5 + 5x^4 + 4x^3 - x^2 - 2x + 5, \\
\sigma_3 & = & -x^3 - x + 3, \\
\sigma_4 & = & 3.
\end{array} \tag{9.126}$$

Computing the remainders in Equation (9.122), we obtain

$$r_1 = -2x + 4, \quad r_2 = 0, \quad r_3 = 0, \quad r_4 = -2. \tag{9.127}$$

$\square$

The next theorem describes the Hensel lifting process.

Theorem 9.52. *Let u be a monic polynomial in $\mathbf{Z}[x]$ with positive degree, and let $p \geq 3$ be a prime number. Suppose that*

$$T_p(u) = v_1 \otimes_p \cdots \otimes_p v_s$$

where $v_1, \ldots, v_s$ are monic, relatively prime polynomials in $\mathbf{Z}_p[x]$. Then, for each positive integer k, there are monic, relatively prime polynomials $v_1^{(k)}, \ldots, v_s^{(k)}$ in $\mathbf{Z}_{p^k}[x]$ such that

1. $T_{p^k}(u) = v_1^{(k)} \otimes_{p^k} \cdots \otimes_{p^k} v_s^{(k)}$,

2. $T_p\left(v_i^{(k)}\right) = v_i$ *for* $i = 1, \ldots, s$.

Proof: The theorem holds for both the non-negative and symmetric representation of $\mathbf{Z}_p$ and $\mathbf{Z}_{p^k}$. Since the proof is quite involved, we prove the theorem for $s = 2$ using the symmetric representation. The proof, which uses mathematical induction, gives an algorithm for computing $v_1^{(k)}$ and $v_2^{(k)}$.

First, for the base case $k = 1$, $v_1^{(1)} = v_1$ and $v_2^{(1)} = v_2$ satisfy properties (1) and (2) of the theorem.

Next, suppose that $k \geq 2$, and let's assume that we have found $v_1^{(k-1)}$ and $v_2^{(k-1)}$, and show how to lift one more step to find $v_1^{(k)}$ and $v_2^{(k)}$. By the induction hypothesis, property (1) of the theorem gives the factorization in $\mathbf{Z}_{p^{k-1}}[x]$

$$T_{p^{k-1}}(u) = v_1^{(k-1)} \otimes_{p^{k-1}} v_2^{(k-1)}. \tag{9.128}$$

However, the product $v_1^{(k-1)} v_2^{(k-1)}$ in $\mathbf{Z}[x]$ may not be a factorization of u, and so we define the "error" in $\mathbf{Z}[x]$ between u and this factorization as

$$E = u - v_1^{(k-1)} v_2^{(k-1)}. \tag{9.129}$$

This error polynomial E is used to lift the factorization from $\mathbf{Z}_{p^{k-1}}[x]$ to $\mathbf{Z}_{p^k}[x]$.

First, let's show that E satisfies the degree condition

$$\deg(E) < \deg(u). \tag{9.130}$$

Since u is monic,

$$\deg(u) = \deg\left(T_{p^{k-1}}(u)\right) = \deg\left(v_1^{(k-1)}\right) + \deg\left(v_2^{(k-1)}\right),$$

and since $v_1^{(k-1)}$ and $v_2^{(k-1)}$ are also monic, the subtraction in Equation (9.129) eliminates the highest degree monomial, which gives the inequality (9.130). In addition, since

$$\deg(u) = \deg\left(T_p(u)\right) = \deg\left(v_1\right) + \deg\left(v_2\right),$$

we have

$$\deg(E) < \deg\left(v_1\right) + \deg\left(v_2\right). \tag{9.131}$$

By Equations (9.128) and (9.129),

$$T_{p^{k-1}}(E) = 0,$$

and so Theorem 9.39(2) implies that p^{k-1} divides the coefficient of each monomial in E.

On the other hand, the projection $T_{p^k}(E)$ in $\mathbf{Z}_{p^k}[x]$ may not be 0, although the coefficient of each monomial in this projection has the form

$r p^{k-1}$ where r is in $\mathbf{Z}_p$. To see why, since each coefficient of E has the form $q p^{k-1}$, dividing q by p we obtain

$$q p^{k-1} = (a p + s) p^{k-1} = a p^k + s p^{k-1}$$

where $0 \leq s \leq p - 1$. From Equation (9.97), we have

$$
\begin{aligned}
T_{p^k}\left(q p^{k-1}\right) &= S_{p^k}\left(\text{irem}\left(q p^{k-1}, p^k\right)\right) \\
&= S_{p^k}\left(\text{irem}\left(a p^k + s p^{k-1}, p^k\right)\right) \\
&= S_{p^k}\left(s p^{k-1}\right) \\
&= \begin{cases} s p^{k-1}, & \text{if } 0 \leq s p^{k-1} \leq \text{iquot}\left(p^k, 2\right), \\ s p^{k-1} - p^k, & \text{if } \text{iquot}\left(p^k, 2\right) < s p^{k-1} < p^k \end{cases} \\
&= r p^{k-1}
\end{aligned}
$$

where

$$
r = \begin{cases} s, & \text{if } 0 \leq s \leq \text{iquot}(p, 2), \\ s - p, & \text{if } \text{iquot}(p, 2) < s < p. \end{cases}
$$

Therefore, the coefficient of each monomial in $T_{p^k}(E)$ has the form $r p^{k-1}$ with r in $\mathbf{Z}_p$.

Next, define

$$F = T_{p^k}(E)/p^{k-1} \tag{9.132}$$

where the above discussion shows that F is in $\mathbf{Z}_p[x]$. In addition, from the inequality (9.131), we have

$$\deg(F) < \deg(v_1) + \deg(v_2),$$

and so Theorem 9.49 implies that there are polynomials r_1 and r_2 in $\mathbf{Z}_p[x]$ such that

$$(r_2 \otimes_p v_1) \oplus_p (r_1 \otimes_p v_2) = F \tag{9.133}$$

with

$$\deg(r_1) < \deg(v_1), \quad \deg(r_2) < \deg(v_2). \tag{9.134}$$

The polynomials r_1 and r_2 are used below to define the polynomials $v_1^{(k)}$ and $v_2^{(k)}$.

However, before we do this, we need to mention a technical point. Suppose that b is in $\mathbf{Z}_{p^{k-1}}$ and c is in $\mathbf{Z}_p$. Then, since the prime $p \geq 3$ is odd, Equation (9.96) implies that

$$-\frac{p^{k-1} - 1}{2} \leq b \leq \frac{p^{k-1} - 1}{2}, \quad -\frac{p - 1}{2} \leq c \leq \frac{p - 1}{2}.$$

Applying these inequalities to the expression $b + p^{k-1} c$, we have

$$-\frac{p^{k-1} - 1}{2} - p^{k-1}\left(\frac{p-1}{2}\right) \leq b + p^{k-1} c \leq \frac{p^{k-1} - 1}{2} + p^{k-1}\left(\frac{p-1}{2}\right),$$

and, therefore,

$$-\frac{p^k - 1}{2} \leq b + p^{k-1} c \leq \frac{p^k - 1}{2} \tag{9.135}$$

which implies $b + p^{k-1} c$ is in $\mathbf{Z}_{p^k}$.

Define the polynomials at the next step k with

$$v_1^{(k)} = v_1^{(k-1)} + p^{k-1} r_1, \qquad v_2^{(k)} = v_2^{(k-1)} + p^{k-1} r_2. \tag{9.136}$$

We show that $v_1^{(k)}$ and $v_2^{(k)}$ satisfy the properties the theorem. First, applying the inequality (9.135) to the coefficients of the monomials in $v_1^{(k)}$ and $v_2^{(k)}$, it follows that both polynomials are in $\mathbf{Z}_{p^k}[x]$. Next, since the polynomials $v_1^{(1)}$, $v_1^{(k-1)}$, $v_2^{(1)}$, and $v_2^{(k-1)}$ are all monic, property (2) of the induction hypothesis implies that

$$\deg\left(v_1^{(k-1)}\right) = \deg(v_1), \qquad \deg\left(v_2^{(k-1)}\right) = \deg(v_2). \tag{9.137}$$

Therefore, the inequality (9.134) and Equations (9.136) and (9.137) imply that $v_1^{(k)}$ and $v_2^{(k)}$ are also monic, and

$$\deg\left(v_1^{(k)}\right) = \deg(v_1), \qquad \deg\left(v_2^{(k)}\right) = \deg(v_2).$$

To show that $v_1^{(k)}$ and $v_2^{(k)}$ satisfy property (1), first observe, using property (2) in the induction hypothesis and Theorem 9.41, there are polynomials M and N in $\mathbf{Z}[x]$ such that

$$v_1^{(k-1)} = v_1 + p M, \qquad v_2^{(k-1)} = v_2 + p N. \tag{9.138}$$

In addition, since, $k \geq 2$, we have

$$2(k - 1) \geq k,$$

and this inequality, Equations (9.136) and (9.138), and the properties of the projection operator (Theorem 9.39) imply that

$$\begin{aligned}
v_1^{(k)} \otimes_{p^k} v_2^{(k)} &= \left(v_1^{(k-1)} + p^{k-1} r_1\right) \otimes_{p^k} \left(v_2^{(k-1)} + p^{k-1} r_2\right) \\
&= T_{p^k}\left(\left(v_1^{(k-1)} + p^{k-1} r_1\right)\left(v_2^{(k-1)} + p^{k-1} r_2\right)\right)
\end{aligned}$$

$$= T_{p^k}\left(v_1^{(k-1)}v_2^{(k-1)} + p^{k-1}\left(r_2v_1 + r_1v_2\right)\right.$$
$$\left. + p^k\left(r_2M + r_1N\right) + p^{2(k-1)}r_1r_2\right)$$
$$= T_{p^k}\left(v_1^{(k-1)}v_2^{(k-1)} + p^{k-1}\left(r_2v_1 + r_1v_2\right)\right). \quad (9.139)$$

Now, by Theorem 9.41 and Equation (9.133), there is a polynomial f in $\mathbf{Z}[x]$ such that

$$F = r_2v_1 + r_1v_2 + p\,f.$$

Therefore, using this relation, Equations (9.129), (9.132), and (9.139) together with the properties of the projection operator (Theorem 9.39), we have

$$\begin{aligned}
v_1^{(k)} \otimes_{p^k} v_2^{(k)} &= T_{p^k}\left(v_1^{(k-1)}v_2^{(k-1)} + p^{k-1}(F - p\,f)\right) \\
&= T_{p^k}\left(v_1^{(k-1)}v_2^{(k-1)}\right) + T_{p^k}\left(p^{k-1}F\right) - T_{p^k}\left(p^k f\right) \\
&= T_{p^k}\left(v_1^{(k-1)}v_2^{(k-1)} + E\right) \\
&= T_{p^k}(u),
\end{aligned}$$

which verifies property (1).

To show property (2), using the representation (9.136), Theorem 9.39(2),(5), and the induction hypothesis, we have

$$\begin{aligned}
T_p\left(v_i^{(k)}\right) &= T_p\left(v_i^{(k-1)} + p^{k-1}r_i\right) \\
&= T_p\left(v_i^{(k-1)}\right) \oplus_p T_p\left(p^{k-1}r_1\right) \\
&= v_i.
\end{aligned}$$

Finally, we leave to the reader the proof that $v_1^{(k)}$ and $v_2^{(k)}$ are relatively prime (Exercise 9). □

In the next example, we illustrate the algorithm in the proof of the lifting theorem.

Example 9.53. Let $p = 3$ and

$$u = x^2 + 11x + 30.$$

We apply two levels of the Hensel lifting process using the symmetric representation for $\mathbf{Z}_3[x]$. First, since

$$T_3(u) = x^2 - x = x \otimes_3 (x - 1),$$

we have

$$v_1^{(1)} = v_1 = x, \qquad v_2^{(1)} = v_2 = x - 1.$$

The preliminary calculations for this example were done in Example 9.50, where we obtained $\sigma_1 = -1$ and $\sigma_2 = 1$. We compute these polynomials at the beginning of the process since they are used at each of the lifting steps.

To obtain $v_1^{(2)}$ and $v_2^{(2)}$, we have

$$E = u - v_1^{(1)} v_2^{(1)} = 12\,x + 30$$

and

$$F = T_{3^2}(E)/3 = (3\,x + 3)/3 = x + 1.$$

Again, from Example 9.50, we have $r_1 = -1$ and $r_2 = -1$. Therefore, in $\mathbf{Z}_{3^2}[x]$,

$$v_1^{(2)} = T_{3^2}(v_1^{(1)} + 3\,r_1) = x - 3, \qquad v_2^{(2)} = T_{3^2}(v_2^{(1)} + 3\,r_2) = x - 4.$$

To obtain $v_1^{(3)}$ and $v_2^{(3)}$, we have

$$E = u - v_1^{(2)} v_2^{(2)} = 18\,x + 18$$

and

$$F = T_{3^3}(E)/3^2 = -x - 1.$$

At this step, by Equation (9.122),

$$r_1 = \mathrm{rem}(F \otimes_p \sigma_1,\ v_1) = -1, \qquad r_2 = \mathrm{rem}(F \otimes_p \sigma_2,\ v_2) = -1,$$

and, therefore,

$$v_1^{(3)} = T_{3^3}(v_1^{(2)} + 3^2\,r_1) = x + 6, \qquad v_2^{(3)} = T_{3^3}(v_2^{(2)} + 3^2\,r_2) = x + 5.$$

Since $u = v_1^{(3)} v_2^{(3)}$ is in $\mathbf{Z}[x]$, the lifting process terminates. (If we continue, the next $E = 0$, and therefore $v_1^{(4)} = v_1^{(3)}$ and $v_2^{(4)} = v_2^{(3)}$.)   □

Hensel Lifting Process. The Hensel lifting process is based on the following calculations. First, using Equations (9.113) and (9.119), we obtain a sequence of polynomials $\sigma_1, \dots, \sigma_n$ in $\mathbf{Z}_p[x]$ such that

$$(\sigma_1 \otimes_p g_1) \oplus_p \cdots \oplus_p (\sigma_n \otimes_p g_s) = 1, \tag{9.140}$$

where

$$v = v_1 \otimes_p \cdots \otimes_p v_s$$

(in unexpanded form) and $g_i = v/v_i$. The polynomials σ_i are computed at the beginning of the process because they are used at each lifting step. Next, let $v_i^{(1)} = v_1$, $i = 1, \ldots, s$, and to obtain $v_i^{(j)}, \ldots, v_s^{(j)}$ (for $2 \leq j \leq k$), let

$$E = u - v_1^{(j-1)} \cdots v_s^{(j-1)}, \tag{9.141}$$
$$F = T_{p^j}(E)/p^j, \tag{9.142}$$

where the algebraic operations (expansion and automatic simplification) are performed in $\mathbf{Z}[x]$. Following this, we obtain the coefficient polynomials r_i (for $1 \leq i \leq s$), defined by Equation (9.122) with the operation (in $\mathbf{Z}_p[x]$)

$$r_i = \text{rem}(F \otimes_p \sigma_i, \; v_i), \quad i = 1, \ldots, s. \tag{9.143}$$

Finally, we obtain the new lifted factors with

$$v_i^{(j)} = T_{p^j}\left(v_i^{(j-1)} + p^{j-1} r_i\right), \quad i = 1, \ldots, s, \tag{9.144}$$

where the algebraic operation on the expression in parentheses is performed in $\mathbf{Z}[x]$.

Example 9.54. Consider the polynomial

$$u = x^5 - 48\,x^4 + 720\,x^3 - 5760\,x^2 + 59904\,x - 110592,$$

and let $p = 11$. By projecting u into $\mathbf{Z}_{11}[x]$ and factoring the (square-free) projection using Berlekamp's algorithm, we have (in the symmetric representation)

$$\begin{aligned}
T_{11}(u) = v &= v_1 \otimes_{11} v_2 \otimes_{11} v_3 \otimes_{11} v_4 \\
&= \left(x^2 - 2\,x + 4\right) \otimes_{11} (x+5) \otimes_{11} (x-5) \tag{9.145} \\
&\quad \otimes_{11} (x-2) \\
&= v_1^{(1)} \otimes_{11} v_2^{(1)} \otimes_{11} v_3^{(1)} \otimes_{11} v_4^{(1)}.
\end{aligned}$$

The preliminary calculations for this example were done in Example 9.51, where the σ_i are given in Equation (9.126). Then, from Equation (9.141) with $j = 2$,

$$\begin{aligned}
E &= u - \left(x^2 - 2\,x + 4\right)(x+5)(x-5)(x-2) \\
&= -44\,x^4 + 737\,x^3 - 5852\,x^2 + 60104\,x - 110792,
\end{aligned}$$

and therefore, by Equation (9.142),

$$F = T_{11^2}(E)/11 = -4\,x^4 + x^3 - 4\,x^2 - 3\,x + 4.$$

Again, from Example 9.51, the polynomials r_i are given in Equation (9.127). Therefore, from Equation (9.144),

$$v_1^{(2)} = x^2 - 24\,x + 48, \quad v_2^{(2)} = x + 5, \quad v_3^{(2)} = x - 5, \quad v_4^{(2)} = x - 24.$$

In a similar way, for $j = 3$,

$$E = 121\,x^3 - 5808\,x^2 + 75504\,x - 139392$$

and

$$F = x^3 - 4\,x^2 - 3\,x + 3.$$

From Equation (9.143), we have

$$r_1 = 0, \quad r_2 = 1, \quad r_3 = -1, \quad r_4 = 0,$$

and from Equation (9.144), we have

$$v_1^{(3)} = x^2 - 24\,x + 48, \quad v_2^{(3)} = x + 126, \quad v_3^{(3)} = x - 126, \quad v_4^{(3)} = x - 24.$$

This example is continued in Examples 9.56 and 9.57 below.   □

Choosing a Prime Number p.   In order to apply the Hensel lifting process, the irreducible factorization of $T_p(u)$ in $\mathbf{Z}_p[x]$ must have relatively prime (and therefore distinct) factors. In other words, $T_p(u)$ must be a square-free polynomial. However, even though u is square-free, $T_p(u)$ may not have this property. (This happens in Example 9.43.) The next theorem shows, however, that it is always possible to find some prime p where $T_p(u)$ is square-free.

Theorem 9.55. *Let p be a prime number, and let u be a square-free polynomial in $\mathbf{Z}[x]$ such that $p \nmid \mathrm{lc}(u)$. Then $T_p(u)$ is not square-free for at most a finite number of p.*

Proof: Recall that $T_p(u)$ is square-free if and only if

$$\mathrm{res}(T_p(u),\ T_p(u)') \neq 0.$$

(See Theorem 7.9 (page 276) and Theorem 9.24 (page 372).) In addition, using Exercise 12,

$$\mathrm{res}(T_p(u), T_p(u)') = T_p(\mathrm{res}(u, u'))$$

and since the integer $\mathrm{res}(u, u')$ is divisible by at most a finite number of primes p, the theorem follows.   □

Example 9.56. Consider the polynomial

$$u = x^5 - 48\,x^4 + 720\,x^3 - 5760\,x^2 + 59904\,x - 110592.$$

Then

$$\mathrm{res}(u, u') = -509012486930992988160000 = -2^{48}\,3^{10}\,5^4\,7^2.$$

Since 2, 3, 5, and 7 are divisors of the resultant, $p = 11$ is the first prime for which $T_p(u)$ is square-free. □

Termination of the Hensel Lifting Process. There are two ways the Hensel lifting process can terminate. First, if at some point in the lifting process the error polynomial $E = 0$, the most recently lifted factors $v_1^{(j-1)}, \ldots, v_s^{(j-1)}$ are the true factors of u in $\mathbf{Z}[x]$. This situation, which arises when u and $T_p(u)$ have the same number of irreducible factors, occurs in Example 9.53. On the other hand, if $T_p(u)$ has more factors than u, the error polynomial E will never be 0. In this case, the lifting process continues until we are guaranteed that the symmetric representation of $\mathbf{Z}_{p^k}$ contains the coefficients of u and all of its irreducible factors. This situation arises in Example 9.57 below. Using the inequality (9.107), we choose this maximum lifting step to be the smallest integer k such that

$$2\,B \le p^k, \tag{9.146}$$

where B is defined in Equation (9.106). In this case, some of the lifted factors may not be factors of u in $\mathbf{Z}[x]$. This point is illustrated in the next example.

Example 9.57. In this example, we obtain the irreducible factorization of the polynomial

$$u = x^5 - 48\,x^4 + 720\,x^3 - 5760\,x^2 + 59904\,x - 110592.$$

First, in Example 9.56 we determined that $p = 11$, and in Example 9.48 we determined that $B = 2^5\sqrt{6} \cdot 110592$. Using the inequality (9.146), we have

$$k \ge \ln(2\,B)/\ln(p) \approx 6.951,$$

and so $k = 7$. The factorization of $T_{11}(u)$ and the first two lifting steps are described in Example 9.54. The lifting steps for $j = 3, 4, 5, 6,$ and 7 are similar with

$$v_1^{(7)} = x^2 - 24\,x + 48, \quad v_2^{(7)} = x - 2868179, \quad v_3^{(7)} = x + 2868179, \quad v_4^{(7)} = x - 24.$$

This is an instance where there are more lifted factors than irreducible factors of u. By polynomial division in $\mathbf{Z}[x]$, $v_1^{(7)}$ and $v_4^{(7)}$ are irreducible factors of u, while $v_2^{(7)}$ and $v_3^{(7)}$ are not. However, the polynomial $v_2^{(7)} \otimes_{11^7} v_3^{(7)} = x^2 - 96$ divides u, and so the irreducible factorization in $\mathbf{Z}[x]$ is

$$u = (x^2 - 24x + 48)(x^2 + 96)(x - 24).$$ □

Procedures

The *Irreducible_factor* Procedure. The procedure *Irreducible_factor*, which obtains the irreducible factorization of a primitive, square-free polynomial, is shown in Figure 9.8. The global symbol y is a mathematical symbol introduced at line 3. We have used a global symbol to avoid using a local variable that would not be assigned before it was used (see page 3).

In lines 1–3, we obtain a monic polynomial V with the transformation in Equation (9.92). At line 4, the *Find_prime* operator obtains a suitable prime number p that is guaranteed by Theorem 9.55 (Exercise 13). At line 5, the *Berlekamp_factor* operator (page 384) obtains the set of irreducible factors of the projection of V in $\mathbf{Z}_p[y]$. Since *Berlekamp_factor* is defined in terms of the non-negative representation of $\mathbf{Z}_p[x]$, the operator *Tnn* projects V into the non-negative representation (Exercise 2).

At line 6, if there is only one factor in S, the irreducible polynomial u is returned. Otherwise, we find the value for k defined by the inequality (9.146) with the *Find_k* operator (Exercise 17), and then, at line 9, find the set of irreducible factors with the *Hensel_lift* operator (defined below). The *Map* operator in this statement transforms the factors in S to the symmetric representation using the projection operator *Ts* described in Exercise 2. Finally, in lines 10–15, we invert the monic transformation from line 3 (see Equation (9.94)).

The *Hensel_lift* Procedure. A procedure that performs the Hensel lifting process is given in Figure 9.9. In lines 1–2, when $k = 1$ there is no lifting to be done. In this case, we recover the factors of u by applying the *True_factors* operator (described below) to the set S.

The case $k \geq 2$ is handled in lines 4–19. Since it is more convenient to work with a list of (distinct) factors than with a set, we obtain a list V at line 4. (The operator *Operand _list* obtains the list of operands of the main operator of an expression (see page 14).) In line 5, we obtain the coefficient polynomials σ_i defined by Equation (9.140) using the operator *Gen_extend_sigma_p* described in Exercise 11.

The actual lifting process is obtained in lines 6–18. At line 7, we form a product Vp of the factors in the list V. (The *Construct* operator, which

Procedure *Irreducible_factor(u ,x)*;
Input
> u : a primitive, square-free polynomial in $\mathbf{Z}[x]$ with positive degree;
> x : a symbol;

Output
> The factored form of u;

Local Variables
> $n, l, V, p, S, k, W, M, i, w$;

Global y;
Begin
```
1     n := Degree_gpe(u, x);
2     l := Leading_coefficient_gpe(u, x);
3     V := Algebraic_expand(Substitute(l^(n-1) * u, x = y/l));
4     p := Find_prime(V, y);
5     S := Berlekamp_factor(Tnn(V, y, p), y, p);
6     if Number_of_operands(S) = 1 then Return(u)
7     else
8        k := Find_k(V, y, p);
9        W := Hensel_lift(V, Map(Ts, S, y, p), y, p, k);
10       W := Substitute(W, y = l * x);
11       M := 1;
12       for i := 1 to Number_of_operands(W) do
13          w := Operand(W, i);
14          M := M * Algebraic_expand(w / Polynomial_content(w, x, [ ], Z));
15       Return(M)
```
End

Figure 9.8. An MPL procedure that obtains the irreducible factorization of a primitive, square-free polynomial in $\mathbf{Z}[x]$. (Implementation: Maple (txt), Mathematica (txt), MuPAD (txt).)

creates the product, is described on page 9.) At line 8, we obtain the error expression E defined in Equation (9.129). In lines 9–10, if $E = 0$, we have found the true factors in $\mathbf{Z}[x]$, which terminates the process, and so we return the set of factors. Otherwise, we perform the next lifting step by computing F (see Equation (9.142)) and the coefficient polynomials r_i (see Equation (9.143)) using the operator *Gen_extend_R_p* described in Exercise 11. In lines 14–18, we form the new list *Vnew* of lifted factors (see Equation (9.144)) and assign this to V in preparation for the next lifting step. If we complete the **for** loop that begins at line 6, V may have too many factors, which are combined at line 19 by the *True_factors* operator.

Procedure *Hensel_lift(u, S, x, p, k)*;
Input
     u : a monic, square-free polynomial in $\mathbf{Z}[x]$ with positive degree;
     S : a set of relatively prime monic polynomials in $\mathbf{Z}_p[x]$
         where $T_p(u)$ is the product in $\mathbf{Z}_p[x]$ of members of S.
         (The members of S use the symmetric representation of $\mathbf{Z}_p$);
     x : a symbol;
     p : a prime ≥ 3;
     k : a positive integer;
Output
     the set of irreducible factors of u;
Local Variables
     $i, j, R, Vp, Vnew, V, E, F, v\_lift, G$;
Begin

```
1     if  k = 1 then
2         Return( True_factors(u, S, x, p, k))
3     else
4         V := Operand _list(S);
5         G := Gen_extend_sigma_p(V, x, p);
6         for  j := 2 to  k do
7             Vp := Algebraic_expand(Construct(" * ", V));
8             E := u − Vp;
9             if  E = 0 then
10                Return( Construct(set, V))
11            else
```

12                 $F := Algebraic\_expand\left(Ts\left(E, x, p^{j}\right) / p^{j-1}\right);$
13                 $R := Gen\_extend\_R\_p(V, G, F, x, p);$
14                 $Vnew := [\,];$

```
15                for  i := 1 to  Number_of_operands(V) do
```

16                     $v\_lift := Algebraic\_expand\Big($
                        $Operand(V, i) + p^{j-1} * Operand(R, i)\Big);$

```
17                    Vnew := Join( Vnew, [v_lift])
18                V := Vnew;
19        Return( True_factors(u, Construct(set, V), x, p, k))
```

End

Figure 9.9. An MPL procedure that performs the Hensel lifting process. (Implementation: Maple (txt), Mathematica (txt), MuPAD (txt).)

The *True_factors* Procedure. The last step in the algorithm uses the lifted factors $v_1^{(k)}, \ldots, v_s^{(k)}$ to determine the irreducible factors of u. This step is based on the observation that each factor of u is either one of the lifted polynomials or the product (in $\mathbf{Z}_{p^k}[x]$) of lifted polynomials. We improve the efficiency of the algorithm by only testing trial factors that are the product of m lifted polynomials, where $m \leq s/2$. Indeed, there is at most one factor constructed with $m > s/2$, and this factor can be obtained by dividing all factors obtained with $m < s/2$ from u.

The *True_factors* procedure shown in Figure 9.10 tests all trial factors until the complete factorization is found. The input to the procedure includes the polynomial u and a set l of lifted factors in $\mathbf{Z}_{p^k}[x]$, and the output is the set of irreducible factors of u in $\mathbf{Z}[x]$.

In lines 1–2, we assign the values of u and l to the local variables U and L, which change in the course of the computation. At each point in the computation, U represents the divisor of u that remains to be factored, and L is the set of lifted polynomials that are used to factor U. The variable *factors* initialized at line 3 contains the set of factors found so far, and the variable m in line 4 represents the number of polynomials from L in a trial factor. Since m is initialized to 1, the procedure first checks if each lifted polynomial is a factor.

The operator *Comb* at line 6 obtains the set C of all m element subsets of L (Exercise 15). The loop that begins at line 7 checks if a member t of C leads to a factor of U. In lines 9–10, we compute a trial factor T by expanding in $\mathbf{Z}_{p^k}[x]$ the product of the members of t. At line 11, we divide U by T, and, if the remainder (*Operand*$(D, 2)$) is 0, we have obtained a factor of U which is added to *factors* (line 13). Since none of the polynomials in t can be used again in another factor, the members of t are removed from L (line 15), and any remaining m element sets in C that contain members of t are removed from C (line 16). (The operator *Clean_up* is described in Exercise 14.) On the other hand, if the remainder *Operand*$(D, 2)$ is not 0, we remove t from C (line 18) and return to the beginning of the loop (line 7). Once all members of C have been eliminated, we increment m (line 19) and return to the beginning of the loop (line 5).

When the loop in lines 5–19 is done, there are two possibilities. If $U = 1$ then all the factors of u have been found. On the other hand, if $U \neq 1$, then this polynomial is also an irreducible factor of u, and so it is added to *factors*. In either case, the set of factors is returned at line 22.

We have given a basic version of the Berlekamp-Hensel polynomial factorization algorithm for polynomials in $\mathbf{Q}[x]$. The polynomial factorization problem for single variable and multivariate polynomials with large coefficients and many variables of high degree is a difficult computational problem. There are, however, more sophisticated versions of the Berlekamp-Hensel algorithm and other algorithms that handle this problem more efficiently than the algorithm given here. The references cited at the end of the chapter survey some of the recent developments for this problem.

Procedure *True_factors*(u, l, x, p, k);

Input

    u : a monic, square-free polynomial in $\mathbf{Z}[x]$ with positive degree;

    l : the set of lifted factors in $\mathbf{Z}_{p^k}[x]$;

    x : a symbol;

    p : a prime number;

    k : a positive integer;

Output

    the set of irreducible factors of u;

Local Variables

    $U, L, factors, m, C, t, T, D$;

Begin

```
1     U := u;
2     L := l;
3     factors := ∅;
4     m := 1;
5     while  m ≤ Number_of_operands(L)/2 do
6        C := Comb(L, m);
7        while  C ≠ ∅ do
8           t := Operand(C, 1);
9           T := Construct(" * ", t);
10          T := Ts(Algebraic_expand(T), x, pᵏ);
11          D := Polynomial_division(U, T, x);
12          if  Operand(D, 2) = 0 then
13             factors := factors ∪ {T};
14             U := Operand(D, 1);
15             L := L ∼ t;
16             C := Clean_up(C, t);
17          else
18             C := C ∼ {t};
19        m := m + 1;
20     if  U ≠ 1 then
21        factors := factors ∪ {U};
22     Return(factors)
End
```

Figure 9.10. The MPL procedure that finds the irreducible factors of u in $\mathbf{Z}[x]$ using the lifted polynomials. (Implementation: Maple (txt), Mathematica (txt), MuPAD (txt).)

Exercises

1. Let b be in the non-negative representation of $\mathbf{Z}_m$. Show that $S_m(b)$ is in the symmetric representation of $\mathbf{Z}_m$. *Hint:* See Exercise 3, page 33.

2. Let u be a polynomial in $\mathbf{Z}[x]$.

   (a) Give a procedure $Tnn(u,\ x,\ m)$ that obtains the projection $T_m(u)$ using the non-negative representation of $\mathbf{Z}_m$.

   (b) Give a procedure $Ts(u,\ x,\ m)$ that obtains the projection $T_m(u)$ using the symmetric representation of $\mathbf{Z}_m$.

3. Prove Theorem 9.39 when $\mathbf{Z}_m$ has the symmetric representation.

4. Let m and n be integers which are ≥ 2, and suppose that $m \mid n$. Show that $T_m(T_n(u)) = T_m(u)$.

5. Suppose that u and v are in $\mathbf{Z}[x]$ where $v \mid u$.

   (a) If $\deg(u) = 2$, show that $||v|| \leq ||u||$.

   (b) Find an example with $\deg(u) = 3$ where $||u|| < ||v||$.

   (c) In Equation (9.105) we give an example where $\deg(u) = 4$ and $||u|| < ||v||$. Find another example.

6. Suppose that u and v are polynomials in $\mathbf{Z}[x]$ where $v \mid u$ and $\deg(v) = 1$. Show that $||v|| \leq ||u||$.

7. Suppose that u and v are polynomials in $\mathbf{Z}[x]$ where $v \mid u$ and u, v and $\text{quot}(u, v)$ have non-negative coefficients. Show that $||v|| \leq ||u||$.

8. Prove Theorem 9.41.

9. Show that the polynomials $v_1^{(k)}$ and $v_2^{(k)}$ in the proof of Theorem 9.52 are relatively prime.

10. (a) Factor $x^2 + 5x + 6$ using the approach in this section.

    (b) Show that $x^4 + 1$ is irreducible in $\mathbf{Z}[x]$ using the approach in this section.

11. Let V be a list of two or more relatively prime, monic polynomials in $\mathbf{Z}_p[x]$, where p is a prime number and $\mathbf{Z}_p$ is represented with the symmetric representation.

    (a) Give a procedure

$$Gen\_extend\_sigma\_p(V, x, p)$$

that returns the coefficient list $[\sigma_1, \ldots, \sigma_s]$ defined by Equation (9.112), where σ_i is represented with the symmetric representation for $\mathbf{Z}_p$.

    (b) Let S be the list obtained by Part (a), and let F be a polynomial in $\mathbf{Z}_p[x]$ that satisfies the inequality (9.108). Give a procedure

$$Gen\_extend\_R\_p(V, S, F, x, p)$$

that returns the list $[r_1, \ldots, r_s]$ defined by Equation (9.121), where r_i is represented with the symmetric representation for $\mathbf{Z}_p$.

Note: This exercise uses operators for polynomial division (Exercise 9, page 125) and the extended Euclidean algorithm (Exercise 4, page 142) that must be modified to operate with the symmetric representation for $\mathbf{Z}_p[x]$.

12. Let p be a prime number and suppose that $p \nmid \text{lc}(v)$ and $p \nmid \text{lc}(w)$. Show that $T_m(\text{res}(v, w)) = \text{res}(T_m(v), T_m(w))$, where the resultant on the right is computed in $\mathbf{Z}_p[x]$.

13. Let u be a square-free polynomial in $\mathbf{Z}[x]$. Give a procedure

$$Find\_prime(u,\ x)$$

that returns a suitable prime number guaranteed by Theorem 9.55. This procedure requires a list of primes. You can either provide a list of primes or use a CAS operator that generates prime numbers such as `ithprime(n)` in Maple, `Prime[n]` in Mathematica, and `ithprime(n)` in MuPAD (Implementation: Maple (mws), Mathematica (nb), MuPAD (mnb).)

14. Let C be a set whose members are also sets, and let t be another set. Give a procedure $Clean\_up(C, t)$ that returns a new set which contains those members s of C such that $s \cap t = \emptyset$. For example, if

$$C = \{\{1, 2\}, \{1, 3\}, \{1, 4\}, \{1, 5\}, \{2, 3\}, \{2, 4\}, \{2, 5\}, \{3, 4\}, \{3, 5\}, \{4, 5\}\}$$

and $t = \{1, 2\}$, then $Clean\_up(C, t) \rightarrow \{\{3, 4\}, \{3, 5\}, \{4, 5\}\}$.

15. Let S be a set of mathematical expressions and let k be an integer with $0 \leq k \leq Number\_of\_operands(S)$. Give a procedure $Comb(S, k)$ that returns the set of all k element subsets of S. For example, if $S = \{a, b, c, d\}$, then

$$Comb(S, 2) \rightarrow \{\{a, b\}, \{a, c\}, \{a, d\}, \{b, c\}, \{b, d\}, \{c, d\}\}.$$

The procedure can be defined by the following recursive transformation rule sequence.

(a) If $k = Number\_of\_operands(S)$, then $Comb(S, k) \rightarrow \{S\}$.

(b) $Comb(S, 0) \rightarrow \{\emptyset\}$.

(c) Let $x = Operand(S, 1)$, $T = S \sim \{x\}$, and $D = Comb(T, k - 1)$. For $D = \{S_1, \ldots, S_n\}$, let $E = \{S_1 \cup \{x\}, \ldots, S_n \cup \{x\}\}$. Then

$$Comb(S, k) \rightarrow Comb(T, k) \cup E.$$

16. The *height* of a polynomial is the maximum of the absolute values of its coefficients. Let u be an algebraic expression. Give a procedure

$$Polynomial\_height(u, x)$$

that returns the height of a polynomial. If u is not a polynomial in x, return the global symbol **Undefined**.

17. Let u be a polynomial in $\mathbf{Z}[x]$, and let p be a prime number. Give a procedure $Find\_k(u, x, p)$ that finds the integer k defined by the inequality (9.146). (See Exercise 16 for the computation of $||u||$.)

18. Modify the procedure $Factor\_sv$ (see Exercise 8, page 370) that factors polynomials in $\mathbf{Q}[x]$ so that operator $Irreducible\_factor$ replaces $Kronecker$.

Further Reading

9.1 Square-Free Polynomials and Factorization. The algorithm in this section is given in Musser [73]. Yun [106] describes a number of algorithms for square-free factorization of single variable and multivariable polynomials, including the algorithm in Exercise 11 on page 359. Other square-free factorization algorithms are given in Wang and Trager [97].

9.3 Factorization in $\mathbf{Z}_p[x]$. The approach here is similar to the approach in Knuth [55] and Akritas [2]. See Dean [31] and Mignotte [66] for a discussion of irreducible factorization in $\mathbf{Z}_p[x]$. Berlekamp's algorithm is also described Davenport, Siret, and Tournier [29], Mignotte and Ştefănescu [67], Geddes, Czapor, and Labahn [39], Yap [105], Winkler [101], and Zippel [108].

9.4 Irreducible Factorization in $\mathbf{Q}[x]$, A Modern Approach. The approach here is similar to the approach in Knuth [55] and Akritas [2]. Modern polynomial factorization algorithms are also described in Davenport, Siret, and Tournier [29], Geddes, Czapor, and Labahn [39], Mignotte and Ştefănescu [67], von zur Gathen and Gerhard [96], Yap [105], Winkler [101] and Zippel [108]. Zippel [108] gives a theory on bounds of coefficients of divisors of a polynomial. Theorem 9.47 is based on Proposition 87 in this book. Coefficient bounds are also discussed in Akritas [2], Cohen [22], and Mignotte [66].

Bibliography

[1] Williams W. Adams and Philippe Loustaunau. *An Introduction to Gröbner Bases*. Graduate Studies in Mathematics, Volume 3. American Mathematical Society, Providence, RI, 1994.

[2] Alkiviadis G. Akritas. *Elements of Computer Algebra with Applications*. John Wiley & Sons, New York, 1989.

[3] Michael Artin. *Algebra*. Prentice Hall, Inc., Englewood Cliffs, NJ, 1991.

[4] E. J. Barbeau. *Polynomials*. Springer-Verlag, New York, 1989.

[5] David Barton and Richard Zippel. Polynomial decomposition algorithms. *Journal of Symbolic Computation*, 1(1):159–168, 1985.

[6] Thomas Becker, Volker Weispfenning, and Heinz Kredel. *Gröbner bases, A Computational Approach to Commutative Algebra*. Springer-Verlag, New York, 1993.

[7] Laurent Bernardin. A review of symbolic solvers. *SIGSAM Bulletin*, 30(1):9–20, March 1996.

[8] Laurent Bernardin. A review of symbolic solvers. In Michael J. Wester, editor, *Computer Algebra Systems, A Practical Guide*, pages 101–120. John Wiley & Sons, Ltd., New York, 1999.

[9] A. S. Besicovitch. On the linear independence of fractional powers of integers. *J. London Math. Soc.*, 15:3–6, 1940.

[10] Garrett Birkhoff and Saunders Mac Lane. *A Survey of Modern Algebra*. A K Peters, Ltd., Natick, MA, 1997.

[11] E. Bond, M. Auslander, S. Grisoff, R. Kenney, M. Myszewski, J. Sammet, R. Tobey, and S. Zilles. Formac–An experimental formula manipulation compiler. In *Proc. 19th ACM National Conference*, pages K2.1-1–K2.1-11, August 1964.

[12] William E. Boyce and Richard C. DiPrima. *Elementary Differential Equations*. Sixth Edition. John Wiley & Sons, New York, 1997.

[13] Manuel Bronstein. *Symbolic Integration I, Transcendental Functions*. Springer-Verlag, New York, 1997.

[14] W. S. Brown. On Euclid's algorithm and the computation of polynomial greatest common divisors. *Journal of the Association for Computing Machinery*, 18(4):478–504, October 1971.

[15] W. S. Brown. The subresultant prs algorithm. *ACM Transactions on Math. Software*, 4(3):237–249, September 1978.

[16] W. S. Brown and J. F. Traub. On Euclid's algorithm and the theory of subresultants. *Journal of the Association for Computing Machinery*, 18(4):505–514, October 1971.

[17] B. Buchberger, G. E. Collins, R. Loos, and R. Albrecht. *Computer Algebra, Symbolic and Algebraic Computation*. Second Edition. Springer-Verlag, New York, 1983.

[18] Bruno Buchberger. Gröbner bases: An algorithmic method in polynomial ideal theory. In N. K. Bose, editor, *Recent Trends in Multidimensional Systems Theory*, pages 184–232. D. Reidel Publishing Company, Dordrecht, Holland, 1985.

[19] Ronald Calinger. *Classics of Mathematics*. Moore Publishing Company Inc., Oak Park, IL, 1982.

[20] Shang-Ching Chou. *Mechanical Geometry Theorem Proving*. D. Reidel Publishing Company, Boston, 1988.

[21] Barry A. Cipra. Do mathematicians still do math. *Science*, 244:769–770, May 19, 1989.

[22] Henri Cohen. *A Course in Computational Algebraic Number Theory*. Springer-Verlag, New York, 1993.

[23] J. S. Cohen, L. Haskins, and J. P. Marchand. Geometry of equilibrium configurations in the ising model. *Journal of Statistical Physics*, 31(3):671–678, June 1983.

[24] Joel S. Cohen. *Computer Algebra and Symbolic Computation: Elementary Algorithms*. A K Peters, Natick, MA, 2002.

[25] George Collins. Subresultants and reduced polynomial remainder sequences. *J. ACM*, 14:128–142, January 1967.

[26] George Collins. The calculation of multivariate polynomial resultants. *J.ACM*, 18(4):515–532, October 1971.

[27] George Collins. Computer algebra of polynomials and rational functions. *American Mathematical Monthly*, 80(7):725–754, September 1973.

[28] Thomas H. Cormen, Charles E. Leiserson, and Ronald L. Rivest. *Introduction to Algorithms*. McGraw-Hill, New York, 1989.

[29] J. H. Davenport, Y. Siret, and E. Tournier. *Computer Algebra, Systems and Algorithms for Algebraic Computation*. Academic Press, New York, 1988.

[30] P. J. Davis and R. Hersh. *The Mathematical Experience*. Birkhäuser, Boston, MA, 1981.

[31] Richard A. Dean. *Classical Abstract Algebra*. Harper and Row, New York, 1990.

[32] William R. Derrick and Stanley I. Grossman. *Differential Equations with Applications*. Third Edition. West Publishing Company, St. Paul, MN, 1987.

[33] F. Dorey and G. Whaples. Prime and composite polynomials. *Journal of Algebra*, 28:88–101, 1974.

[34] H. T. Engstrom. Polynomial substitutions. *Amer. J. of Mathematics*, 63:249–255, 1941.

[35] James F. Epperson. *An Introduction to Numerical Methods and Analysis*. John Wiley & Sons, New York, 2002.

[36] R. J. Fateman. Macsyma's general simplifier. philosophy and operation. In V.E. Lewis, editor, *Proceedings of MACSYMA's Users' Conference*, Washington, D.C., June 20–22 1979, pages 563–582. MIT Laboratory for Computer Science, Cambridge, MA, 1979.

[37] Richard J. Fateman. Symbolic mathematics system evaluators. In Michael J. Wester, editor, *Computer Algebra Systems, A Practical Guide*, 255–284. John Wiley & Sons, Ltd., New York, 1999.

[38] Richard J. Gaylord, N. Kamin, Samuel, and Paul R. Wellin. *An Introduction to Programming with Mathematica, Second Edition*. Springer-Verlag, New York, 1996.

[39] K.O. Geddes, S.R. Czapor, and G. Labahn. *Algorithms for Computer Algebra*. Kluwer Academic Publishers, Boston, 1992.

[40] Jürgen Gerhard, Walter Oevel, Frank Postel, and Stefan Wehmeier. *MuPAD Tutorial, English Edition*. Springer-Verlag, New York, 2000.

[41] John W. Gray. *Mastering Mathematica, Programming Methods and Applications*. Second Edition. Academic Press, New York, 1997.

[42] J. Gutierrez and T. Recio. Advances on the simplification of sine-cosine equations. *Journal of Symbolic Computation*, 26(1):31–70, July 1998.

[43] G. H. Hardy and E. M. Wright. *An Introduction To The Theory of Numbers*. Oxford at The Clarendon Press, London, 1960.

[44] K. M. Heal, M. L. Hansen, and K. M. Rickard. *Maple 6 Learning Guide*. Waterloo Maple Inc., Waterloo, ON, Canada, 2000.

[45] André Heck. *Introduction to Maple*. Second Edition. Springer-Verlag, New York, 1996.

[46] I. N. Herstein. *Topics in Algebra*. Second Edition. Xerox Publishing Company, Lexington, MA, 1975.

[47] E. W. Hobson. *Treatise on Plane and Advanced Trigonometry*. Seventh Edition. Dover Publications, Inc., New York, 1957.

[48] Douglas R. Hofstadter. *Gödel, Escher, Bach: An Eternal Golden Braid*. Random House Inc., New York, 1980.

[49] David J. Jeffrey and Albert D. Rich. Simplifying square roots of square roots by denesting. In Michael J. Wester, editor, *Computer Algebra Systems, A Practical Guide*, pages 61–72. John Wiley & Sons, Ltd., New York, 1999.

[50] Richard D. Jenks and Robert S. Sutor. *Axiom, The Scientific Computation System*. Springer-Verlag, New York, 1992.

[51] N. Kajler, editor. *Computer-Human Interaction in Symbolic Computation.* Springer-Verlag, New York, 1998.

[52] Israel Kleiner. Field theory, from equations to axiomatization. part i. *American Mathematical Monthly,* 106(7):677–684, August-September 1999.

[53] Israel Kleiner. Field theory, from equations to axiomatization. part ii. *American Mathematical Monthly,* 106(9):859–863, November 1999.

[54] Morris Kline. *Mathematics and The Search for Knowledge.* Oxford University Press, New York, 1985.

[55] D. Knuth. *The Art of Computer Programming,* volume 2. Second Edition. Addison-Wesley, Reading, MA, 1981.

[56] Donald Knuth, Ronald Graham, and Oren Patashnik. *Concrete Mathematics, A Foundation For Computer Science.* Addison-Wesley, Reading, MA, 1989.

[57] K. Korsvold. An on-line algebraic simplify program. Technical report, Stanford University, 1965. Stanford University Artificial Intelligence Project, Memorandum 37.

[58] J. S. Kowalik, editor. *Coupling Symbolic and Numerical Computing in Expert Systems.* Elsevier Science Publishers, New York, 1986.

[59] Dexter Kozen and Susan Landau. Polynomial decomposition algorithms. *Journal of Symbolic Computation,* 7:445–456, 1989.

[60] Susan Landau. Simplification of nested radicals. *SIAM J. Comput.,* 21(1):85–110, February 1992.

[61] Ulrich Libbrecht. *Chinese Mathematics in the Thirteenth Century.* MIT Press, Cambridge, MA, 1973.

[62] R. Lidl and H. Niederreiter. *Introduction to Finite Fields and their Applications.* Revised Edition. Cambridge University Press, New York, 1994.

[63] C. C. Lin and L. A. Segel. *Mathematics Applied to Deterministic Problems in the Natural Sciences.* Classics in Applied Mathematics 1. Society for Industrial and Applied Mathematics, Philadelphia, 1988.

[64] John D. Lipson. *Elements of Algebra and Algebraic Computing.* Benjamin/Cummings, Menlo Park, CA, 1981.

[65] Stephen B. Maurer and Anthony Ralston. *Discrete Algorithmic Mathematics.* A K Peters, Ltd., Natick, MA, 1998.

[66] Maurice Mignotte. *Mathematics for Computer Algebra.* Springer-Verlag, New York, 1991.

[67] Maurice Mignotte and Doru Ştefănescu. *Polynomials, An Algorithmic Approach.* Springer-Verlag, New York, 1999.

[68] Bhubaneswar Mishra. *Algorithmic Algebra.* Springer-Verlag, New York, 1993.

[69] M. B. Monagan, K. O. Geddes, K. M. Heal, G. Labahn, S. M. Vorkoetter, and J. McCarron. *Maple 6 Programming Guide.* Waterloo Maple Inc., Waterloo, ON, Canada, 2000.

[70] Joel Moses. *Symbolic Integration.* PhD thesis, MIT, September 1967.

[71] Joel Moses. Algebraic simplification: A guide for the perplexed. *Communications of the ACM*, 14(8):527–537, August 1971.

[72] George M. Murphy. *Ordinary Differential Equations and Their Solutions.* D. Van Nostrand, New York, 1960.

[73] David Musser. *Algorithms for Polynomial Factorization.* PhD thesis, Department of Computer Science, University of Wisconsin, 1971.

[74] Paul J. Nahin. *An Imaginary Tale, The Story of $\sqrt{-1}$.* Princeton University Press, Princeton, NJ, 1998.

[75] Jurg Nievergelt, J. Craig Farrar, and Edward M. Reingold. *Computer Approaches to Mathematical Problems.* Prentice-Hall, Englewood Cliffs, NJ, 1974.

[76] F. S. Nowlan. Objectives in the teaching college mathematics. *American Mathematical Monthly*, 57(1):73–82, February 1950.

[77] R. Pavelle, M. Rothstein, and J. P. Fitch. Computer algebra. *Scientific American*, 245:136–152, 1981.

[78] Louis L. Pennisi. *Elements of Complex Variables.* Holt, Rinehart and Winston, New York, 1963.

[79] Charles Pinter. *A Book of Abstract Algebra.* Second Edition. McGraw-Hill, New York, 1990.

[80] Marcelo Polezzi. A geometrical method for finding an explicit formula for the greatest common divisor. *American Mathematical Monthly*, 104(5):445–446, May 1997.

[81] Frank Postel and Paul Zimmermann. Solving ordinary differential equations. In Michael J. Wester, editor, *Computer Algebra Systems, A Practical Guide*, pages 191–209. John Wiley & Sons, Ltd., New York, 1999.

[82] T. W. Pratt. *Programming Languages, Design and Implementation*. Second Edition. Prentice Hall, Englewood Cliffs, NJ, 1984.

[83] Gerhard Rayna. *Reduce, Software for Algebraic Computation*. Springer-Verlag, New York, 1987.

[84] D. Richardson. Some undecidable problems involving elementary functions of a real variable. *Journal of Symbolic Logic*, 33(4):511–520, December 1968.

[85] J. F. Ritt. Prime and composite polynomials. *Trans. Am. Math. Soc.*, 23:51–66, 1922.

[86] P. Sconzo, A. LeSchack, and R. Tobey. Symbolic computation of f and g series by computer. *The Astronomical Journal*, 70(1329):269–271, May 1965.

[87] George F. Simmons. *Differential Equations with Applications and Historical Notes*. Second Edition. McGraw-Hill, New York, 1991.

[88] George F. Simmons. *Calculus with Analytic Geometry*. Second Edition. McGraw-Hill, New York, 1996.

[89] Barry Simon. Symbolic magic. In Michael J. Wester, editor, *Computer Algebra Systems, A Practical Guide*, pages 21–24. John Wiley & Sons, Ltd., New York, 1999.

[90] Barry Simon. Symbolic math powerhouses revisited. In Michael J. Wester, editor, *Computer Algebra Systems, A Practical Guide*. John Wiley & Sons, Ltd., New York, 1999.

[91] Trevor J. Smedley. Fast methods for computation with algebraic numbers. Research Report CS-90-12, Department of Computer Science, University of Waterloo, May 1990.

[92] Jerome Spanier and Keith B. Oldham. *An Atlas of Functions*. Hemisphere Publishing Corporation, New York, 1987.

[93] Frederick W. Stevenson. *Exploring the Real Numbers*. Prentice Hall, Upper Saddle River, NJ, 2000.

[94] David R. Stoutemyer. Crimes and misdemeanors in the computer algebra trade. *Notices of the American Mathematical Society*, 38(7):778–785, September 1991.

[95] R. Tobey, R. Bobrow, and S. Zilles. Automatic simplification in Formac. In *Proc. AFIPS 1965 Fall Joint Computer Conference*, volume 27, pages 37–52. Spartan Books, Washington, DC, November 1965. Part 1.

[96] Joachim von zur Gathen and Jürgen Gerhard. *Modern Computer Algebra*. Cambridge University Press, New York, 1999.

[97] Paul Wang and Barry Trager. New algorithms for polynomial square free decomposition over the integers. *SIAM Journal of Comp.*, 8:300–305, 1979.

[98] Mark Allen Weiss. *Data Structures and Problem Solving Using C++*. Second Edition. Addison-Wesley, Reading, MA, 2000.

[99] Clark Weissman. *Lisp 1.5 Primer*. Dickenson Publishing Company, Belmont, CA, 1967.

[100] Michael J. Wester. *Computer Algebra Systems, A Practical Guide*. John Wiley & Sons, Ltd., New York, 1999.

[101] F. Winkler. *Polynomial Algorithms in Computer Algebra*. Springer-Verlag, New York, 1996.

[102] Stephen Wolfram. *The Mathematica Book*. Fourth Edition. Cambridge University Press., New York, 1999.

[103] D. Wooldridge. An algebraic simplify program in Lisp. Technical report, Stanford University, December 1965. Artificial Intelligence Project, Memo 11.

[104] W. A. Wulf, M. Shaw, P. Hilfinger, and L. Flon. *Fundamental Structures of Computer Science*. Addison-Wesley, Reading, MA, 1981.

[105] Chee Keng Yap. *Fundamental Problems of Algorithmic Algebra*. Oxford University Press, New York, 2000.

[106] David Y. Y. Yun. On square-free decomposition algorithms. In R. D. Jenks, *Proceedings of the 1976 ACM Symposium of Symbolic and Algebraic Computation*, pages 26–35. ACM, New York, 1976.

[107] David Y. Y. Yun and David R. Stoutemyer. Symbolic mathematical computation. In J. Belzer, A.G. Holzman, and A. Kent, editors, *Encyclopedia of Computer Science and Technology*, volume 15, pages 235–310. M. Dekker, New York, 1980.

[108] Richard Zippel. *Effective Polynomial Computation*. Kluwer Academic Publishers, Boston, 1993.

[109] Daniel Zwillinger. *Handbook of Differential Equations*. Academic Press, Boston, MA, 1989.

Index

Printed and bound by CPI Group (UK) Ltd, Croydon, CR0 4YY

24/10/2024

01778284-0002